Expansions and Asymptotics for Statistics

MONOGRAPHS ON STATISTICS AND APPLIED PROBABILITY

General Editors

F. Bunea, V. Isham, N. Keiding, T. Louis, R. L. Smith, and H. Tong

1 Stochastic Population Models in Ecology and Epidemiology *M.S. Barlett* (1960)
2 Queues *D.R. Cox and W.L. Smith* (1961)
3 Monte Carlo Methods *J.M. Hammersley and D.C. Handscomb* (1964)
4 The Statistical Analysis of Series of Events *D.R. Cox and P.A.W. Lewis* (1966)
5 Population Genetics *W.J. Ewens* (1969)
6 Probability, Statistics and Time *M.S. Barlett* (1975)
7 Statistical Inference *S.D. Silvey* (1975)
8 The Analysis of Contingency Tables *B.S. Everitt* (1977)
9 Multivariate Analysis in Behavioural Research *A.E. Maxwell* (1977)
10 Stochastic Abundance Models *S. Engen* (1978)
11 Some Basic Theory for Statistical Inference *E.J.G. Pitman* (1979)
12 Point Processes *D.R. Cox and V. Isham* (1980)
13 Identification of Outliers *D.M. Hawkins* (1980)
14 Optimal Design *S.D. Silvey* (1980)
15 Finite Mixture Distributions *B.S. Everitt and D.J. Hand* (1981)
16 Classification *A.D. Gordon* (1981)
17 Distribution-Free Statistical Methods, 2nd edition *J.S. Maritz* (1995)
18 Residuals and Influence in Regression *R.D. Cook and S. Weisberg* (1982)
19 Applications of Queueing Theory, 2nd edition *G.F. Newell* (1982)
20 Risk Theory, 3rd edition *R.E. Beard, T. Pentikäinen and E. Pesonen* (1984)
21 Analysis of Survival Data *D.R. Cox and D. Oakes* (1984)
22 An Introduction to Latent Variable Models *B.S. Everitt* (1984)
23 Bandit Problems *D.A. Berry and B. Fristedt* (1985)
24 Stochastic Modelling and Control *M.H.A. Davis and R. Vinter* (1985)
25 The Statistical Analysis of Composition Data *J. Aitchison* (1986)
26 Density Estimation for Statistics and Data Analysis *B.W. Silverman* (1986)
27 Regression Analysis with Applications *G.B. Wetherill* (1986)
28 Sequential Methods in Statistics, 3rd edition
G.B. Wetherill and K.D. Glazebrook (1986)
29 Tensor Methods in Statistics *P. McCullagh* (1987)
30 Transformation and Weighting in Regression
R.J. Carroll and D. Ruppert (1988)
31 Asymptotic Techniques for Use in Statistics
O.E. Bandorff-Nielsen and D.R. Cox (1989)
32 Analysis of Binary Data, 2nd edition *D.R. Cox and E.J. Snell* (1989)
33 Analysis of Infectious Disease Data *N.G. Becker* (1989)
34 Design and Analysis of Cross-Over Trials *B. Jones and M.G. Kenward* (1989)
35 Empirical Bayes Methods, 2nd edition *J.S. Maritz and T. Lwin* (1989)
36 Symmetric Multivariate and Related Distributions
K.T. Fang, S. Kotz and K.W. Ng (1990)
37 Generalized Linear Models, 2nd edition *P. McCullagh and J.A. Nelder* (1989)
38 Cyclic and Computer Generated Designs, 2nd edition
J.A. John and E.R. Williams (1995)
39 Analog Estimation Methods in Econometrics *C.F. Manski* (1988)
40 Subset Selection in Regression *A.J. Miller* (1990)
41 Analysis of Repeated Measures *M.J. Crowder and D.J. Hand* (1990)
42 Statistical Reasoning with Imprecise Probabilities *P. Walley* (1991)
43 Generalized Additive Models *T.J. Hastie and R.J. Tibshirani* (1990)

44 Inspection Errors for Attributes in Quality Control
N.L. Johnson, S. Kotz and X. Wu (1991)

45 The Analysis of Contingency Tables, 2nd edition *B.S. Everitt* (1992)

46 The Analysis of Quantal Response Data *B.J.T. Morgan* (1992)

47 Longitudinal Data with Serial Correlation—A State-Space Approach
R.H. Jones (1993)

48 Differential Geometry and Statistics *M.K. Murray and J.W. Rice* (1993)

49 Markov Models and Optimization *M.H.A. Davis* (1993)

50 Networks and Chaos—Statistical and Probabilistic Aspects
O.E. Barndorff-Nielsen, J.L. Jensen and W.S. Kendall (1993)

51 Number-Theoretic Methods in Statistics *K.-T. Fang and Y. Wang* (1994)

52 Inference and Asymptotics *O.E. Barndorff-Nielsen and D.R. Cox* (1994)

53 Practical Risk Theory for Actuaries
C.D. Daykin, T. Pentikäinen and M. Pesonen (1994)

54 Biplots *J.C. Gower and D.J. Hand* (1996)

55 Predictive Inference—An Introduction *S. Geisser* (1993)

56 Model-Free Curve Estimation *M.E. Tarter and M.D. Lock* (1993)

57 An Introduction to the Bootstrap *B. Efron and R.J. Tibshirani* (1993)

58 Nonparametric Regression and Generalized Linear Models
P.J. Green and B.W. Silverman (1994)

59 Multidimensional Scaling *T.F. Cox and M.A.A. Cox* (1994)

60 Kernel Smoothing *M.P. Wand and M.C. Jones* (1995)

61 Statistics for Long Memory Processes *J. Beran* (1995)

62 Nonlinear Models for Repeated Measurement Data
M. Davidian and D.M. Giltinan (1995)

63 Measurement Error in Nonlinear Models
R.J. Carroll, D. Rupert and L.A. Stefanski (1995)

64 Analyzing and Modeling Rank Data *J.J. Marden* (1995)

65 Time Series Models—In Econometrics, Finance and Other Fields
D.R. Cox, D.V. Hinkley and O.E. Barndorff-Nielsen (1996)

66 Local Polynomial Modeling and its Applications *J. Fan and I. Gijbels* (1996)

67 Multivariate Dependencies—Models, Analysis and Interpretation
D.R. Cox and N. Wermuth (1996)

68 Statistical Inference—Based on the Likelihood *A. Azzalini* (1996)

69 Bayes and Empirical Bayes Methods for Data Analysis
B.P. Carlin and T.A Louis (1996)

70 Hidden Markov and Other Models for Discrete-Valued Time Series
I.L. MacDonald and W. Zucchini (1997)

71 Statistical Evidence—A Likelihood Paradigm *R. Royall* (1997)

72 Analysis of Incomplete Multivariate Data *J.L. Schafer* (1997)

73 Multivariate Models and Dependence Concepts *H. Joe* (1997)

74 Theory of Sample Surveys *M.E. Thompson* (1997)

75 Retrial Queues *G. Falin and J.G.C. Templeton* (1997)

76 Theory of Dispersion Models *B. Jørgensen* (1997)

77 Mixed Poisson Processes *J. Grandell* (1997)

78 Variance Components Estimation—Mixed Models, Methodologies and Applications *P.S.R.S. Rao* (1997)

79 Bayesian Methods for Finite Population Sampling
G. Meeden and M. Ghosh (1997)

80 Stochastic Geometry—Likelihood and computation
O.E. Barndorff-Nielsen, W.S. Kendall and M.N.M. van Lieshout (1998)

81 Computer-Assisted Analysis of Mixtures and Applications—
Meta-analysis, Disease Mapping and Others *D. Böhning* (1999)

82 Classification, 2nd edition *A.D. Gordon* (1999)

83 Semimartingales and their Statistical Inference *B.L.S. Prakasa Rao* (1999)
84 Statistical Aspects of BSE and vCJD—Models for Epidemics
C.A. Donnelly and N.M. Ferguson (1999)
85 Set-Indexed Martingales *G. Ivanoff and E. Merzbach* (2000)
86 The Theory of the Design of Experiments *D.R. Cox and N. Reid* (2000)
87 Complex Stochastic Systems
O.E. Barndorff-Nielsen, D.R. Cox and C. Klüppelberg (2001)
88 Multidimensional Scaling, 2nd edition *T.F. Cox and M.A.A. Cox* (2001)
89 Algebraic Statistics—Computational Commutative Algebra in Statistics
G. Pistone, E. Riccomagno and H.P. Wynn (2001)
90 Analysis of Time Series Structure—SSA and Related Techniques
N. Golyandina, V. Nekrutkin and A.A. Zhigljavsky (2001)
91 Subjective Probability Models for Lifetimes
Fabio Spizzichino (2001)
92 Empirical Likelihood *Art B. Owen* (2001)
93 Statistics in the 21st Century
Adrian E. Raftery, Martin A. Tanner, and Martin T. Wells (2001)
94 Accelerated Life Models: Modeling and Statistical Analysis
Vilijandas Bagdonavicius and Mikhail Nikulin (2001)
95 Subset Selection in Regression, Second Edition *Alan Miller* (2002)
96 Topics in Modelling of Clustered Data
Marc Aerts, Helena Geys, Geert Molenberghs, and Louise M. Ryan (2002)
97 Components of Variance *D.R. Cox and P.J. Solomon* (2002)
98 Design and Analysis of Cross-Over Trials, 2nd Edition
Byron Jones and Michael G. Kenward (2003)
99 Extreme Values in Finance, Telecommunications, and the Environment
Bärbel Finkenstädt and Holger Rootzén (2003)
100 Statistical Inference and Simulation for Spatial Point Processes
Jesper Møller and Rasmus Plenge Waagepetersen (2004)
101 Hierarchical Modeling and Analysis for Spatial Data
Sudipto Banerjee, Bradley P. Carlin, and Alan E. Gelfand (2004)
102 Diagnostic Checks in Time Series *Wai Keung Li* (2004)
103 Stereology for Statisticians *Adrian Baddeley and Eva B. Vedel Jensen* (2004)
104 Gaussian Markov Random Fields: Theory and Applications
Håvard Rue and Leonhard Held (2005)
105 Measurement Error in Nonlinear Models: A Modern Perspective, Second Edition
Raymond J. Carroll, David Ruppert, Leonard A. Stefanski,
and Ciprian M. Crainiceanu (2006)
106 Generalized Linear Models with Random Effects: Unified Analysis via H-likelihood
Youngjo Lee, John A. Nelder, and Yudi Pawitan (2006)
107 Statistical Methods for Spatio-Temporal Systems
Bärbel Finkenstädt, Leonhard Held, and Valerie Isham (2007)
108 Nonlinear Time Series: Semiparametric and Nonparametric Methods
Jiti Gao (2007)
109 Missing Data in Longitudinal Studies: Strategies for Bayesian Modeling and Sensitivity Analysis
Michael J. Daniels and Joseph W. Hogan (2008)
110 Hidden Markov Models for Time Series: An Introduction Using R
Walter Zucchini and Iain L. MacDonald (2009)
111 ROC Curves for Continuous Data
Wojtek J. Krzanowski and David J. Hand (2009)
112 Antedependence Models for Longitudinal Data
Dale L. Zimmerman and Vicente A. Núñez-Antón (2009)
113 Mixed Effects Models for Complex Data
Lang Wu (2010)
114 Intoduction to Time Series Modeling
Genshiro Kitagawa (2010)
115 Expansions and Asymptotics for Statistics
Christopher G. Small (2010)

Monographs on Statistics and Applied Probability 115

Expansions and Asymptotics for Statistics

Christopher G. Small
University of Waterloo
Waterloo, Ontario, Canada

CRC Press
Taylor & Francis Group
Boca Raton London New York

CRC Press is an imprint of the
Taylor & Francis Group an **informa** business
A CHAPMAN & HALL BOOK

Chapman & Hall/CRC
Taylor & Francis Group
6000 Broken Sound Parkway NW, Suite 300
Boca Raton, FL 33487-2742

Printed in the United States of America on acid-free paper
10 9 8 7 6 5 4 3 2 1

International Standard Book Number: 978-1-58488-590-0 (Hardback)

Library of Congress Cataloging-in-Publication Data

Small, Christopher G.
Expansions and asymptotics for statistics / Christopher G. Small.
p. cm. -- (Monographs on statistics and applied probability ; 115)
Includes bibliographical references and index.
ISBN 978-1-58488-590-0 (hardcover : alk. paper)
1. Asymptotic distribution (Probability theory) 2. Asymptotic expansions. I. Title. II. Series.

QA273.6.S63 2010
519.5--dc22 2010010969

Visit the Taylor & Francis Web site at
http://www.taylorandfrancis.com

and the CRC Press Web site at
http://www.crcpress.com

Contents

Preface **xi**

1 Introduction **1**

1.1 Expansions and approximations 1

1.2 The role of asymptotics 3

1.3 Mathematical preliminaries 4

1.4 Two complementary approaches 16

1.5 Problems 18

2 General series methods **23**

2.1 A quick overview 23

2.2 Power series 24

2.3 Enveloping series 40

2.4 Asymptotic series 47

2.5 Superasymptotic and hyperasymptotic series 63

2.6 Asymptotic series for large samples 66

2.7 Generalised asymptotic expansions 68

2.8 Notes 69

2.9 Problems 69

3 Padé approximants and continued fractions **75**

3.1 The Padé table 75

3.2 Padé approximations for the exponential function 79

3.3 Two applications 81

3.4 Continued fraction expansions 85

3.5 A continued fraction for the normal distribution 88

3.6 Approximating transforms and other integrals 90

3.7 Multivariate extensions 92

3.8 Notes 93

3.9 Problems 94

4 The delta method and its extensions 99

4.1 Introduction to the delta method 99

4.2 Preliminary results 100

4.3 The delta method for moments 103

4.4 Using the delta method in Maple 108

4.5 Asymptotic bias 109

4.6 Variance stabilising transformations 111

4.7 Normalising transformations 114

4.8 Parameter transformations 116

4.9 Functions of several variables 119

4.10 Ratios of averages 119

4.11 The delta method for distributions 121

4.12 The von Mises calculus 123

4.13 Obstacles and opportunities: robustness 134

4.14 Problems 137

5 Optimality and likelihood asymptotics 143

5.1 Historical overview 143

5.2 The organisation of this chapter 151

5.3 The likelihood function and its properties 152

5.4 Consistency of maximum likelihood 159

5.5 Asymptotic normality of maximum likelihood 161

5.6 Asymptotic comparison of estimators 164
5.7 Local asymptotics 171
5.8 Local asymptotic normality 177
5.9 Local asymptotic minimaxity 181
5.10 Various extensions 185
5.11 Problems 187

6 The Laplace approximation and series 193
6.1 A simple example 193
6.2 The basic approximation 195
6.3 The Stirling series for factorials 200
6.4 Laplace expansions in Maple 201
6.5 Asymptotic bias of the median 202
6.6 Recurrence properties of random walks 205
6.7 Proofs of the main propositions 207
6.8 Integrals with the maximum on the boundary 211
6.9 Integrals of higher dimension 212
6.10 Integrals with product integrands 215
6.11 Applications to statistical inference 219
6.12 Estimating location parameters 220
6.13 Asymptotic analysis of Bayes estimators 222
6.14 Notes 223
6.15 Problems 223

7 The saddle-point method 227
7.1 The principle of stationary phase 227
7.2 Perron's saddle-point method 229
7.3 Harmonic functions and saddle-point geometry 234
7.4 Daniels' saddle-point approximation 238
7.5 Towards the Barndorff-Nielsen formula 241

7.6 Saddle-point method for distribution functions 251
7.7 Saddle-point method for discrete variables 253
7.8 Ratios of sums of random variables 254
7.9 Distributions of M-estimators 256
7.10 The Edgeworth expansion 258
7.11 Mean, median and mode 262
7.12 Hayman's saddle-point approximation 263
7.13 The method of Darboux 268
7.14 Applications to common distributions 269
7.15 Problems 274

8 Summation of series 279
8.1 Advanced tests for series convergence 279
8.2 Convergence of random series 285
8.3 Applications in probability and statistics 286
8.4 Euler-Maclaurin sum formula 291
8.5 Applications of the Euler-Maclaurin formula 295
8.6 Accelerating series convergence 297
8.7 Applications of acceleration methods 309
8.8 Comparing acceleration techniques 313
8.9 Divergent series 314
8.10 Problems 316

9 Glossary of symbols 321

10 Useful limits, series and products 325

References 327

Index 331

Preface

The genesis for this book was a set of lectures given to graduate students in statistics at the University of Waterloo. Many of these students were enrolled in the Ph.D. program and needed some analytical tools to support their thesis work. Very few of these students were doing theoretical work as the principal focus of their research. In most cases, the theory was intended to support a research activity with an applied focus. This book was born from a belief that the toolkit of methods needs to be broad rather than particularly deep for such students. The book is also written for researchers who are not specialists in asymptotics, and who wish to learn more.

The statistical background required for this book should include basic material from mathematical statistics. The reader should be thoroughly familiar with the basic distributions, their properties, and their generating functions. The characteristic function of a distribution will also be discussed in the following chapters. So, a knowledge of its basic properties would be very helpful. The mathematical background required for this book varies depending on the module. For many chapters, a good course in analysis is helpful but not essential. Those who have a background in calculus equivalent to say that in Spivak (1994) will have more than enough. Chapters which use complex analysis will find that an introductory course or text on this subject is more than sufficient as well.

I have tried as much as possible to use a unified notation that is common to all chapters. This has not always been easy. However, the notation that is used in each case is fairly standard for that application. At the end of the book, the reader will find a list of the symbols and notation common to all chapters of the book. Also included is a list of common series and products. The reader who wishes to expand an expression or to simplify an expansion should check here first.

The book is meant to be accessible to a reader who wishes to browse a particular topic. Therefore the structure of the book is modular. Chapters 1–3 form a module on methods for expansions of functions arising in probability and statistics. Chapter 1 discusses the role of expansions and asymptotics in statistics, and provides some background material necessary for the rest of the book. Basic results on limits of random variables are stated, and some of the notation, including order notation, limit superior and limit inferior, etc., are explained in detail.

Chapter 2 also serves as preparation for the chapters which follow. Some basic properties of power series are reviewed and some examples given for calculating cumulants and moments of distributions. Enveloping series are introduced because they appear quite commonly in expansions of distributions and integrals. Many enveloping series are also asymptotic series. So a section of Chapter 2 is devoted to defining and discussing the basic properties of asymptotic series. As the name suggests, asymptotic series appear quite commonly in asymptotic theory.

The partial sums of power series and asymptotic series are both rational functions. So, it is natural to generalise the discussion from power series and asymptotic series to the study of rational approximations to functions. This is the subject of Chapter 3. The rational analogue of a Taylor polynomial is known as a Padé approximant. The class of Padé approximants includes various continued fraction expansions as a special case. Padé approximations are not widely used by statisticians. But many of the functions that statisticians use, such as densities, distribution functions and likelihoods, are often better approximated by rational functions than by polynomials.

Chapters 4 and 5 form a module in their own right. Together they describe core ideas in statistical asymptotics, namely the asymptotic normality and asymptotic efficiency of standard estimators as the sample size goes to infinity. Both the delta method for moments and the delta method for distributions are explained in detail. Various applications are given, including the use of the delta method for bias reduction, variance stabilisation, and the construction of normalising transformations. It is natural to place the von Mises calculus in a chapter on the delta method because the von Mises calculus is an extension of the delta method to statistical functionals.

The results in Chapter 5 can be studied independently of Chapter 4, but are more naturally understood as the application of the delta method to the likelihood. Here, the reader will find much of the standard theory that derives from the work of R. A. Fisher, H. Cramér, L. Le Cam and others. Properties of the likelihood function, its logarithm and derivatives are described. The consistency of the maximum likelihood estimator is sketched, and its asymptotic normality proved under standard regularity. The concept of asymptotic efficiency, due to R. A. Fisher, is also explained and proved for the maximum likelihood estimator. Le Cam's critique of this theory, and his work on local asymptotic normality and minimaxity, are briefly sketched, although the more challenging technical aspects of this work are omitted.

Chapters 6 and 7 form yet another module on the Laplace approximation and the saddle-point method. In statistics, the term "saddle-point approximation" is taken to be synonymous with "tilted Edgeworth expansion." However, such an identification does not do justice to the full power of the saddle-point method, which is an extension of the Laplace method to contour integrals in the complex plane. Applied mathematicians often recognise the close connection between the saddle-point approximation and the Laplace method by using the former term to cover both techniques. In the broadest sense used in applied mathematics, the central limit theorem and the Edgeworth expansion are both saddle-point methods.

Finally, Chapter 8, on the summation of series, forms a module in its own right. Nowadays, Monte Carlo techniques are often the methods of choice for numerical work by both statisticians and probablists. However, the alternatives to Monte Carlo are often missed. For example, a simple approach to computing anything that can be written as a series is simply to sum the series. This will work provided that the series converges reasonably fast. Unfortunately, many series do not. Nevertheless, a large amount of work has been done on the problem of transforming series so that they converge faster, and many of these techniques are not widely known. When researchers complain about the slow convergence of their algorithms, they sometimes ignore simple remedies which accelerate the convergence. The topics of series convergence and the acceleration of that convergence are the main ideas to be found in Chapter 8.

Another feature of the book is that I have supplemented some topics with a discussion of the relevant Maple* commands that implement the ideas on that topic. Maple is a powerful symbolic computation package that takes much of the tedium out of the difficult work of doing the expansions. I have tried to strike a balance here between theory and computation. Those readers who are not interested in Maple will have no trouble if they simply skip the Maple material. Those readers who use, or who wish to use Maple, will need to have a little bit of background in symbolic computation as this book is not a self-contained introduction to the subject. Although the Maple commands described in this book will work on recent versions of Maple, the reader is warned that the precise format of the output from Maple will vary from version to version.

Scattered throughout the book are a number of vignettes of various people in statistics and mathematics whose ideas have been instrumental in the development of the subject. For readers who are only interested in the results and formulas, these vignettes may seem unnecessary. However, I include these vignettes in the hope that readers who find an idea interesting will ponder the larger contributions of those who developed the idea.

Finally, I am most grateful to Melissa Smith of Graphic Services at the University of Waterloo, who produced the pictures. Thanks are also due to Ferdous Ahmed, Zhenyu Cui, Robin Huang, Vahed Maroufy, Michael McIsaac, Kimihiro Noguchi, Reza Ramezan and Ying Yan, who proofread parts of the text. Any errors which remain after their valuable assistance are entirely my responsibility.

CHAPTER 1

Introduction

1.1 Expansions and approximations

We begin with the observation that any finite probability distribution is a partition of unity. For example, for $p+q=1$, the binomial distribution may be obtained from the binomial expansion

$$\begin{aligned} 1 &= (p+q)^n \\ &= \binom{n}{0} p^n + \binom{n}{1} p^{n-1} q + \binom{n}{2} p^{n-2} q^2 + \cdots + \binom{n}{n} q^n . \end{aligned}$$

In this expansion, the terms are the probabilities for the values of a binomial random variable. For this reason, the theory of sums or series has always been closely tied to probability. By extension, the theory of infinite series arises when studying random variables that take values in some denumerable range.

Series involving partitions go back to some of the earliest work in mathematics. For example, the ancient Egyptians worked with geometric series in practical problems of partitions. Evidence for this can be found in the Rhind papyrus, which is dated to 1650 BCE. Problem 64 of that papyrus states the following.

> Divide ten heqats of barley among ten men so that the common difference is one eighth of a heqat of barley.

Put in more modern terms, this problem asks us to partition ten heqats* into an arithmetic series

$$10 = a + \left(a + \frac{1}{8}\right) + \left(a + \frac{2}{8}\right) + \ldots + \left(a + \frac{9}{8}\right) .$$

That is, to find the value of a in this partition. The easiest way to solve this problem is to use a formula for the sum of a finite arithmetic series.

* The heqat was an ancient Egyptian unit of volume corresponding to about 4.8 litres.

A student in a modern course in introductory probability has to do much the same sort of thing when asked to compute the normalising constant for a probability function of given form. If we look at the solutions to such problems in the Rhind papyrus, we see that the ancient Egyptians well understood the standard formula for simple finite series.

However the theory of infinite series remained problematic throughout classical antiquity and into more modern times until differential and integral calculus were placed on a firm foundation using the modern theory of analysis. Isaac Newton, who with Gottfried Leibniz developed calculus, is credited with the discovery of the binomial expansion for general exponents, namely

$$(1+x)^y = 1 + \binom{y}{1} x + \binom{y}{2} x^2 + \binom{y}{3} x^3 + \cdots$$

where the binomial coefficient

$$\binom{y}{n} = \frac{y\,(y-1)\,(y-2)\,\cdots\,(y-n+1)}{n!}$$

is defined for any real value y. The series converges when $|x| < 1$. Note that when $y = -1$ the binomial coefficients become $(-1)^n$ so the expansion is the usual formula for an infinite geometric series.

In 1730, a very powerful tool was added to the arsenal of mathematicians when James Stirling discovered his famous approximation to the factorial function. It was this approximation which formed the basis for De Moivre's version of the central limit theorem, which in its earliest form was a normal approximation to the binomial probability function. The result we know today as Stirling's approximation emerged from the work and correspondence of Abraham De Moivre and James Stirling. It was De Moivre who found the basic form of the approximation, and the numerical value of the constant in the approximation. Stirling evaluated this constant precisely.† The computation of $n!$ becomes a finite series when logarithms are taken. Thus

$$\ln(n!) = \ln 1 + \ln 2 + \cdots + \ln n\,. \tag{1.1}$$

De Moivre first showed that

$$\frac{n!}{\sqrt{n}\, n^n\, e^{-n}} \to \text{constant} \qquad \text{as } n \to \infty\,. \tag{1.2}$$

Then Stirling's work showed that this constant is $\sqrt{2\,\pi}$.

† Gibson (1927, p. 78) wrote of Stirling that "next to Newton I would place Stirling as the man whose work is specially valuable where series are in question."

With this result in hand, combinatorial objects such as binomial coefficients can be approximated by smooth functions. See Problem 2 at the end of the chapter. By approximating binomial coefficients, De Moivre was able to obtain his celebrated normal approximation to the binomial distribution. Informally, this can be written as

$$\mathcal{B}(n, p) \approx \mathcal{N}(n\,p,\; n\,p\,q)$$

as $n \to \infty$. We state the precise form of this approximation later when we consider a more general statement of the central limit theorem.

1.2 The role of asymptotics

For statisticians, the word "asymptotics" usually refers to an investigation into the behaviour of a statistic as the sample size gets large. In conventional usage, the word is often limited to arguments claiming that a statistic is "asymptotically normal" or that a particular statistical method is "asymptotically optimal." However, the study of asymptotics is much broader than just the investigation of asymptotic normality or asymptotic optimality alone.

Many such investigations begin with a study of the limiting behaviour of a sequence of statistics $\{W_n\}$ as a function of sample size n. Typically, an asymptotic result of this form can be expressed as

$$F(t) = \lim_{n\to\infty} F_n(t)\,.$$

The functions $F_n(t)$, $n = 1, 2, 3, \ldots$ could be distribution functions as the notation suggests, or moment generating functions, and so on. For example, the asymptotic normality of the sample average $\bar{X}_n$ for a random sample $X_1, \ldots, X_n$ from some distribution can be expressed using a limit of standardised distribution functions.

Such a limiting result is the natural thing to derive when we are proving asymptotic normality. However, when we speak of asymptotics generally, we often mean something more than this. In many cases, it is possible to expand $F_n(t)$ to obtain (at least formally) the series

$$F_n(t) \;\sim\; F(t)\left\{1 + \frac{a_1(t)}{n} + \frac{a_2(t)}{n^2} + \frac{a_3(t)}{n^3} + \cdots\right\}.$$

We shall call such a series an *asymptotic series* for $F_n(t)$ in the variable n.

Stirling's approximation, which we encountered above, is an asymptotic result. Expressed as a limit, this approximation states that

$$\sqrt{2\,\pi} = \lim_{n\to\infty} \frac{n!}{n^{n+1/2}\,e^{-n}}\,.$$

This is better known in the form

$$n! \sim \sqrt{2\pi n}\, n^n e^{-n} \tag{1.3}$$

for large n. When put into the form of a series, Stirling's approximation can be sharpened to

$$\frac{n!}{n^{n+1/2} e^{-n}} \sim \sqrt{2\pi}\left\{1 + \frac{1}{12n} + \frac{1}{288n^2} - \cdots\right\} \tag{1.4}$$

as $n \to \infty$. We shall also speak of *k-th order* asymptotic results, where k denotes the number of terms of the asymptotic series that are used in the approximation.

The idea of expanding a function into a series in order to study its properties has been around for a long time. Newton developed some of the standard formulas we use today, Euler gave us some powerful tools for summing series, and Augustin-Louis Cauchy provided the theoretical framework to make the study of series a respectable discipline. Thus series expansions are certainly older than the subject of statistics itself if, by that, we mean statistics as a recognisable discipline. So it is not surprising to find series expansions used as an analytical tool in many areas of statistics. For many people, the subject is almost synonymous with the theory of asymptotics. However, series expansions arise in many contexts in both probability and statistics which are not usually called asymptotics, per se. Nevertheless, if we define asymptotics in the broad sense to be the study of functions or processes when certain variables take limiting values, then all series expansions are essentially asymptotic investigations.

1.3 Mathematical preliminaries

1.3.1 Supremum and infimum

Let A be any set of real numbers. We say that A is bounded above if there exists some real number u such that $x \leq u$ for all $x \in A$. Similarly, we say that A is bounded below if there exists a real number b such that $x \geq b$ for all $x \in A$. The numbers u and b are called an *upper bound* and a *lower bound*, respectively.

Upper and lower bounds for infinite sequences are defined in much the same way. A number u is an upper bound for the sequence

$$x_1, x_2, x_3, \ldots$$

if $u \geq x_n$ for all $n \geq 1$. The number b is a lower bound for the sequence if $b \leq x_n$ for all n.

Isaac Newton (1642–1727)

Co-founder of the calculus, Isaac Newton also pioneered many of the techniques of series expansions including the binomial theorem.

"And from my pillow, looking forth by light
Of moon or favouring stars, I could behold
The antechapel where the statue stood
Of Newton with his prism and silent face,
The marble index of a mind for ever
Voyaging through strange seas of Thought, alone."

William Wordsworth, The Prelude, Book 3, lines 58–63.

Definition 1. *A real number u is called a* least upper bound *or* supremum *of any set A if u is an upper bound for A and is the smallest in the sense that $c \geq u$ whenever c is any upper bound for A.*

A real number b is called a greatest lower bound *or* infimum *of any set A if b is a lower bound for A and is the greatest in the sense that $c \leq b$ whenever c is any lower bound for A.*

It is easy to see that a supremum or infimum of A is unique. Therefore, we write $\sup A$ for the unique supremum of A, and $\inf A$ for the unique infimum.

Similar definitions hold for sequences. We shall define the supremum or infimum of a sequence of real numbers as follows.

Definition 2. *Let x_n, $n \geq 1$ be a infinite sequence of real numbers. We define $\sup x_n$, the supremum of x_n, to be the upper bound which is smallest, in the sense that $u \geq \sup x_n$ for every upper bound u. The infimum of the sequence is defined correspondingly, and written as $\inf x_n$.*

In order for a set or a sequence to have a supremum or infimum, it is necessary and sufficient that it be bounded above or below, respectively. This is summarised in the following proposition.

Proposition 1. *If A (respectively x_n) is bounded above, then A (respectively x_n) has a supremum. Similarly, if A (respectively x_n) is bounded below, then A (respectively x_n) has an infimum.*

This proposition follows from the completeness property of the real numbers. We omit the proof. For those sets which do not have an upper bound the collection of all upper bounds is empty. For such situations, it is useful to adopt the fiction that the smallest element of the empty set $\emptyset$ is ∞ and the largest element of $\emptyset$ is $-\infty$. With this fiction, we adopt the convention that $\sup A = \infty$ when A has no upper bound. Similarly, when A has no lower bound we set $\inf A = -\infty$. For sequences, these conventions work correspondingly. If x_n, $n \geq 1$ is not bounded above, then $\sup x_n = \infty$, and if not bounded below then $\inf x_n = -\infty$.

1.3.2 Limit superior and limit inferior

A real number u is called an *almost upper bound* for A if there are only finitely many $x \in A$ such that $x \geq u$. The *almost lower bound* is defined correspondingly. Any infinite set that is bounded (both above and below) will have almost upper bounds, and almost lower bounds.

Let B be the set of almost upper bounds of any infinite bounded set A. Then B is bounded below. Similarly, let C be the set of almost lower bounds of A. Then C is bounded above. See Problem 3. It follows from Proposition 1 that B has an infimum.

Definition 3. *Let A be an infinite bounded set, and let B be the set of almost upper bounds of A. The infimum of B is called the limit superior of A. We write* $\limsup A$ *for this real number. Let C be the set of almost lower bounds of A. The supremum of C is called the limit inferior of A. We write the limit inferior of A as* $\liminf A$.

We can extend these definitions to the cases where A has no upper bound or no lower bound. If A has no upper bound, then the set of almost upper bounds will be empty. Since $B = \emptyset$ we can define $\inf \emptyset = \infty$ so that $\limsup A = \infty$ as well. Similarly, if A has no lower bound, we set $\sup \emptyset = -\infty$ so that $\liminf A = -\infty$.

The definitions of limit superior and limit inferior extend to sequences with a minor modification. Let x_n, $n \geq 1$ be a sequence of real numbers. For each $n \geq 1$ define

$$\begin{aligned} y_n &= \sup_{k \geq n} x_k \\ &= \sup \{x_n, x_{n+1}, x_{n+2}, \ldots\}, \end{aligned}$$

and

$$\begin{aligned} z_n &= \inf_{k \geq n} x_k \\ &= \inf \{x_n, x_{n+1}, x_{n+2}, \ldots\}. \end{aligned}$$

It can be checked that y_n is monotone decreasing, and z_n is monotone increasing. We then define

$$\limsup x_n = \inf y_n \qquad \liminf x_n = \sup z_n . \tag{1.5}$$

The limit superior of a sequence can be shown to be the largest cluster point of the sequence, and the limit inferior the smallest cluster point.

To illustrate the definitions of limits superior and inferior, let us consider two examples. Define $x_n = (-1)^n + n^{-1}$, so that

$$x_1 = 0, \qquad x_2 = \frac{3}{2}, \qquad x_3 = -\frac{2}{3},$$

and so on. The reader can check that $\limsup x_n = 1$ and $\liminf x_n =$

-1. As another example, consider $x_n = n$ for all $n \geq 1$. In this case $\limsup x_n = \liminf x_n = \infty$.

Proposition 2. *Limits superior and inferior of sequences* x_n, $n \geq 1$ *satisfy the following list of properties.*

1. *In general we have*

$$\inf x_n \leq \liminf x_n \leq \limsup x_n \leq \sup x_n .$$

2. *Moreover, when* $\limsup x_n < \sup x_n$, *then the sequence* x_n, $n \geq 1$ *has a maximum (i.e., a largest element). Similarly, when* $\liminf x_n > \inf x_n$, *then* x_n, $n \geq 1$ *has a minimum.*
3. *The limits superior and inferior are related by the identities*

$$\liminf x_n = -\limsup(-x_n), \qquad \limsup x_n = -\liminf(-x_n).$$

The proof of this proposition is left as Problem 5 at the end of the chapter.

1.3.3 The O-notation

The handling of errors and remainder terms in asymptotics is greatly enhanced by the use of the Bachmann-Landau O-notation.‡ When used with care, this order notation allows the quick manipulation of vanishingly small terms with the need to display their asymptotic behaviour explicitly with limits.

Definition 4. *Suppose* $f(x)$ *and* $g(x)$ *are two functions of some variable* $x \in S$. *We shall write*

$$f(x) = O[\,g(x)\,]$$

if there exists some constant $\alpha > 0$ *such that* $|\,f(x)\,| \leq \alpha\,|\,g(x)\,|$ *for all* $x \in S$.

Equivalently, when $g(x) \neq 0$ for all $x \in S$, then $f(x) = O[\,g(x)\,]$ provided that $f(x)/g(x)$ is a bounded function on the set S.

‡ Commonly known as the Landau notation. See P. Bachmann. *Die Analytische Zahlentheorie. Zahlentheorie.* 2, Teubner, Leipzig 1894, and E. Landau. *Handbuch der Lehre von der Verteilung der Primzahlen.* Vol 2, Teubner, Leipzig 1909, pp. 3–5.

For example, on $S = (-\infty, \infty)$, we have $\sin 2x = O(x)$, because

$$|\sin 2x| \leq 2\,|x|$$

for all real x.

In many cases, we are only interested in the properties of a function on some region of a set S such as a neighbourhood of some point x_0. We shall write

$$f(x) = O[\,g(x)\,], \qquad \text{as } x \to x_0$$

provided that there exists $\alpha > 0$ such that $|\,f(x)\,| \leq \alpha\,|\,g(x)\,|$ for all x in some punctuated neighbourhood of x_0. We shall be particularly interested in the cases where $x_0 = \pm\infty$ and $x_0 = 0$. For example, the expression

$$\sin(x^{-1}) = O[\,x^{-1}\,], \qquad \text{as } x \to \infty$$

is equivalent to saying that there exists positive constants c and α such that $|\sin(x^{-1})| \leq \alpha\,|\,x^{-1}|$ for all $x > c$.

The virtue of this O-notation is that $O[g(x)]$ can be introduced into a formula in place of $f(x)$ and treated as if it were a function. This is particularly useful when we wish to carry a term in subsequent calculations, but are only interested in its size and not its exact value. Algebraic manipulations using order terms become simpler if $g(x)$ is algebraically simpler to work with than $f(x)$, particularly when $g(x) = x^k$.

Of course, $O[g(x)]$ can represent many functions. So, the use of an equals sign is an abuse of terminology. This can lead to confusion. For example

$$\sin x = O(x) \qquad \text{and} \qquad \sin x = O(1)$$

as $x \to 0$. However, it is not true that the substitution $O(x) = O(1)$ can be made in any calculation. The confusion can be avoided if we recall that $O[\,g(x)\,]$ represents functions including those of smaller order than $g(x)$ itself. So the ease and flexibility of the Landau O-notation can also be its greatest danger for the unwary.[§]

Nevertheless, the notation makes many arguments easier. The advantage of the notation is particularly apparent when used with Taylor expansions of functions. For example as $x \to 0$ we have

$$e^x = 1 + x + O(x^2) \qquad \text{and} \qquad \ln(1+x) = x + O(x^2)\,.$$

Therefore

$$e^x \ln(1+x) \quad = \quad [\,1 + x + O(x^2)\,] \cdot [\,x + O(x^2)\,]$$

[§] A more precise notation is to consider $O[\,g(x)\,]$ more properly as a class of functions and to write $f(x) \in O[\,g(x)\,]$. However, this conceptual precision comes at the expense of algebraic convenience.

$$\begin{aligned} &= [1+x+O(x^2)]\cdot x+[1+x+O(x^2)]\cdot O(x^2) \\ &= x+x^2+O(x^3)+O(x^2)+O(x^3)+O(x^4) \\ &= x+O(x^2)\,, \end{aligned}$$

as $x \to 0$.

The O-notation is also useful for sequences, which are functions defined on the domain of natural numbers. When $S = \{1, 2, 3, \ldots\}$, then we will write

$$f(n) = O[\,g(n)\,] \qquad \text{or} \qquad f_n = O[\,g_n\,]$$

as the context suggests. It is left to the reader to see that when $g(n)$ vanishes nowhere, then a sequence $f(n)$ is of order $O[\,g(n)\,]$ if and only if $f(n)$ is of order $O[\,g(n)\,]$ as $n \to \infty$.

In Maple, the Landau order symbol is not invoked directly as a function. However, it does appear as Maple output to requests for series representations of functions. For example, a request in Maple for the Taylor expansion of $\sin x$ about $x = 0$, namely

$> \mathit{taylor}(\sin(x), x)$

yields the output

$$x - \frac{1}{6}\,x^3 + \frac{1}{120}\,x^5 + O(x^6)$$

where the order notation implicitly assumes that $x \to 0$. Of course, the order term can be replaced by $O(x^7)$ by explicitly requesting the expansion to that order–the default is $O(x^6)$, namely

$> \mathit{taylor}(\sin(x), x, 7)$

which yields

$$x - \frac{1}{6}\,x^3 + \frac{1}{120}\,x^5 + O(x^7)$$

with the coefficient on x^6 explicitly evaluated as zero. The default value of the order in *taylor* when the degree is not specified is given by the *Order* variable. This may be redefined to n using the command

$> \mathit{Order} := n$

or by setting, as above for $n = 7$, the value of n within the *taylor* command.

1.3.4 Asymptotic equivalence

Definition 5. *Let $f(x)$ and $g(x)$ be nonzero in some neighbourhood* *of x_0. Then $f(x)$ and $g(x)$ are said to be asymptotically equivalent as $x \to x_0$ provided that $f(x)\,/\,g(x) \to 1$ as $x \to x_0$.*

Again, x_0 can be—and often is—equal to $\pm\infty$. We shall write

$$f(x) \sim g(x) \qquad \text{as } x \to x_0\,.$$

Also as before, x can be restricted to the natural numbers, for which we write

$$f(n) \sim g(n) \qquad \text{or} \qquad f_n \sim g_n\,,$$

as $n \to \infty$. For example, we may write

$$n \sim n + \ln n \qquad \text{as } n \to \infty\,.$$

We also encountered the concept of asymptotic equivalence earlier with Stirling's approximation in (1.3) which was

$$n! \sim \sqrt{2\,\pi\,n}\,n^n\,e^{-n} \qquad \text{as } n \to \infty\,.$$

Using $\sim$ we can find necessary and sufficient conditions for the convergence of series using the comparison test. As an example, for which values of k does the infinite sum

$$\sum_{n=1}^{\infty} \binom{2\,n}{n}^k 2^{-(2\,k\,n)}$$

converge? To determine this, it is sufficient to use Stirling's approximation to show that

$$\binom{2\,n}{n}^k 2^{-(2\,k\,n)} \sim \left(\frac{1}{\sqrt{\pi\,n}}\right)^k$$

as $n \to \infty$. The comparison test tells us that the series will converge if $\sum n^{-k/2}$ converges and diverge otherwise. This converges if $k > 2$ and diverges otherwise.

1.3.5 The o-notation

Definition 6. *Let $f(x)$ and $g(x)$ be defined in some neighbourhood of x_0, with $g(x)$ nonzero. We write*

$$f(x) = o[\,g(x)\,] \qquad \text{as } x \to x_0$$

whenever $f(x)\,/\,g(x) \to 0$ as $x \to x_0$.

Typically again $x_0 = 0$ or $\pm\infty$, and x may be restricted to the natural numbers.

The o-notation can be used to express asymptotic equivalence. Suppose $f(x)$ and $g(x)$ are nonzero. Then

$$f(x) \sim g(x) \qquad \text{if and only if} \qquad f(x) = g(x)\,[1 + o(1)]\,.$$

It is sometimes useful to write

$$1 + o(1) = e^{o(1)}\,.$$

For example, Stirling's approximation in (1.3) can be written in the alternative form

$$n! = \sqrt{2\,\pi\,n}\,n^n\,e^{-n+o(1)}\,.$$

Two useful formulas are

$$o[\,f(x)\,g(x)\,] = o[\,f(x)\,]\,O[\,g(x)\,] \tag{1.6}$$

and

$$o[\,f(x)\,g(x)\,] = f(x)\,o[\,g(x)\,] \tag{1.7}$$

See Problem 6.

The o-notation is often used in situations where we cannot be or do not wish to be as precise as the O-notation allows. For example, as $x \to 0$ the statements

$$e^x = 1 + x + O(x^2) \qquad \text{and} \qquad e^x = 1 + x + o(x)$$

are both true. However, the first statement is stronger, and implies the second. Nevertheless, to determine a linear approximation to e^x around $x = 0$, the second statement is sufficient for the purpose. While both statements are true for the exponential function, the second statement can be proved more easily, as its verification only requires the value of e^x and the first derivative of e^x at $x = 0$.

For sequences $f(n)$ and $g(n)$, where $n = 1, 2, \ldots$, we may define the o-notation for $n \to \infty$. In this case, we write

$$f(n) = o[\,g(n)\,] \qquad \text{or} \qquad f_n = o[\,g_n\,]$$

provided $f(n)/g(n) \to 0$ (respectively, $f_n/g_n \to 0$) as $n \to \infty$.

1.3.6 The O_p-notation

For stochastic functions and sequences, the Landau notation is limited in its utility because random functions are often not bounded with probability one. So we must replace the concept of boundedness with the concept of boundedness in probability.

Let X_s, $s \in S$ be a family of random variables indexed by $s \in S$. We say that X_s, $s \in S$ is *bounded in probability* if for all $\epsilon > 0$ there exists some $\alpha > 0$ such that

$$P(|\,X_s\,| \le \alpha) \ge 1 - \epsilon$$

for all $s \in S$.

Definition 7. *If X_s, Y_s, $s \in S$ are two indexed families of random variables, with $P(Y_s = 0) = 0$ for all s. We write*

$$X_s = O_p(Y_s) \qquad \text{for } s \in S$$

when the ratio X_s/Y_s is bounded in probability.

In particular, if $g(s)$ is a deterministic nonvanishing function, we shall write

$$X_s = O_p[\,g(s)\,] \qquad \text{for } s \in S$$

provided $X_s/g(s)$ is bounded in probability.

Our most important application is to a sequence X_n of random variables. An infinite sequence of random variables is bounded in probability if it is bounded in probability at infinity. See Problem 7. Therefore, we write

$$X_n = O_p[\,g(n)\,] \qquad \text{as } n \to \infty$$

provided $X_n/g(n)$ is bounded in probability.

1.3.7 The o_p-notation

There is also a stochastic version of the o-notation.

Definition 8. *We write*

$$X_n = o_p(Y_n) \qquad \text{as } n \to \infty$$

whenever, for all $\epsilon > 0$,

$$P\left(\left|\frac{X_n}{Y_n}\right| \ge \epsilon\right) \to 0$$

as $n \to \infty$.

This notation can be applied when Y_n is replaced by a nonrandom function $g(n)$. In this case, we write $X_n = o[\,g(n)\,]$. In particular, $X_n = o_p(1)$ if and only if $P(|X_n| \ge \epsilon) \to 0$ for all $\epsilon > 0$. This is a special case of convergence in probability, as defined below.

1.3.8 Modes of convergence

Some of the main modes of convergence for a sequence of random variables are listed in the following definition.

Definition 9. *Let $X_n, n \geq 1$ be a sequence of random variables.*

1. *The sequence X_n, $n \geq 1$ converges to a random variable X almost surely if*
$$P\left(\lim_{n\to\infty} X_n = X\right) = 1\,.$$
We write $X_n \stackrel{a.s.}{\to} X$.
2. *The sequence X_n, $n \geq 1$ converges to a random variable X in probability if for all $\epsilon > 0$, $P(|X_n - X| \geq \epsilon) \to 0$ as $n \to \infty$. We write $X_n \stackrel{P}{\to} X$. A sequence X_n converges in probability to X if and only if $X_n = X + o_p(1)$.*
3. *Let p be any positive real number. We say that X_n converges in L_p to X whenever*
$$E\,(|X_n - X|^p) \to 0$$
as $n \to \infty$. We write $X_n \stackrel{L_p}{\to} X$.
4. *Let X_n have distribution function $F_n(t) = P(X_n \leq t)$, and let X have distribution function $F(t)$. Then X_n converges to X in distribution provided $F_n(t) \to F(t)$ for every value t at which F is continuous. When X_n converges in distribution to X we write $X_n \stackrel{d}{\Longrightarrow} X$.*

Various implications can be drawn between these modes of convergence.

Proposition 3. *The following results can be proved.*

1. *If $X_n \stackrel{a.s.}{\to} X$ then $X_n \stackrel{P}{\to} X$.*
2. *If $X_n \stackrel{P}{\to} X$ then $X_n \stackrel{d}{\Longrightarrow} X$.*
3. *If $X_n \stackrel{L_p}{\to} X$ then $X_n \stackrel{P}{\to} X$.*
4. *It follows from the statements above that either convergence almost surely or convergence in L_p to X implies convergence in distribution.*
5. *In the case of a constant limit, convergence in probability and convergence in distribution are equivalent. That is, $X_n \stackrel{d}{\Longrightarrow} c$ if and only if $X_n \stackrel{P}{\to} c$.*

The proofs of these statements are omitted. Two useful results about convergence in distribution are the following, which we state without proof.

Proposition 4. *Let $g(x)$ be a continous real-valued function of a real variable. Then $X_n \stackrel{d}{\Longrightarrow} X$ implies that $g(X_n) \stackrel{d}{\Longrightarrow} g(X)$.*

Proposition 5 (Slutsky's theorem). *Suppose $X_n \stackrel{d}{\Longrightarrow} X$ and $Y_n \stackrel{P}{\to} c$. Then*

1. $X_n + Y_n \stackrel{d}{\Longrightarrow} X + c$, *and*
2. $X_n Y_n \stackrel{d}{\Longrightarrow} cX$.

Slutsky's theorem is particularly useful when combined with the central limit theorem, which is stated in Section 1.3.10 below in a version due to Lindeberg and Feller.

1.3.9 The law of large numbers

Laws of large numbers are often divided into strong and weak forms. We begin with a standard version of the *strong law of large numbers.*

Proposition 6. *Let $X_1, X_2, \ldots$ be independent, identically distributed random variables with mean $E(X_j) = \mu$. Let $\overline{X}_n = n^{-1}(X_1 + \cdots + X_n)$. Then $\overline{X}_n$ converges almost surely to the mean μ as $n \to \infty$:*

$$\overline{X}_n \stackrel{a.s.}{\to} \mu \tag{1.8}$$

as $n \to \infty$.

Convergence almost surely implies convergence in probability. Therefore, we may also conclude that

$$\overline{X}_n \stackrel{P}{\to} \mu \tag{1.9}$$

This is the *weak law of large numbers.* This conclusion can be obtained by assumptions that may hold when the assumptions of the strong law fail. For example, the weak law of large numbers will be true whenever $\text{Var}(\overline{X}_n) \to 0$. The weak law comes in handy when random variables are either dependent or not identically distributed. The most basic version of the weak law of large numbers is proved in Problems 9–11.

1.3.10 The Lindeberg-Feller central limit theorem

Let X_1, X_2, ... be independent random variables with distribution functions F_1, F_2, ..., respectively. Suppose that

$$E(X_j) = 0, \qquad \mathrm{Var}(X_j) = \sigma_j^2 .$$

Let $s_n^2 = \sum_{j=1}^n \sigma_j^2$.

Proposition 7. *Assume the* Lindeberg condition, *which states that for every* $t > 0$,

$$s_n^{-2} \sum_{j=1}^n \int_{\{\,|y| \ge t\, s_n\,\}} y^2 \, dF_j(y) \;\to\; 0 \tag{1.10}$$

as $n \to \infty$. *Then the standardised partial sums converge to normal, namely*

$$\frac{X_1 + X_2 + \cdots + X_n}{s_n} \overset{d}{\Longrightarrow} \mathcal{N}(0,\, 1)\,. \tag{1.11}$$

as $n \to \infty$. *That is,*

$$P\,(X_1 + \cdots + X_n \le t\, s_n) \to \int_{-\infty}^{t} \frac{1}{\sqrt{2\,\pi}}\, e^{-z^2/2}\, dz$$

as $n \to \infty$.

The best known special case where the Lindeberg condition is satisfied occurs when the random variables are identically distributed so that $F_j = F$, and $\sigma_j^2 = \sigma^2$ for all j. Then $s_n^2 = n\,\sigma^2$, and the Lindeberg condition reduces to

$$\int_{\{\,|y| \ge t\,\sigma\,\sqrt{n}\,\}} y^2 \, dF(y) \;\to\; 0$$

which is satisfied for all $t > 0$. This special case is often proved on its own using generating functions. See Problems 12–15.

1.4 Two complementary approaches

With the advent of modern computing, the analyst has often been on the defensive, and has had to justify the relevance of his or her discipline in the face of the escalating power of successive generations of computers. Does a statistician need to compute an asymptotic property of a statistic if a quick simulation can provide an excellent approximation? The traditional answer to this question is that analysis fills in the gaps where the computer has trouble. For example, in his excellent 1958 monograph on asymptotic methods, N. G. de Bruijn considered an imaginary dialogue between a numerical analyst (NA) and an asymptotic analyst (AA).

- The NA wishes to know the value of $f(100)$ with an error of at most 1%.
- The AA responds that $f(x) = x^{-1} + O(x^{-2})$ as $x \to \infty$.
- But the NA questions the error term in this result. Exactly what kind of error is implied in the term $O(x^{-2})$? Can we be sure that this error is small for $x = 100$? The AA provides a bound on the error term, which turns out to be far bigger than the 1% error desired by the NA.
- In frustration, the NA turns to the computer, and computes the value of $f(100)$ to 20 decimal places!
- However, the next day, she wishes to compute the value of $f(1000)$, and finds that the resulting computation will require a month of work at top speed on her computer! She returns to the AA and "gets a satisfactory reply."

For all the virtues of this argument, it cannot be accepted as sufficient justification for the use of asymptotics in statistics or elsewhere. Rather, our working principle shall be the following.

> A primary goal of asymptotic analysis is to obtain a deeper *qualitative* understanding of *quantitative* tools. The conclusions of an asymptotic analysis often supplement the conclusions which can be obtained by numerical methods.

Thus numerical and asymptotic analysis are partners, not antagonists. Indeed, many numerical techniques, such as Monte Carlo, are motivated and justified by theoretical tools in analysis, including asymptotic results such as the law of large numbers and the central limit theorem. When coupled with numerical methods, asymptotics becomes a powerful way to obtain a better understanding of the functions which arise in probability and statistics. Asymptotic answers to questions will usually provide incomplete descriptions of the behaviour of functions, be they estimators, tests or functionals on distributions. But they are part of the picture, with an indispensable role in understanding the nature of statistical tools.

With the advent of computer algebra software (CAS), the relationship between the computer on one side and the human being on the other side has changed. Previously, the human being excelled at analysis and the computer at number crunching. The fact that computers can now manipulate complex formulas with greater ease than humans is not to be seen as a threat but rather as an invaluable assistance with the more tedious parts of any analysis. I have chosen Maple as the CAS of this book. But another choice of CAS might well have been made, with only a minor modification of the coding of the examples.

1.5 Problems

1. Solve Problem 64 from the Rhind papyrus as stated in Section 1.

2.(a) Use Stirling's approximation to prove that

$$\binom{n}{\frac{n}{2} + x\,\frac{\sqrt{n}}{2}} \sim 2^n \sqrt{\frac{2}{\pi\, n}}\; e^{-x^2/2}\,.$$

(b) Prove that as $n \to \infty$,

$$\binom{2\,n}{n} \sim \frac{4^n}{\sqrt{n\,\pi}} \cdot \left[1 - \frac{1}{8\,n} + \frac{1}{128\,n^2} + O\left(\frac{1}{n^3}\right)\right].$$

3. Suppose that A is an infinite set of real numbers that is bounded above and below. Prove that the set B of almost upper bounds of A is nonempty and bounded below. Similarly, prove that the set C of almost lower bounds of A is nonempty and bounded above.

4. For the sequence $x_n = n^{-1}$, find $\liminf x_n$ and $\limsup x_n$.

5. Prove Proposition 2.

6. Prove (1.6) and (1.7).

7. Suppose S is a finite set, and that X_s, $s \in S$ is a family of random variables indexed by the elements of S.

   (a) Prove that X_s, $s \in S$ is bounded in probability.

   (b) Prove that a sequence X_n, $n \geq 1$ is bounded in probability if and only if it is bounded in probability at infinity. That is, there is some n_0 such that X_n, $n \geq n_0$ is bounded in probability.

8. Let $X_n \stackrel{d}{\Longrightarrow} X$ and $Y_n \stackrel{P}{\rightarrow} c$, where $c \neq 0$. Use Propositions 4 and 5 to prove that $X_n/Y_n \stackrel{d}{\Longrightarrow} X/c$.

9. *The following three questions are concerned with a proof of the weak law of large numbers using Markov's and Chebyshev's inequalities.* Suppose $P(X \geq 0) = 1$. Prove Markov's inequality, which states that

$$P(X \geq \epsilon) \leq \frac{E(X)}{\epsilon}$$

for all $\epsilon > 0$. (Hint: write $X = X\,Y + X\,(1 - Y)$, where $Y = 1$ when $X \geq \epsilon$ and $Y = 0$ when $X < \epsilon$. Then prove that $E(X) \geq E(X\,Y) \geq \epsilon\,P(Y = 1)$.)

10. Suppose X has mean μ and variance σ^2. Replace X by $(X - \mu)^2$ in Markov's inequality to prove Chebyshev's inequality, which states that when $a > 0$,

$$P\left(|X - \mu| \geq a\right) \leq \frac{\sigma^2}{a^2}\,.$$

11. Let X_n, $n \geq 1$ be independent, identically distributed random variables with mean μ and variance σ^2. Let

$$\overline{X}_n = \frac{X_1 + \cdots + X_n}{n}\,.$$

Use Chebyshev's inequality to prove that $\overline{X}_n \stackrel{P}{\rightarrow} \mu$ as $n \rightarrow \infty$.

12. *The next two questions are concerned with a proof of the most basic form of the central limit theorem using moment generating functions.* Let $X_1, \ldots, X_n$ be a random sample from a distribution with mean μ, variance σ^2 and moment generating function $M(t) = E\,e^{t\,X_1}$. Let

$$\bar{X}_n = \frac{X_1 + \cdots + X_n}{n}$$

denote the sample average. Show that the moment generating function of $\sqrt{n}\,(\bar{X}_n - \mu)$ is

$$e^{-\sqrt{n}\,\mu\,t}\,[M(t/\sqrt{n})]^n.$$

13. Suppose that the moment generating function $M(t)$ from the previous question is finite in an open interval about $t = 0$. Use the fact that $M(t)$ is differentiable at zero to all orders to expand the function

$K(t) = \ln[M(t)]$ about $t = 0$. Show that the logarithm of moment generating function of $\sqrt{n}\,(\bar{X}_n - \mu)$ can be expressed in the form

$$-\sqrt{n}\,t + n\,K(t/\sqrt{n}) = \frac{\sigma^2 t^2}{2} + \frac{K'''(\xi)\,t^3}{6\,\sqrt{n}},$$

where ξ lies strictly between 0 and t. Use this formula to verify that $\sqrt{n}\,(\bar{X}_n - \mu)$ converges in distribution to $\mathcal{N}(0, \sigma^2)$ as $n \to \infty$.

14. In the notation of Lindeberg-Feller central limit theorem, suppose that the random variables X_n are uniformly bounded in the sense that there exists a c such that $P(-c \le X_n \le c) = 1$ for all $n \ge 1$. Suppose also that $s_n \to \infty$. Prove that the Lindeberg condition is satisfied.

15. Suppose that Y_n, $n \ge 1$ are independent, identically distributed random variables with mean zero and variance σ^2. Let $X_n = n\,Y_n$. Prove that the Lindeberg condition is satisfied for X_n, $n \ge 1$.

16. One of the most famous limits in mathematics is

$$\lim_{n\to\infty}\left(1 + \frac{x}{n}\right)^n = e^x.$$

Verify the asymptotic formula

$$\left(1 + \frac{x}{n}\right)^n = e^x\left\{1 - \frac{x^2}{2n} + \frac{x^3(3x+8)}{24n^2} - \frac{x^4(x+2)(x+6)}{48n^3} + O\left(\frac{1}{n^4}\right)\right\}$$

out to the order shown. To check this expansion in Maple, try the command

> $asympt\left(\left(1 + \frac{x}{n}\right)^n, n, 4\right)$

17. Let $A(t) = E(t^X)$ denote the probability generating function for a random variable X with distribution $\mathcal{P}(\mu)$, *i.e.*, Poisson with mean μ. Let $A_n(t)$ denote the probability generating function for a random variable X_n whose distribution is $\mathcal{B}(n, \mu/n)$, *i.e.*, binomial with parameters n and μ/n (where clearly $n \ge \mu$).

(a) Prove that

$$A_n(t) = A(t)\left\{1 - \frac{\mu^2(t-1)^2}{2n} + O\left(\frac{1}{n^2}\right)\right\}.$$

as $n \to \infty$.

(b) Show that

$$P(X_n = k) = P(X = k)\left\{1 + \frac{k - (\mu - k)^2}{2\,n} + O\left(\frac{1}{n^2}\right)\right\}.$$

(c) Argue that $\mathcal{B}(n, \mu/n)$ converges in distribution to $\mathcal{P}(\mu)$ as $n \to \infty$.

(d) Using the next term of order n^{-1} in the expansion on the right-hand side, argue that as $n \to \infty$,

$$P(X_n = k) > P(X = k)$$

when k is close to the mean μ, and that

$$P(X_n = k) < P(X = k)$$

for values of k further away from the mean.

18. In their textbook on mathematical statistics, Peter Bickel and Kjell Doksum¶ declare that

> Asymptotics has another important function beyond suggesting numerical approximations .... If they are simple, asymptotic formulae suggest qualitative properties that may hold even if the approximation itself is not adequate.

What is meant by this remark?

¶ See Bickel and Doksum, *Mathematical Statistics, Vol. 1*, 2nd edition, Prentice Hall, Upper Saddle River 2001, p. 300.

CHAPTER 2

General series methods

2.1 A quick overview

By an infinite series we shall mean an expression of the form

$$a_0 + a_1 + a_2 + a_3 + a_4 + \cdots ,$$

where each term a_j is a real (or occasionally a complex) number. We also write the series in the form $\sum_{j=0}^{\infty} a_j$ or as $\sum a_j$ when the limits of the summation are clear from the context. (The range for the index j may also be the strictly positive integers rather than the nonnegative integers. The context dictates which choice or any other for that matter is the most natural.) The sequence

$$s_0 = a_0,\ s_1 = a_0 + a_1,\ s_2 = a_0 + a_1 + a_2,\ \ldots$$

is called the sequence of partial sums of the series. When the sequence of partial sums has a limit $s = \lim_n s_n$ as $n \to \infty$, we say that the infinite series is *convergent* or *summable*, that its sum is s, or that it converges to s. If the partial sums do not have a limit, the series is said to be *divergent*. Divergent series are often subdivided into two types, namely *properly divergent* series where $\lim_n s_n = \infty$ or $\lim_n s_n = -\infty$, and *oscillatory* series where $\lim_n s_n$ does not exist, even among the extended reals.

In most cases, the series that we shall consider will arise as formal expansions of functions, and will be infinite sums of functions. By a *function series* we mean an infinite series of the form

$$\sum_{j=0}^{\infty} a_j(x) = a_0(x) + a_1(x) + a_2(x) + \cdots$$

for some sequence of functions $\{a_j(x)\}$. Letting $s_n(x) = \sum_{j=0}^{n} a_j(x)$, then the function series converges to $s(x)$ where $s(x) = \lim_n s_n(x)$ for all x for which the limit is defined. The set of all x such that the limit exists is called the *domain of convergence* of the series. Typically, the terms of the series will be drawn from a family of functions of a particular type. For example, series of the form

$$a_0 + a_1\,(x-c) + a_2\,(x-c)^2 + a_3\,(x-c)^3 + \cdots$$

are *power series* about c. An extended family of mathematical relatives of power series are the series expansions into *orthogonal polynomials* which have the form

$$a_0 + a_1\,p_1(x) + a_2\,p_2(x) + a_3\,p_3(x) + \cdots$$

where $p_j(x)$ is a polynomial of degree j. The particular orthogonality condition used to define the polynomials will vary from application to application, but can often be written in the form $E\,[p_j(X)\,p_k(X)] = \delta_{jk}$ where X is a random variable with given distribution, and δ_{jk} is the Dirac delta which equals one or zero as $j = k$ or $j \neq k$, respectively.

Another important type of series is the asymptotic series, which has the form

$$a_0 + \frac{a_1}{x} + \frac{a_2}{x^2} + \frac{a_3}{x^3} + \cdots$$

which is like a power series, but written out in terms of nonpositive powers of x. We shall encounter the full definition of an asymptotic series in Section 2.4.

Both asymptotic series and power series provide a sequence of approximations where each approximant is a rational function. The sequence $s_n(x)$ of approximants in each case can be thought of as a family of functions indexed by a nonnegative integer value n. The family of rational approximants to a function is naturally doubly indexed in the sense that for any pair of nonnegative integers (m, n) we might seek a rational approximant of the form $p(x)/q(x)$, where deg $p = m$ and deg $q = n$. The theory of Padé approximants provides such an approximation with the property that the resulting rational function, upon expansion in ascending powers of x, agrees with the power series to $m + n + 1$ terms or more. This will be the subject of the next chapter.

2.2 Power series

2.2.1 Basic properties

One of the basic ideas of a function series is to expand a function whose properties may be complex or unknown in terms of functions which are well understood and have desirable properties. For example, power series are ways of writing functions as combinations of functions of the form $(x-c)^n$. Such powers have algebraically attractive properties. Since $(x-c)^n\,(x-c)^m = (x-c)^{n+m}$ we may quickly organise the terms of the product of two such series into a series of the same kind. Moreover, the term-by-term derivative of a power series or the term-by-term integral is also a power series.

Suppose that $\sum_{n=0}^{\infty} a_n\,(x-c)^n$ is a power series. We define the *radius of convergence* to be

$$r = \lim_{n\to\infty} \left| \frac{a_n}{a_{n+1}} \right| \tag{2.1}$$

when this limit exists, or, more generally, by the *Cauchy-Hadamard formula*

$$r = \frac{1}{\limsup_{n\to\infty} \sqrt[n]{|\,a_n\,|}}\,. \tag{2.2}$$

when the limit does not exist.* Note that both $r = 0$ and $r = \infty$ are acceptable values. We observe the following.

1. When $|x-c| < r$ the power series converges. When $|x-c| > r$, the series diverges. At $x = c \pm r$, no general conclusion can be made. For this reason, the open interval $(c-r,\, c+r)$ is called the *interval of convergence*.

2. Inside the interval of convergence, the power series can be differentiated or integrated term by term. That is

$$\frac{d}{dx}\sum_{n=0}^{\infty} a_n\,(x-c)^n = \sum_{n=1}^{\infty} n\,a_n\,(x-c)^{n-1}\,, \tag{2.3}$$

   and

$$\int_c^y \sum_{n=0}^{\infty} a_n\,(x-c)^n\,dx = \sum_{n=0}^{\infty} \frac{a_n}{n+1}\,(y-c)^{n+1}\,. \tag{2.4}$$

3. The interval of convergence of the power series that results from differentiating or integrating is the same as that of the original series.

4. Within the interval of convergence, let the power series converge to a function $f(x)$. Then $a_n = f^{(n)}(c)/n!$ for all n. With the coefficients expressed in this form, the resulting series is called the *Taylor series* for the function $f(x)$. The Taylor series representation of a power series is named after the mathematician *Brook Taylor*.

5. If $\sum a_n\,(x-c)^n = \sum b_n\,(x-c)^n$ for all x in some nonempty open interval, then $a_n = b_n$ for all n.

* While the limit of a sequence of numbers may not exist, the limit superior–the largest cluster point of the sequence–always exists. See Chapter 1 for a brief introduction to limits superior and inferior. For more information, the reader is referred to Spivak (1994).

6. Any function $f(x)$ which can be represented as a power series about the value c for all x in some open interval containing c, is said to be *analytic* at c. It can be shown that a function which is analytic at some point is also analytic in some open interval that contains that point.

7. A function which is analytic at every real value c is said to be an *entire* function.

8. However, to be analytic–*i.e.*, to have such a power series representation at a point–requires more than the existence of a Taylor series about that point. If $f(x)$ is infinitely differentiable at c, the Taylor series for $f(x)$ is defined about the point c. Nevertheless, this series may not converge except for the trivial case where $x = c$. Even if the Taylor series converges for all x in some open interval containing c, then it may not converge to the function $f(x)$, but rather to some other function $g(x)$, which is analytic at c. At $x = c$, we will have $f^{(n)}(c) = g^{(n)}(c)$, for all n. However, the derivatives need not agree elsewhere.

Many standard functions have power series representations. To find the Taylor expansion for a function $f(x)$ in Maple we can use the *taylor* command. For example,

$> taylor(f(x), x = c, n)$

produces the Taylor expansion of $f(x)$ about $x = c$, out to order n. Thus

$> taylor(exp(x), x = 0, 5)$

yields

$$1 + x + \frac{1}{2}x^2 + \frac{1}{6}x^3 + \frac{1}{24}x^4 + O(x^5)$$

where the order term at the end is for the limit as $x \to 0$. More generally for expansions about $x = c$, the final order term will be of the form $O((x - c)^n)$ as $x \to c$. It is possible to obtain a coefficient from a power series directly using the *coeftayl* command. For example, the command

$> coeftayl(f(x), x = c, n)$

provides the coefficient on the term with $(x-c)^n$ in the Taylor expansion of a function or expression $f(x)$ about $x = c$.

In addition to this command, which does not require that any special package be invoked, there are a number of commands for the manipulation of power series in the *powseries* package. This package can be called using the command

Brook Taylor (1685–1731)

In addition to his famous series, Brook Taylor worked on finite differences and the mathematical theory of perspective.

> "A study of Brook Taylor's life and work reveals that his contribution to the development of mathematics was substantially greater than the attachment of his name to one theorem would suggest."
>
> Philip S. Jones, *Dictionary of Scientific Biography*, Vol. 13, p. 267.

> *with(powseries)*

whenever such manipulations are required. We shall consider this package in greater detail in Section 2.2.6.

2.2.2 Normal distribution function

To illustrate these ideas, we consider the power series representation for the normal distribution function. Expanding in a power series about zero, we have

$$\begin{aligned}\exp\left(-z^2/2\right) &= 1-\frac{z^2}{2}+\frac{1}{2!}\left(\frac{z^2}{2}\right)^2-\frac{1}{3!}\left(\frac{z^2}{2}\right)^3+\cdots \\ &= 1-\frac{z^2}{2}+\frac{z^4}{8}-\frac{z^6}{48}+\cdots. \qquad (2.5)\end{aligned}$$

Here, the coefficients have the form

$$a_{2n}=(-1)^n[n!\,2^n]^{-1} \text{ and } a_{2n+1}=0\,.$$

Applying the Cauchy-Hadamard formula, we find that the radius of convergence of this series is

$$\begin{aligned}r &= \lim_{n\to\infty} \sqrt[2n]{n!\,2^n} \\ &= \infty\,. \qquad (2.6)\end{aligned}$$

Thus the series converges for all z.

As mentioned above in item 2 of Section 2.2.1, we may freely integrate term by term within the radius of convergence. Integrating term by term, we obtain an expansion for the distribution function, namely

$$\begin{aligned}\Phi(x) &= \frac{1}{2}+\int_0^x \frac{1}{\sqrt{2\,\pi}}\exp\left(-\frac{z^2}{2}\right)\,dz \\ &= \frac{1}{2}+\frac{1}{\sqrt{2\pi}}\left(x-\frac{x^3}{3\cdot 2}+\frac{x^5}{5\cdot 2!\cdot 2^2}-\frac{x^7}{7\cdot 3!\cdot 2^3}+\cdots\right) \qquad (2.7)\end{aligned}$$

As we are integrating within the radius of convergence about zero, this series also converges for all real values. For small values of $|x|$, particularly those less than one, the convergence of this series is quite rapid with a small number of terms. However, when $|x|$ is large, the convergence can be quite slow, at least initially. Later in this chapter, we will consider a different expansion which, in contrast, is particularly accurate when $|x|$ is large, and poor when $|x|$ is small.

2.2.3 Generating functions and power families

Let X be a nonnegative integer-valued random variable. For each $n = 0, 1, 2, \ldots$, define $\pi_n = P(X = n)$. We define the *probability generating function* of X to be

$$\begin{aligned} A(\theta) &= E\left(\theta^X\right) \\ &= \pi_0 + \pi_1\,\theta + \pi_2\,\theta^2 + \pi_3\,\theta^3 + \cdots, \end{aligned} \tag{2.8}$$

for all $-r < \theta < r$, where r is the radius of convergence of the power series. Since $\{\pi_n\,;\ n = 0, 1, 2, \ldots\}$ are probability weights, it can be seen from the Cauchy-Hadamard formula that $r \geq 1$.

Many of the standard properties of probability generating functions follow easily from their representations as power series in θ. For example, we may differentiate term by term at $\theta = 0$ within the radius of convergence to obtain

$$\pi_n = \frac{A^{(n)}(0)}{n!}\,.$$

Provided $r > 1$, we may differentiate the power series term by term at $\theta = 1$ to obtain $E(X) = A'(1)$. This result generalises to the case where $r = 1$, for which the more general formula

$$E(X) = \lim_{\theta \to 1-} A'(\theta)$$

can be shown. This equation is always valid provided that infinite values are admitted on both sides.

It is possible to extend the distribution of X to a family of distributions indexed by $\theta \in [0, r)$ called a *power family* as follows. For $0 \leq \theta < r$, we define

$$\begin{aligned} \pi_n(\theta) &= \frac{P(X = n)\,\theta^n}{A(\theta)} \\ &= \frac{\pi_n\,\theta^n}{A(\theta)}\,, \qquad n = 0, 1, 2, \ldots. \end{aligned}$$

It is easy to check that $\{\pi_n(\theta)\}$ is a set of probability weights for every $0 < \theta < r$. The family of distributions on the nonnegative integers so obtained is indexed by the parameter θ, and is called the *power family* of distributions associated with X. The original distribution of X can be recovered as the distribution of the power family with the parameter value $\theta = 1$.

To understand the geometrical features of such power family distributions, it is helpful to introduce a new set of coefficients. Let us write

$$A(\theta) = \pi_0 \left[1 + \frac{\theta}{\rho_1} + \frac{\theta^2}{\rho_1\,\rho_2} + \cdots + \frac{\theta^n}{\rho_1\,\rho_2 \cdots \rho_n} + \cdots\right]. \tag{2.9}$$

With this representation, a number of geometric and parametric features of the power family distributions can be determined.

1. The Cauchy-Hadamard formula for the radius of convergence of $A(\theta)$ tells us that this is determined by the limiting behaviour of the geometric mean of ρ_n, $n \geq 1$. In particular, if $\rho_n \to r$, then r is the radius of convergence. More generally,

$$r = \liminf_{n \to \infty} \sqrt[n]{\rho_1 \, \rho_2 \, \cdots \, \rho_n} \, .$$

2. If $\rho_n = c^{-1} n, n = 1, 2, \ldots$, for some value of $c > 0$, then the distribution is *Poisson* with mean $\mu = c\,\theta$.

3. If $\rho_n = c^{-1}, n = 1, 2, \ldots$, where $c < 1$, then the distribution is *geometric* with mean $c\theta/(1 - c\theta)$.

4. If ρ_n is a strictly increasing function of n, then the distribution is *unimodal*, with mode at $\nu = \nu(\theta)$ such that $\rho_\nu < \theta < \rho_{\nu+1}$. In the special case where $0 < \theta < \rho_1$, the mode is at $\nu = 0$.

5. Under the conditions above, the mode ν is a nondecreasing function of θ.

6. Certain results relate the position $\nu = \nu(\theta)$ of the mode to the probability weight $\widehat{\pi}(\theta) = \pi_\nu(\theta)$ of the mode. In particular, under the conditions of remark 3, above, we find that

$$\ln \widehat{\pi}(\theta) = \int_0^\theta \frac{\nu(\theta)}{t} \, dt + \ln \pi_0(\theta) \, , \tag{2.10}$$

provided that $\pi_0(\theta) \neq 0$. Under the conditions of remark 3, and the additional condition that the radius of convergence of $A(\theta)$ is infinite, we have

$$\limsup_{\theta \to \infty} \frac{\ln \nu(\theta)}{\ln \theta} = \limsup_{\theta \to \infty} \frac{\ln \ln \widehat{\pi}(\theta)}{\ln \theta} \, . \tag{2.11}$$

2.2.4 The noncentral chi-square and F distributions

Let $X_1, \ldots, X_n$ be independent normal random variables, all with unit variance, but with nonzero means $\mu_1, \ldots, \mu_n$, respectively. Define $\delta = \sum_{j=1}^n \mu_j^2$. Define also

$$U = \sum_{j=1}^n X_j^2 \, . \tag{2.12}$$

We seek the distribution of U.

Consider an orthogonal $n \times n$ matrix (a_{jk}), and the transformation

$$Y_j = \sum_{k=1}^{n} a_{jk}\, X_k\,, \quad j = 1, \ldots, n\,.$$

Since the X_j are independent normal random variables, and the linear transformation is orthogonal, it follows that $Y_1, \ldots, Y_n$ are also independent normal random variables, also having unit variances.

Of course, there is considerable choice in the coefficients of the orthogonal transformation. Let us choose a transformation such that

$$E(Y_1) = \cdots = E(Y_{n-1}) = 0\,.$$

Let us set $\nu = E(Y_n)$. Since the linear transformation is orthogonal, it follows that $\sum_j X_j^2 = \sum_j Y_j^2$. It is left to the reader to check that

$$\nu = \sqrt{\mu_1^2 + \mu_2^2 + \cdots + \mu_n^2}\,. \tag{2.13}$$

If we define

$$V = \sum_{j=1}^{n-1} Y_j^2\,, \quad W = Y_n^2\,, \tag{2.14}$$

then V has a chi-square distribution with $n-1$ degrees of freedom. So V has density function

$$g(v) \propto v^{(n-3)/2} \exp(-v/2)\,, \tag{2.15}$$

namely the $\mathcal{G}((n-1)/2,\, 1/2)$ density. Here the constant of integration is ignored for the time being. Since Y_n is $\mathcal{N}(\nu,\, 1)$, we can easily determine the density function of $W = Y_n^2$ to be

$$h(w) \propto \frac{1}{2\sqrt{w}} \left\{ \exp\left[-\frac{1}{2}\,(\sqrt{w} - \sqrt{\nu})^2\right] + \exp\left[-\frac{1}{2}\,(-\sqrt{w} - \sqrt{\nu})^2\right]\right\}, \tag{2.16}$$

for $w > 0$, again ignoring the constant of integration for the moment.

To find the density function for U, we introduce the change of variables $U = V + W$ and $R = V/(V + W)$, or

$$\begin{aligned} V &= U\,R\,, \\ W &= U\,(1 - R)\,. \end{aligned} \tag{2.17}$$

The absolute value of the Jacobian of the latter transformation (2.17) is U, so that the desired marginal density f(u) of U is found by integrating the joint density of U and R to obtain

$$f(u) \propto \int_0^1 g\,(u\,r)\; h\,(u - u\,r)\; u\,dr\,. \tag{2.18}$$

Unfortunately, the integral in (2.18) cannot be evaluated in simple form. Therefore, we turn to an alternative power series representation of $h(w)$.

Returning to formula (2.16) for the density of W, we can expand the exponentials in power series to obtain

$$h(w) \propto w^{-\frac{1}{2}} \exp\left[-\frac{1}{2}(w+\nu)\right] \sum_{j=0}^{\infty} \frac{(w\,\nu)^j}{(2\,j)!}. \tag{2.19}$$

Inserting (2.19) into (2.18) gives us

$$\begin{aligned} f(u) &\propto \int_0^1 \exp\left(-\frac{u+\nu}{2}\right) u^{\frac{n}{2}-1} r^{\frac{n}{2}-\frac{3}{2}} (1-r)^{-\frac{1}{2}} \sum_{j=0}^{\infty} \frac{u^j (1-r)^j \nu^j}{(2\,j)!}\, dr \\ &= \exp\left(-\frac{u+\nu}{2}\right) u^{\frac{n}{2}-1} \sum_{j=0}^{\infty} \frac{u^j \nu^j}{(2\,j)!} \int_0^1 r^{\frac{n}{2}-\frac{3}{2}} (1-r)^{j-\frac{1}{2}}\, dr \\ &= \exp\left(-\frac{u+\nu}{2}\right) u^{\frac{n}{2}-1} \sum_{j=0}^{\infty} \frac{u^j \nu^j}{(2\,j)!} \frac{\Gamma\left(\frac{n-1}{2}\right) \Gamma\left(j+\frac{1}{2}\right)}{\Gamma\left(\frac{n}{2}+j\right)}. \end{aligned} \tag{2.20}$$

It remains to determine the constant of proportionality. Note that the parameter ν only appears in the formula for $f(u)$ through the density $h(w)$ given in (2.16). However, in (2.16) the constant of proportionality does not depend upon ν. It follows therefore that the constant of proportionality in (2.20) also does not depend upon ν. So we can determine the constant by setting $\nu = 0$, where the distribution of U is known to be precisely chi-square with n degrees of freedom. Plugging this constant in, we obtain

$$\begin{aligned} f(u) &= 2^{-\frac{n}{2}} \exp\left(-\frac{u+\nu}{2}\right) u^{\frac{n}{2}-1} \sum_{j=0}^{\infty} \frac{u^j \nu^j}{(2\,j)!} \frac{\Gamma\left(j+\frac{1}{2}\right)}{\Gamma\left(\frac{1}{2}\right) \Gamma\left(\frac{n}{2}+j\right)} \\ &= 2^{-\frac{n}{2}} \exp\left(-\frac{u+\nu}{2}\right) u^{\frac{n}{2}-1} \sum_{j=0}^{\infty} \frac{(u\,\nu)^j}{2^{2\,j}\, j!\, \Gamma\left(\frac{n}{2}+j\right)}. \end{aligned} \tag{2.21}$$

It can be checked that the power series in $u\,\nu$ in formula (2.21) converges for all values of u and ν.

The random variable U is said to have a *noncentral chi-square distribution* with n degrees of freedom and noncentrality parameter ν. We shall write this distribution using the notation $\mathcal{X}(n, \nu)$. In the particular case $\nu = 0$, formula (2.21) reduces to the usual formula for the density function of the chi-square distribution with n degrees of freedom, which we write as $\mathcal{X}(n)$. The noncentral chi-square distribution can arise in hypothesis testing when the asymptotic distribution of a test statistic is governed by some general parameter value under the alternative hypothesis. In this case, the null hypothesis is usually associated with the value $\nu = 0$, where the test statistic has an asymptotic chi-square distribution.

Also of fundamental importance in hypothesis testing is the noncentral version of the F distribution which we shall define next. Suppose that U_1 and U_2 are independent noncentral chi-square random variables with degrees of freedom n_1 and n_2, and noncentrality parameters ν_1 and ν_2, respectively. We say that the random variable

$$S = \frac{U_1/n_1}{U_2/n_2} \tag{2.22}$$

has a *noncentral F distribution* with (n_1, n_2) degrees of freedom, and noncentrality parameters (ν_1, ν_2).

Using methods similar to those leading to formula (2.21), it can be shown that the random variable S has density function

$$f(s) = (n_2 + n_1 s)^{j+k-(n_1+n_2)/2} \times$$

$$\sum_{j=0}^{\infty} \sum_{k=0}^{\infty} \frac{\Gamma\left(\frac{n_1+n_2}{2} + j + k\right) n_1^{n_1/2+j} \, n_2^{n_2/2+k} \, \nu_1^j \, \nu_2^k \, s^{n_1/2+j-1}}{\exp\left(\frac{\nu_1+\nu_2}{2}\right) 2^{j+k} \, \Gamma\left(\frac{n_1}{2} + j\right) \Gamma\left(\frac{n_2}{2} + k\right)}.$$

Like the chi-square distributions, the F distributions arise as the distributions of test statistics. For example, when testing for the equality of variances between two different normal populations, it is natural to obtain two samples and to compare the two sample variances by taking their ratio. The family of F distributions is even more important in experimental design, where the test statistic

$$F = \frac{\mathrm{MS}_{\mathrm{Treatment}}}{\mathrm{MS}_{\mathrm{Error}}},$$

the ratio of the treatment mean square to the error mean square, has an F distribution under the null hypothesis of no treatment effect. Under the alternative hypothesis, the treatment sum of squares has a noncentral chi-square distribution, making the ratio a noncentral F.

2.2.5 Moments and cumulants

Our next example illustrates how power series can be used to calculate the moments of a distribution as functions of its cumulants. Let X be a random variable with moment generating function

$$M(t) = E\left(e^{tX}\right)$$

and cumulant generating function

$$K(t) = \ln[\,M(t)\,]\,.$$

For $r = 1, 2, 3, \ldots$, the r-th moment μ_r and cumulant κ_r can be defined through the series expansions

$$\begin{aligned} M(t) &= 1 + \mu_1\, t + \mu_2\, \frac{t^2}{2!} + \mu_3\, \frac{t^3}{3!} + \cdots + \mu_r\, \frac{t^r}{r!} + \cdots \\ K(t) &= \kappa_1\, t + \kappa_2\, \frac{t^2}{2!} + \kappa_3\, \frac{t^3}{3!} + \cdots + \kappa_r\, \frac{t^r}{r!} + \cdots \end{aligned} \qquad (2.23)$$

While the moments of a random variable have a simpler interpretation than the cumulants, the cumulants are often mathematically easier to work with. In particular, the cumulants have an additive property for sums of independent random variables: if $X_1, \ldots, X_n$ are independent with r-th cumulants $\kappa_{1r}, \ldots, \kappa_{nr}$, respectively, then $X_1 + \cdots + X_n$ has r-th cumulant $\sum_{j=1}^{n} \kappa_{jr}$. This property makes the calculation of cumulants for sums of random variables particularly easy. So to calculate the moments of random variables, it is often useful to calculate the cumulants first, and then to convert these cumulants into moments. To find the relationship between cumulants and moments, we write $M(t) = \exp[K(t)]$, or

$$\begin{aligned} 1 + \mu_1 t + \frac{\mu_2 t^2}{2} + \cdots &= \exp\left(\kappa_1 \frac{t}{1!} + \kappa_2 \frac{t^2}{2!} + \kappa_3 \frac{t^3}{3!} + \cdots\right) \\ &= \exp\left(\kappa_1 \frac{t}{1!}\right) \exp\left(\kappa_2 \frac{t^2}{2!}\right) \exp\left(\kappa_3 \frac{t^3}{3!}\right) \cdots \\ &= \prod_{r=1}^{\infty} \left(1 + \frac{\kappa_r t^r}{r!} + \frac{\kappa_r^2 t^{2r}}{2!(r!)^2} + \cdots\right). \end{aligned} \qquad (2.24)$$

The next step is to expand out the right-hand side of (2.24) and to collect terms with common powers of t. In this way, we can write this expanded expression as a power series by organising the lowest powers of t first in the summation. Our identity becomes

$$1 + \mu_1 t + \frac{\mu_2 t^2}{2!} + \frac{\mu_3 t^3}{3!} \cdots = 1 + \kappa_1 t + \frac{(\kappa_1^2 + \kappa_2) t^2}{2!} + \frac{(\kappa_3 + 3\kappa_1\kappa_2 + \kappa_1^3) t^3}{3!} \cdots.$$

Since these two power series are equal, their coefficients can be equated. Thus $\mu_1 = \kappa_1$, $\mu_2 = \kappa_1^2 + \kappa_2$, and so on. Problem 7 at the end of the chapter asks the reader to derive the formula for μ_r in terms of cumulants for $1 \le r \le 5$. To write the cumulants in terms of the moments, we can invert these equations step by step by hand. Alternatively, we can invert the power series so that

$$\kappa_1 t + \frac{\kappa_2 t^2}{2!} + \frac{\kappa_3 t^3}{3!} \cdots = \ln\left(1 + \frac{\mu_1 t}{1!} + \frac{\mu_2 t^2}{2!} + \frac{\mu_3 t^3}{3!} + \cdots\right)$$

$$= \sum_{j=1}^{\infty} (-1)^{j-1} \frac{\left(\frac{\mu_1 t}{1!} + \frac{\mu_2 t^2}{2!} + \frac{\mu_3 t^3}{3!} \cdots\right)^j}{j}.$$

Once again, we expand the right-hand side and gather terms with common powers of t. Equating coefficients on the right-hand side with coefficients on the left-hand side, we can show that

$$\kappa_1 t + \frac{\kappa_2 t^2}{2!} + \frac{\kappa_3 t^3}{3!} \cdots = \mu_1 t + \frac{(\mu_2 - \mu_1^2)t^2}{2!} + \frac{(\mu_3 - 3\mu_1\mu_2 + 2\mu_1^3)t^3}{3!} \cdots .$$

Therefore $\kappa_1 = \mu_1$, $\kappa_2 = \mu_2 - \mu_1^2$, and so on. Again, see Problem 7.

The general formulas relating moments and cumulants can be found from Faà di Bruno's formula† for the higher derivatives of the composition of two functions. This formula, which is a generalisation of the chain rule, is not particularly convenient for numerical work. It is more practical to calculate moments or cumulants recursively. A formula that is easy to program is

$$\mu_n - \kappa_n = \sum_{j=1}^{n-1} \binom{n-1}{j-1} \kappa_j \, \mu_{n-j} \,, \tag{2.25}$$

which gives the moment μ_n (respectively, κ_n) in terms of the cumulant κ_n (respectively, μ_n) and previous moments and cumulants.

When programming in Maple, it is worth noting that (2.25) need not be coded directly. This is because Maple possesses the *powseries* package, which allows these sorts of manipulations of power series with ease. This package is discussed immediately below. The reader who wishes to see how to apply the *powseries* package to convert cumulants to moments or vice versa should go to Section 4 of Chapter 4, where the method is discussed in the context of the implementation of the delta method in Maple.

2.2.6 *Power series in Maple: a worked example*

Expanding out a series is sometimes a tedious business. Fortunately, this tedious work can be performed in Maple using the *powseries* package. To invoke this package, we use the command

† Francesco Faà di Bruno was an Italian mathematician who was beatified by Pope John Paul II in 1988.

$> with(powseries)$

at the command prompt. The commands of this package allow the quick manipulation of *formal power series*.

In order to understand how the commands in the *powseries* package are to be used, let us work through a particular example that requires power series manipulations. Suppose a fair coin is tossed independently until two heads appear in immediate succession for the first time. It is well known that the number of tosses required to obtain two heads has a negative binomial distribution. However, we typically have to wait longer if we want the heads to be in immediate succession. Let $X + 2$ denote the number of tosses required. For example, if the sequence is

$$\mathrm{H\,T\,T\,H\,T\,H\,H}\cdots$$

then $X = 5$. Clearly, X can take any nonnegative integer value. Let $p_n = P(X = n)$. It is easy to see that $p_0 = \frac{1}{4}$, and that $p_1 = \frac{1}{8}$, by simply enumerating the possible outcomes for two or three tosses. After that, the cases get more difficult. However, it is possible to construct a simple recursion for values of p_n beyond that. We have

$$p_n = \frac{1}{2}\,p_{n-1} + \frac{1}{4}\,p_{n-2}\,. \tag{2.26}$$

See Problem 10 at the end of the chapter. From this recursion, the probability generating function

$$A(\theta) = p_0 + p_1\,\theta + p_2\,\theta^2 + \cdots$$

for X can be created in Maple using the *powcreate* command

$$> powcreate\left(p(n) = \frac{p(n-1)}{2} + \frac{p(n-2)}{4},\ p(0) = \tfrac{1}{4},\ p(1) = \tfrac{1}{8}\right)$$

as displayed. More generally, the coefficients of a formal power series can be generated using a series of equations which form the arguments of *powcreate*. The first of these equations gives a formula for the coefficient on the general term. Subsequent equations provide initial assigned values of the coefficients. In case there is any disagreement between the general formula and the assigned values, the assigned values always override the general formula.

Note that this formal power series has no assigned variable for its powers. To assign this, we have to turn the formal power series into an ordinary expression using the *tpsform* command. Thus

$> tpsform(p,\ \theta,\ 5)$

generates the output

$$\frac{1}{4} + \frac{1}{8}\,\theta + \frac{1}{8}\,\theta^2 + \frac{3}{32}\,\theta^3 + \frac{5}{64}\,\theta^4 + O(\theta^5).$$

The first argument in *tpsform* specifies the coefficient function, the second argument the variable of the power series, and the optional third argument the degree. If the degree is omitted, the default value of $n = 6$ is used.

By inspection, we can conclude that $p_n = 2^{-(n+2)}\, F_{n+1}$, where F_k is the k-th Fibonacci number. We get some insight into the probability generating function for X if we take the reciprocal of the formal series, using

$> tpsform(inverse(p),\, \theta,\, 5)$

which leads to the output

$$4 - 2\,\theta - \theta^2 + O(\theta^5)\,.$$

Playing with the degree a bit, we see that the order term at the end is really zero. So the probability generating function of X is

$$\begin{aligned} A(\theta) &= E(\theta^X) \\ &= (4 - 2\,\theta - \theta^2)^{-1}\,. \end{aligned} \tag{2.27}$$

The command *inverse* is one of many commands in the *powseries* package which act like ordinary functions, but are defined on formal power series instead. Commands such as *inverse*, *powexp*, *powsin*, *powcos*, and so on perform formal operations on power series that correspond to the associated functions, namely inversion, exponentiation, application of a trigonometric function, etc. Rather than returning a real number as an ordinary function does, these commands return another formal power series.

2.2.7 *Power series involving operators*

There is a close relationship between the Taylor expansion of a function and an operation known as the *exponential tilt*. Let ∂ denote the differentiation operator

$$\partial : f \mapsto f'$$

which takes any differentiable function f and returns its derivative. We represent higher order derivatives with a formal power notation as

$$\begin{aligned} (\partial^n f)(x) &= \partial[\partial^{n-1} f](x) \\ &= f^{(n)}(x)\,. \end{aligned}$$

We can also multiply ∂^n by a constant c that does not depend upon x using the formula

$$[(c\,\partial^n)\,f]\,(x) = c\,f^{(n)}(x)\,.$$

Continuing in this way, we can build up linear combinations of such operators, with formulas such as

$$[(c_1\,\partial^n + c_2\,\partial^m)\,f]\,(x) = c_1\,f^{(n)}(x) + c_2\,f^{(m)}(x)\,.$$

By extending such formalism, we can construct polynomials and power series of operators such as ∂.

With this in mind, let us consider an operator whose formal representation is $T_1^y = \exp(y\,\partial)$, where we interpret the exponential function as a power series about zero in $y\,\partial$. So we can write

$$\exp(y\,\partial) = 1 + y\,\partial + \frac{y^2\partial^2}{2!} + \frac{y^3\partial^3}{3!} + \cdots.$$

This expansion can be interpreted by applying each side to some function. Applying this operator formally to a function f and evaluating at x, we get

$$\begin{aligned}(T_1^y f)\,(x) &= \left[\left(1 + y\,\partial + \frac{y^2\partial^2}{2!} + \frac{y^3\partial^3}{3!} + \cdots\right) f\right]\,(x) \\ &= f(x) + y\,f'(x) + \frac{y^2\,f''(x)}{2!} + \frac{y^3\,f'''(x)}{3!} + \cdots,\end{aligned}$$

which is simply the Taylor expansion for $f(x+y)$. Thus the exponential of the differentiation operator is the shift operator, and we may write

$$\exp(y\,\partial)f(x) = f(x+y)\,. \tag{2.28}$$

Suppose $f(x)$ is a probability density function. If we think in terms of statistical models, then we can see that this exponential operator generates a location model.

Note that the exponent y which appears in the notation T_1^y is formally justified by the identity

$$T_1^{y+z} = T_1^y\,T_1^z\,.$$

The multiplication of operators on the right-hand side is to be understood as operator composition.

More generally, the operator

$$T_n^y = \exp\left(\frac{y\,\partial^n}{n!}\right)$$

is associated with shifts in the n-th cumulant of the distribution. To see that this is the case, let us consider the moment generating function of a transformed density. Suppose a distribution with density $f(x)$ has moment generating function $M(t)$ and cumulant generating function $K(t)$. Correspondingly, let $(T_n^y f)(x)$ have moment generating function $M^*(t)$ and cumulant generating function $K^*(t)$. For notational convenience, let us write $z = y/n!$ so that $T_n^y = \exp(z\,\partial^n)$. We have

$$\begin{aligned}
M^*(t) &= \int_{-\infty}^{+\infty} e^{t\,x}\,(T_n^y f)(x)\,dx \\
&= \int_{-\infty}^{+\infty} e^{t\,x} \sum_{j=0}^{\infty} \frac{z^j\,\partial^{nj}}{j!} f(x)\,dx \\
&= \sum_{j=0}^{\infty} \frac{z^j}{j!} \int_{-\infty}^{+\infty} e^{t\,x} f^{(nj)}(x)\,dx \\
&= \sum_{j=0}^{\infty} \frac{z^j}{j!} (-1)^{nj}\, t^{nj} \int_{-\infty}^{+\infty} e^{t\,x} f(x)\,dx \quad \text{(See Problem 12)} \\
&= \sum_{j=0}^{\infty} \frac{[z\,(-t)^n]^j}{j!} M(t) \\
&= M(t)\exp[z\,(-t)^n]\,. \qquad (2.29)
\end{aligned}$$

One of the steps above is left as Problem 12 at the end of the chapter for the reader to verify when sufficient regularity holds. Now taking the logarithm of both sides of the equation (2.29), we obtain the relationship between the cumulant generating functions $K(t)$ and $K^*(t)$, namely

$$K^*(t) = K(t) + z\,(-t)^n\,.$$

We can write out $K(t) + z\,(-t)^n$ and $K^*(t)$ as power series in t. Since these two series are equal, we can equate their coefficients, which yields a set of equations for the respective cumulants. So

$$\kappa_j^* = \kappa_j \quad \text{for } j \neq n$$

and

$$\begin{aligned}
\kappa_n^* &= \kappa_n + (-1)^n\,n!\,z \\
&= \kappa_n + (-1)^n\,y\,.
\end{aligned}$$

Continuing along these lines, we see that we can write down formal expressions for operators that shift two or more cumulants simultaneously. For example, the operator

$$\exp\left(\frac{y\,\partial^m}{m!} + \frac{z\,\partial^n}{n!}\right) = \exp\left(\frac{y\,\partial^m}{m!}\right)\exp\left(\frac{z\,\partial^n}{n!}\right)$$

is the composition of two shifts on the m-th and n-th cumulants, respectively.

This idea leads to the following general formula. Suppose X and X^* are two random variables having cumulants of all orders κ_j and κ_j^*, respectively. Let X and X^* have respective densities $f(x)$ and $f^*(x)$. Then we obtain the formal expansion

$$f^*(x) = \exp\left[\sum_{j=1}^{\infty} \frac{\kappa_j^* - \kappa_j}{j!} (-\partial)^j\right] f(x) \,. \tag{2.30}$$

See Problem 13. This formula also extends to functions which are not normalised to integrate to one. But in this case the summation operator in the exponential must include the "zero-th cumulant" term where $j = 0$.

It cannot be too strongly emphasised that the expansion in (2.30) is derived by formal manipulations alone. We can make no claims that the expansion on the right-hand side will converge in general. Nor can we conclude that, if it converges, it will converge to the left-hand side. Indeed, it is not even clear in what order we should sum the terms of the exponential upon further expansion. As well shall see, there is an important special case of this expansion due to F. Y. Edgeworth. We will obtain this Edgeworth expansion as an approximation to a saddle-point expansion in Chapter 7.

2.3 Enveloping series

2.3.1 Definition and properties of enveloping series

A series $\sum a_n$ will be said to *envelop* a number t when, for every $n = 0, 1, 2, \ldots$, there exists $0 < \xi_n < 1$ such that the equality

$$t = (a_0 + a_1 + \cdots + a_n) + \xi_n \, a_{n+1} \tag{2.31}$$

is satisfied. We shall denote the fact that the series $\sum a_n$ envelops t by writing this as $\sum a_n \rightsquigarrow t$. Generally, we will say that a series is enveloping if there is some number t which it envelops. The enveloping of a number by a series is closely related to convergence. However, an enveloping series may converge or diverge. The following can be easily verified.

1. If $\sum a_n \rightsquigarrow t$, then t will lie between any two successive partial sums.
2. If the series converges and also envelops a value t, then it must converge to t.

3. An enveloping series is an *alternating series.* That is, the terms alternate in sign.
4. Note that not every alternating series is enveloping.

However, the following holds.

Proposition 1. *If an alternating series is of the form* $a_0 - a_1 + a_2 - a_3 + \cdots$*, where* $a_0 > a_1 > a_2 > \cdots > 0$*, then the series is enveloping. Furthermore, if* $a_n \to 0$*, then the series envelops a unique value* t *to which it converges.*

Proof. See Problem 16. ■

Proposition 2. *Let* $y > 0$*. Suppose the following hold.*

1. *The function* $f(x)$ *has derivatives of all orders for all* x *in the interval* $[x_0, x_0 + y]$.
2. *For all* $n \geq 1$*, the derivatives* $|f^{(n)}(x)|$ *are strictly decreasing in* x *over the interval* $[x_0, x_0 + y]$.

Then for all $x \in [x_0, x_0 + y]$,

$$f(x_0) + \frac{f'(x_0)}{1!}(x - x_0) + \frac{f''(x_0)}{2!}(x - x_0)^2 + \cdots \rightsquigarrow f(x).$$

The conclusion also follows if $f(x)$ *has derivatives of all orders for all* x *in* $[x_0 - y, x_0]$*, and* $|f^{(n)}(x)|$ *are all strictly increasing in the interval* $[x_0 - y, x_0]$.

Proof. This follows from Taylor's theorem with the Lagrange form of the remainder. The proof is left for the reader as Problem 17. ■

While an enveloping series does not necessarily converge, it has some desirable features. When we need to compute a function from an expansion which envelops the function, we can put exact upper and lower bounds on the error, since the function must lie between any two successive partial sums. *Moreover, the error obtained by terminating the expansion at the* n*-th step is bounded in absolute value by the absolute value of the* $(n + 1)$*-st term.* Thus even a diverging series which envelops a function can be used for computation provided that at least one term of the series is sufficiently small.

2.3.2 *Application to the exponential density*

We consider the Taylor expansion of the probability density function $f(x)$ of an *exponential distribution* with mean $\mu > 0$. We can write

$$\begin{aligned} f(x) &= \frac{1}{\mu} \exp\left(-\frac{x}{\mu}\right), \quad x \geq 0, \\ &= \frac{1}{\mu} - \frac{x}{\mu^2} + \frac{x^2}{2\,\mu^3} - \frac{x^3}{6\,\mu^4} + \frac{x^4}{24\,\mu^5} - \cdots \end{aligned} \tag{2.32}$$

This is an alternating series for positive x. Taking successive derivatives, we see that

$$\left|f^{(n)}(x)\right| = \frac{1}{\mu^{n+1}} \exp\left(-\frac{x}{\mu}\right) \tag{2.33}$$

which is a decreasing function of x. It follows from Proposition 2 that

$$\frac{1}{\mu} - \frac{x}{\mu^2} + \frac{x^2}{2\,\mu^3} - \frac{x^3}{6\,\mu^4} + \frac{x^4}{24\,\mu^5} - \cdots \rightsquigarrow \frac{1}{\mu} \exp\left(-\frac{x}{\mu}\right). \tag{2.34}$$

So the series envelops the exponential density.

2.3.3 *Application to the normal density*

Using Proposition 2, it is possible to build enveloping series for a function even when the function does not satisfy the conditions of the proposition. The following example shows how this can be applied to the density function for the normal distribution.

The probability density function of the *normal distribution* with mean μ and variance σ^2 has the form

$$f(x) = \frac{1}{\sqrt{2\,\pi}\,\sigma} \exp\left[-\frac{(x-\mu)^2}{2\,\sigma^2}\right]. \tag{2.35}$$

Expanding $f(x)$ about the point μ we get

$$f(x) = \frac{1}{\sqrt{2\,\pi}\,\sigma} \sum_{n=0}^{\infty} (-1)^n \frac{(x-\mu)^{2n}}{2^n\, n!\, \sigma^{2n}}. \tag{2.36}$$

Figure 2.1 shows the sucessive polynomial approximations to $f(x)$ up to degree four that are obtained from the partial sums of this expansion for the special case $\mu = 0$ and $\sigma^2 = 1$. As can be seen, the Taylor expansion is an enveloping series. In order to apply Proposition 2, we write $f(x) = g[h(x)]$, where

$$g(u) = \frac{1}{\sqrt{2\,\pi}\,\sigma} \exp(-u), \text{ and } h(x) = \frac{(x-\mu)^2}{2\,\sigma^2}.$$

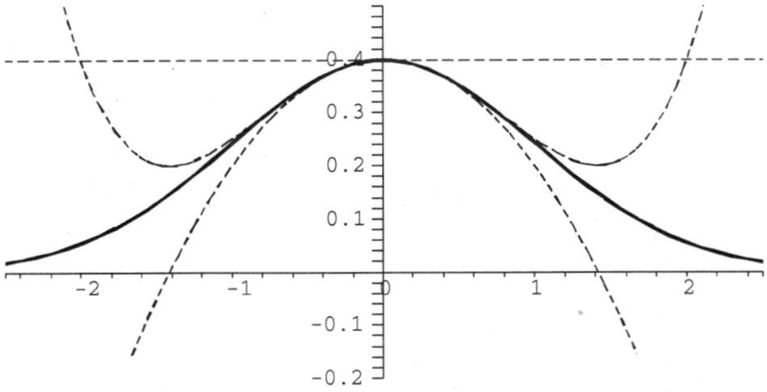

Figure 2.1 *Enveloping polynomials (dashed lines) to the normal density (solid line)*

It can be seen that $g(u)$ satisfies the conditions in part 1 of Proposition 2. So

$$\frac{1}{\sqrt{2\pi}\,\sigma}\sum_{n=0}^{\infty}(-1)^n\frac{(x-\mu)^{2n}}{2^n\,n!\,\sigma^{2n}} \rightsquigarrow g[h(x)] = f(x)\,.$$

2.3.4 Expansions for normal tail probabilities

In our next example, we consider the problem of finding a series expansion for the distribution function of the normal. Let $\phi(x)$ denote the probability density function of the standard normal distribution, and let $\Phi(z)$ denote its cumulative distribution function. We shall restrict attention to the case $z > 0$. We can write

$$1 - \Phi(z) = \int_z^{\infty} \phi(x)\,dx\,. \tag{2.37}$$

Now

$$\begin{aligned}\int_z^{\infty} \phi(x)\,dx &= \int_z^{\infty} \frac{1}{\sqrt{2\pi}}\exp\left(-\frac{x^2}{2}\right)dx \\ &= \int_z^{\infty}\left(x^{-1}\frac{1}{\sqrt{2\pi}}\right)\left[\exp\left(-\frac{x^2}{2}\right)x\,dx\right].\end{aligned}$$

The expression in square brackets can be integrated in closed form. So we apply integration by parts to obtain

$$\int_z^{\infty}\left(x^{-1}\frac{1}{\sqrt{2\pi}}\right)\left[\exp\left(-\frac{x^2}{2}\right)x\,dx\right] =$$

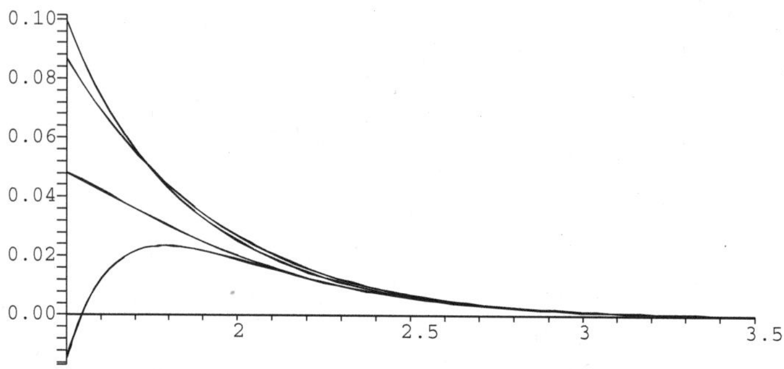

Figure 2.2 *Enveloping approximations for the normal tail probability*

$$\frac{\phi(z)}{z} - \int_z^\infty \left(x^{-3}\,\frac{1}{\sqrt{2\,\pi}}\right)\left[\exp\left(-\frac{x^2}{2}\right)\,x\,dx\right].$$

Since the integral on the right-hand side is positive, the expression $\phi(z)/z$ is an upper bound for the required tail probability. Note that this integral remainder is of the same form as our original integral. Thus we can apply the same trick again using integration by parts to obtain

$$1 - \Phi(z) = \frac{\phi(z)}{z} - \frac{\phi(z)}{z^3} + \int_z^\infty \left(3\,x^{-5}\,\frac{1}{\sqrt{2\,\pi}}\right)\left[\exp\left(-\frac{x^2}{2}\right)\,x\,dx\right].$$

Once again, the integral is positive. So the first two terms sum to a lower bound for the required probability. We can continue indefinitely in this fashion to obtain a formal infinite series. Since the partial sums are alternately upper and lower bounds, the entire series is an enveloping series. Thus

$$\frac{\phi(z)}{z}\left[1 - \frac{1}{z^2} + \frac{3}{z^4} - \frac{3\cdot 5}{z^6} + \frac{3\cdot 5\cdot 7}{z^8} - \cdots\right] \rightsquigarrow 1 - \Phi(z). \qquad (2.38)$$

Figure 2.2 shows the first four partial sums plotted as a function of the variable z.

There are two ways of studying the asymptotic properties of this expansion. The first way is to add up only a finite number of terms in the sum, and see how the partial sum of those terms behaves as $z \to \infty$. For example, suppose we set

$$s_n(z) = \frac{\phi(z)}{z}\left[1 - \frac{1}{z^2} + \frac{3}{z^4} - \cdots + (-1)^{n-1}\frac{3\cdot 5\cdot 7\cdots(2n-3)}{z^{2n-2}}\right].$$

As $z \to \infty$, the behaviour of $s_n(z)$ appears to be quite satisfactory.

This can be seen graphically in Figure 2.2, for $s_n(z)$, $n \leq 4$. To see this analytically, it suffices to use the fact that the series is enveloping. For example, when $n = 1$, we see that $s_1(z) \geq 1 - \Phi(z)$ and $s_2(z) \leq 1 - \Phi(z)$. Therefore,

$$1 - z^{-2} \leq \frac{1 - \Phi(z)}{s_1(z)} \leq 1 .$$

When $z \to \infty$, the left-hand side goes to one. So the ratio in the middle is squeezed between the bounds and must also go to one. Even tighter bounds are obtained when more terms are included.

The second way to study the asymptotic properties of this expansion is to fix the value of z and to consider the limiting behaviour of $s_n = s_n(z)$ $n \to \infty$. When $z \leq 1$ the series clearly diverges because the terms of the series do not go to zero. Suppose $z > 1$. As $n \to \infty$, the denominator of the n-th term goes to infinity exponentially. However, the numerator of the n-th term goes to infinity factorially, which is faster than exponential growth. So the series also diverges when $z > 1$, because, once again, the terms do not go to zero. This strange behaviour can be summarised by stating as follows.

1. As $z \to \infty$ with n fixed, the value of the partial sum $s_n(z)$ is asymptotically "correct."
2. As $n \to \infty$, with z fixed, the series diverges.

In the next section, we shall see that the behaviour of this series is typical of what are called asymptotic series, which usually diverge as more and more terms are added up.

Nevertheless, this enveloping series is useful for computation of tail probabilities when $z \geq 3$. For example, using the enveloping property, we have

$$\max(s_2, s_4, s_6, \cdots) \leq 1 - \Phi(z) \leq \min(s_1, s_3, s_5, \ldots) .$$

So for $z = 3$, say, this reduces to

$$0.001337 \leq 1 - \Phi(3) \leq 0.001361 .$$

We might average these upper and lower bounds to obtain our "best" approximation for the probability, namely $1 - \Phi(3) \approx 0.001349$. Surprisingly, even for a value of z as low as this, the approximation to the tail probability is correct to five decimal places.

From the calculations above, we have seen that the ratio

$$\Psi(z) = \frac{1 - \Phi(z)}{\phi(z)} \tag{2.39}$$

can be written as an enveloping series involving negative powers of z. It is known as *Mills' ratio* although its properties had been investigated earlier than 1926, when it was studied by Mills. As we shall see later in the next chapter, the first tables of the standard normal distribution function were based upon rational approximations of this ratio.

2.3.5 Tail expansions for characteristic functions

The trick that we used in the previous example to generate an enveloping series used integration by parts. It turns out that integration by parts can be used on a wide variety of integrals to produce enveloping series. Often these series will be divergent. But as we have seen, the divergence of a series need not be a complete impediment to its use! Our final example of this section examines an expansion for the characteristic function using integration by parts. Suppose that X is a positive random variable with probability density function $f(x)$ for $x \geq 0$. The characteristic function is given by

$$\begin{aligned} E\left[e^{i\,t\,X}\right] &= \int_0^\infty e^{i\,t\,x}\, f(x)\, dx \\ &= \int_0^\infty \cos(t\,x)\, f(x)\, dx + i \int_0^\infty \sin(t\,x)\, f(x)\, dx\,, \end{aligned}$$

where $i = \sqrt{-1}$. Let $t \neq 0$. Integrating by parts, we see that

$$\int_0^\infty e^{i\,t\,x}\, f(x)\, dx = \left[f(x)\, \frac{e^{i\,t\,x}}{i\,t} \right]_0^\infty - \frac{1}{i\,t} \int_0^\infty e^{i\,t\,x}\, f'(x)\, dx\,.$$

Provided that $\lim_{x\to\infty} f(x) = 0$, the expression $f(x)\, e^{i\,t\,x}$ will also vanish in the limit as $x \to \infty$. Let us assume that this holds, and that additionally

$$\lim_{x\to\infty} f^{(n)}(x) = 0 \text{ for all } n = 0,\, 1,\, 2,\, \ldots.$$

Integrating by parts,

$$\begin{aligned} \int_0^\infty e^{i\,t\,x}\, f(x)\, dx &= \frac{i\, f(0)}{t} + \frac{i}{t} \int_0^\infty e^{i\,t\,x}\, f'(x)\, dx \\ &= \frac{i\, f(0)}{t} + \frac{i}{t} \left[\frac{i\, f'(0)}{t} + \frac{i}{t} \int_0^\infty e^{i\,t\,x}\, f''(x)\, dx \right] \\ &= \frac{i\, f(0)}{t} - \frac{f'(0)}{t^2} - \frac{i\, f''(0)}{t^3} + \frac{f'''(0)}{t^4} + \cdots. \end{aligned}$$

At this stage, the last equality must be treated with suspicion, because the integral remainder obtained at each successive step need not go to zero. Even when the infinite series converges, it may not converge to the characteristic function. However, suppose we break up the expansion into its real and imaginary components. Provided that

- for every $n \geq 0$, the function $|f^{(n)}(x)|$ decreases monotonically to zero as $x \to \infty$,

we find that the resulting series which form the real and imaginary parts are enveloping. This follows from the fact that the integral remainder terms for the real and imaginary components of the series have alternating signs. Thus

$$-\frac{f'(0)}{t^2} + \frac{f'''(0)}{t^4} - \frac{f^{(v)}(0)}{t^6} + \cdots \rightsquigarrow \int_0^\infty \cos(t\,x)\, f(x)\, dx\,, \qquad (2.40)$$

and

$$\frac{f(0)}{t} - \frac{f''(0)}{t^3} + \frac{f^{(iv)}(0)}{t^5} - \cdots \rightsquigarrow \int_0^\infty \sin(t\,x)\, f(x)\, dx \qquad (2.41)$$

A similar expansion is available for the moment generating function of X. Using the integration by parts technique as before, we find that

$$\frac{f(0)}{t} + \frac{f'(0)}{t^2} + \frac{f''(0)}{t^3} + \cdots \rightsquigarrow \int_0^\infty e^{-t\,x}\, f(x)\, dx \qquad (2.42)$$

when $t > 0$. In this case, the derivatives alternate in sign.

This last example suggests that the definition of an enveloping series can be extended to series whose terms are complex numbers. We shall say that a complex series $a_0 + a_1 + a_2 + \cdots$ *weakly envelops* a complex number s if

$$|s - (a_0 + a_1 + \cdots + a_n)| < |a_{n+1}| \qquad (2.43)$$

for all $n \geq 0$. If the real and imaginary parts of the series are enveloping in our original sense, then the complex series is weakly enveloping as defined above. Moreover, a weakly enveloping series with real terms whose terms are strictly monotone decreasing in absolute value will be enveloping.

2.4 Asymptotic series

2.4.1 Definition and properties of asymptotic series

In some previous examples, we have found that the series expansion of the function has the form

$$a_0 + \frac{a_1}{x} + \frac{a_2}{x^2} + \frac{a_3}{x^3} + \cdots + \frac{a_n}{x^n} + \cdots. \qquad (2.44)$$

Let $s_n(x) = \sum_{j=0}^{n} a_j\, x^{-j}$ be the sum of the first $n+1$ terms of this series. We shall say that the given series is an *asymptotic series*[‡] for a function $f(x)$ if, for every $n \geq 0$,

$$\lim_{x\to\infty} x^n [f(x) - s_n(x)] = 0\,. \tag{2.45}$$

In order notation, this can be written as

$$f(x) = a_0 + \frac{a_1}{x} + \frac{a_2}{x^2} + \cdots + \frac{a_n}{x^n} + o(x^{-n})\,, \qquad \text{as } x \to \infty\,,$$

for all $n \geq 0$. When this is the case, we shall write

$$f(x) \quad \sim \quad a_0 + \frac{a_1}{x} + \frac{a_2}{x^2} + \frac{a_3}{x^3} + \cdots\,. \tag{2.46}$$

Suppose we let $r_n(x) = f(x) - s_n(x)$ be the remainder after summing $n+1$ terms. Then for an asymptotic series, we have

$$\lim_{x\to\infty} x^n\, r_n(x) = 0\,, \text{ where } n \text{ is fixed,}$$

even though it is also quite commonly the case that

$$\lim_{n\to\infty} |x^n\, r_n(x)| = \pm\infty\,, \text{ where } x \text{ is fixed,}$$

in addition. The second of these two limits occurs so commonly with asymptotic series that it is occasionally taken as additional part of the definition. However, it is quite unnecessary to assume it when defining an asymptotic series.

When we considered enveloping series in Section 2.3, a number of expansions appeared that looked formally like the asymptotic series defined above. For example, the series for Mills' ratio $\Psi(z)$ that we considered in Section 2.3.4 is asymptotic as well as enveloping. To check that the series is enveloping, it is only necessary to show that the remainder term is alternating in sign. However, to check that the series is asymptotic, it is necessary to study the order of the remainder term more carefully. We will consider this in greater detail in Section 2.4.2 below. At this point, let us simply note that

$$\Psi(z) \quad \sim \quad \frac{1}{z} - \frac{1}{z^3} + \frac{3}{z^5} - \frac{3\cdot 5}{z^7} + \frac{3\cdot 5\cdot 7}{z^9} - \cdots.$$

It is quite common to find that series are both asymptotic and enveloping, although the definitions of each are quite distinct.

When $f(x) = g(x)/h(x)$, we shall also find it convenient to write

$$g(x) \quad \sim \quad h(x)\left[a_0 + \frac{a_1}{x} + \frac{a_2}{x^2} + \frac{a_3}{x^3} + \cdots\right].$$

‡ Although asymptotic series have been studied for centuries by James Stirling, Colin Maclaurin and Leonhard Euler, the basic definition given here is due to Henri Poincaré in *Acta Mathematica*, Vol. 8, (1886), pp. 295–344.

Using this notation, we have

$$1 - \Phi(z) \sim \frac{\phi(z)}{z}\left[1 - \frac{1}{z^2} + \frac{3}{z^4} - \frac{3\cdot 5}{z^6} + \frac{3\cdot 5\cdot 7}{z^8} - \cdots\right].$$

The following observations can be made about asymptotic series:

1. An asymptotic series for $f(x)$ is uniquely determined in the sense that

$$f(x) \sim a_0 + \frac{a_1}{x} + \frac{a_2}{x^2} + \frac{a_3}{x^3} + \cdots \text{ and}$$
$$f(x) \sim b_0 + \frac{b_1}{x} + \frac{b_2}{x^2} + \frac{b_3}{x^3} + \cdots$$

implies that $a_n = b_n$ for all $n \geq 0$.

2. However, the asymptotic series does not uniquely determine $f(x)$ in the sense that it is possible to have

$$f(x) \sim a_0 + \frac{a_1}{x} + \frac{a_2}{x^2} + \frac{a_3}{x^3} + \cdots \text{ and}$$
$$g(x) \sim a_0 + \frac{a_1}{x} + \frac{a_2}{x^2} + \frac{a_3}{x^3} + \cdots$$

for two functions such that $f(x) \neq g(x)$ for all x.

The first of the statements follows from the fact that the coefficients of the asymptotic series can be determined recursively by the behaviour of $f(x)$ at infinity. Thus,

$$a_0 = \lim_{x\to\infty} f(x)$$
$$a_1 = \lim_{x\to\infty} x\,[f(x) - a_0]$$
$$a_2 = \lim_{x\to\infty} x^2\,[f(x) - (a_0 + a_1\,x^{-1})]$$

and so on. To see that a given asymptotic series does not uniquely determine a function, it is sufficient to note that $f(x) = e^{-x}$ and $g(x) \equiv 0$ both have the same asymptotic series.

Asymptotic series behave in some ways like power series with a few important differences. As with power series, the term-by-term sum of two asymptotic series for two respective functions is an asymptotic series for the sum of the two functions. Similarly, if

$$f(x) \sim a_0 + \frac{a_1}{x} + \frac{a_2}{x^2} + \cdots \text{ and } g(x) \sim b_0 + \frac{b_1}{x} + \frac{b_2}{x^2} + \cdots$$

then

$$f(x)\,g(x) \sim a_0\,b_0 + \frac{a_0\,b_1 + b_0\,a_1}{x} + \frac{a_0\,b_2 + a_1\,b_1 + a_2\,b_0}{x^2} + \cdots. \quad (2.47)$$

Additionally we note that the rules for raising a series to a power are similar for asymptotic series to the ones for power series. The proofs of these results are similar to those for power series. Asymptotic series can also be integrated term by term provided certain restrictions are in place. The following proposition illustrates this.

Proposition 3. Let $f(x) \sim a_2 x^{-2} + a_3 x^{-3} + \cdots$. Then for $y > 0$,

$$\int_y^\infty f(x)\, dx \sim \frac{a_2}{y} + \frac{a_3}{2\,y^2} + \frac{a_4}{3\,y^3} + \cdots. \tag{2.48}$$

Proof. Let $s_n(x) = \sum_{k=2}^n a_k\, x^{-k}$. Note that

$$\int_y^\infty s_n(x)\, dx = \frac{a_2}{y} + \frac{a_3}{2\,y^2} + \frac{a_4}{3\,y^3} + \cdots + \frac{a_n}{(n-1)\,y^{n-1}}. \tag{2.49}$$

Now it follows from (2.45) that for all $\epsilon > 0$, there exists an $x_0 > 0$ such that $|f(x) - s_n(x)| < \epsilon\, x^{-n}$, for all $x > x_0$. So when $y > x_0$,

$$\begin{aligned} \left| \int_y^\infty f(x)\, dx - \int_y^\infty s_n(x)\, dx \right| &\le \int_y^\infty |f(x) - s_n(x)|\, dx \\ &< \int_y^\infty \epsilon\, x^{-n}\, dx \\ &= \frac{\epsilon}{(n-1)y^{n-1}}. \end{aligned}$$

Since ϵ can be arbitrarily small, the left-most expression can be made arbitrarily close to zero for y sufficiently large. ■

Although an asymptotic series can be integrated term-by-term, it cannot generally be differentiated term-by-term. Consider the function

$$f(x) = e^{-x}\, \sin(e^x).$$

This function and its derivative are plotted in Figure 2.3. On the left-hand side, the function $f(x)$ is plotted. On the right-hand side, the derivative of the same function is shown. While $f(x)$ has an asymptotic series–its coefficients are all zero–the derivative of $f(x)$ does not.

2.4.2 An application to Mills' ratio

In Section 2.3.4, we derived an enveloping series for the tail of the normal distribution function and for Mills' ratio

$$\Psi(z) = \frac{1 - \Phi(z)}{\phi(z)}$$

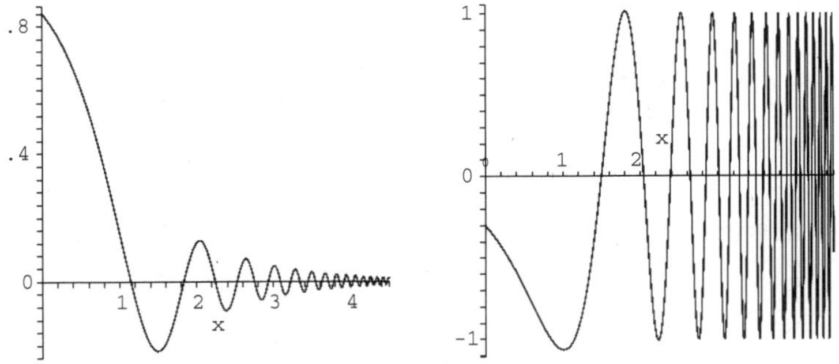

Figure 2.3 *A function whose asymptotic series cannot be differentiated term-by-term. Left: the function. Right: the derivative of the function*

using integration by parts. This series can also be shown to be an asymptotic series, so that

$$\Psi(z) \sim \frac{1}{z} - \frac{1}{z^3} + \frac{3}{z^5} - \frac{3 \cdot 5}{z^7} + \frac{3 \cdot 5 \cdot 7}{z^9} - \cdots \tag{2.50}$$

To prove that this is an asymptotic series for $\Psi(z)$, we can examine the integral remainder terms for the series. We have

$$\begin{aligned} \Psi(z) &= \frac{1}{z} - \frac{1}{z^3} + \cdots + (-1)^{n-1} \frac{3 \cdot 5 \cdots (2n-3)}{z^{2n-1}} \\ &\quad + (-1)^n \frac{3 \cdot 5 \cdots (2n-1)}{\phi(z)} \int_z^\infty x^{-2n} \, \phi(x) \, dx. \end{aligned} \tag{2.51}$$

So it is sufficient to show that for all n,

$$\lim_{z \to \infty} \frac{z^{2n-1}}{\phi(z)} \int_z^\infty x^{-2\,n} \, \phi(x) \, dx = 0\,. \tag{2.52}$$

The verification of this limit is left to Problem 19.

2.4.3 Power series in x^{-1}

The expansion of a function $f(x)$ in an asymptotic series results in a power series in x^{-1}. Are power series in x^{-1} automatically asymptotic series?

Suppose $f(x)$ is defined for all $x > c \geq 0$, and that $f(x)$ is analytic at infinity. So $g(x) = f(1/x)$ is defined for all $0 < x < c^{-1}$, and may be extended to a function which is analytic at zero. So $g(x)$ can be expressed as a convergent power series

$$g(x) = a_0 + a_1\,x + a_2\,x^2 + a_3\,x^3 + \cdots$$

in some neighbourhood of the origin, and

$$g(x) = a_0 + a_1\,x + \cdots + a_n\,x^n + o(x^n) \tag{2.53}$$

as $x \to 0$. We can replace x by x^{-1} in these expansions so that $f(x)$ has series representation

$$f(x) = a_0 + \frac{a_1}{x} + \frac{a_2}{x^2} + \frac{a_3}{x^3} + \cdots$$

for all $0 < x < \infty$, and

$$f(x) = a_0 + \frac{a_1}{x} + \cdots + \frac{a_n}{x^n} + o(x^{-n}) \tag{2.54}$$

as $x \to \infty$. So the power series in x^{-1} is asymptotic. For this reason, if $f(x)$ is analytic at infinity, then it has a convergent power series in x^{-1} which is asymptotic for $f(x)$.

However, when $f(x)$ is not analytic at infinity it may still have an asymptotic expansion. We have seen one such example in Section 2.4.2, where a divergent asymptotic series for the normal tail probability was obtained.§

When a formal series in x^{-k} is obtained using integration by parts, then little can be concluded about the asymptotic nature of the series without futher investigation. Expansions derived in (2.40–2.42) all look like asymptotic series, in the sense that their terms are formally power series in x^{-1}. However, they cannot be shown to be convergent or divergent asymptotic series without additional assumptions on $f(x)$. In (2.40) and (2.41) the easiest approach in practice to prove the asymptotic nature of the series is directly from the definition, i.e., by checking the order condition on the integral remainder term directly. The expansion in (2.42) is often asymptotic in nature.

The following example for an exponential distribution is typical. We consider the characteristic function of an exponential distribution with mean one. The real part of the characteristic function is

$$\begin{aligned}
\int_0^\infty \cos(t\,x)\,e^{-x}\,dx &= \frac{1}{1+t^2} \\
&= \frac{1}{t^2}\left(1+\frac{1}{t^2}\right)^{-1} \\
&\sim \frac{1}{t^2}\left(1-\frac{1}{t^2}+\frac{1}{t^4}-\cdots\right)
\end{aligned}$$

§ Some authors choose to define an asymptotic series as has been done in Section 2.4.1, with the additional assumption that the series diverges.

$$\sim \quad \frac{1}{t^2} - \frac{1}{t^4} + \frac{1}{t^6} - \cdots$$

which converges when $t > 1$. But this is also the expansion that is obtained from formula (2.40) by considering the case $f(x) = e^{-x}$.

2.4.4 Asymptotic series in Maple

Many of the standard asymptotic series for a function can be obtained using the *taylor* command in Maple. A command such as

$> \; taylor(f(x), x = \infty)$

or

$> \; series(f(x), x = \infty)$

will expand some functions $f(x)$ as power series in $1/x$. An additional positive integer parameter specifying the order of the approximation can also be included. Alternatively, we can expand $f(x)$ using the command

$> asympt(f(x), x)$

to the same effect. For example, the command

$$> \; taylor\left(\frac{1}{1+x^2},\; x = \infty,\; 10\right)$$

gives the output

$$\frac{1}{x^2} - \frac{1}{x^4} + \frac{1}{x^6} - \frac{1}{x^8} + O\left(\frac{1}{x^{10}}\right).$$

However, the methods developed in this book for asymptotic series are not identical to those invoked by *asympt* or by the *taylor* command with $x = \infty$ option. The Maple package performs the computation by making the substitution $x = 1/x$ in the function $f(x)$, expanding the resulting function about zero and then substituting $x = 1/x$ once again. The code for the Maple expansion can be written out more fully as

$$> \; subs\left(x = \tfrac{1}{x},\; series\left(subs\left(x = \tfrac{1}{x},\; f(x)\right),\; x = 0,\; n\right)\right)$$

where the *series* command produces a generalised series, akin to Taylor series, but with the possibility of negative or fractional powers of x.

When possible, the *series* command given above will attempt to factor out a function if the asymptotic behaviour of that function does not satisfy a power law. For example, the output from

$$> series\left(\frac{\exp(-x)}{1+x^2},\ x=\infty\right)$$

is

$$\frac{\frac{1}{x^2}-\frac{1}{x^4}+O\left(\frac{1}{x^6}\right)}{e^x}$$

and not the asymptotic series whose coefficients are all zero. However, the expansion may fail, in which case Maple returns a message to that effect.

2.4.5 Watson's Lemma

To verify that a series expansion is asymptotic can be quite difficult. It is not sufficient to verify that it is a formal expansion in powers of x^{-1} as there is no guarantee that the remainder term for any partial sum is of small order. Verification of the necessary order condition can be done using ad hoc methods for particular series. For example, the expansions obtained using integration by parts leave the remainder term in integral form. As we have seen, it is often possible to put a bound on the absolute value of an integral using a bound on the integrand. However, there are many expansions which are asymptotic in form, but not derived using integration by parts. So it is useful to have a general result which can be applied to many expansions. Arguably, the most important result of this kind is Watson's lemma, which provides asymptotic expansions of integrals which are Laplace transforms (or moment generating functions).

Proposition 4. (Watson's Lemma.) Suppose $f(x)$ is a function of a real variable x such that

1. $f(x)$ is analytic for all x in some neighbourhood of the origin $|x| < \delta + \epsilon$, for some $\delta,\ \epsilon > 0$.
2. There exists constants $\alpha,\ K > 0$, such that

   $$|f(x)| \le K\, e^{\alpha\, x}$$

   for all $x \ge \delta$.

Then

$$\int_0^\infty e^{-t\,x}\, f(x)\, dx \;\sim\; \frac{f(0)}{t}+\frac{f'(0)}{t^2}+\frac{f''(0)}{t^3}+\cdots$$

as $t \to \infty$.

Proof. For any given integer m, we can obtain a constant $C > 0$ such that

$$\left| f(x) - \sum_{n=0}^{m} \frac{f^{(n)}(0)}{n!} x^n \right| < C\, x^{m+1}\, e^{\alpha\, x} \tag{2.55}$$

for all $x \geq 0$. Problem 23 asks the reader to verify that a constant C can be found.

Integrating $e^{-t\,x} f(x)$ we get

$$\begin{aligned}
\int_0^\infty e^{-t\,x} f(x)\, dx &= \sum_{n=0}^{m} \frac{f^{(n)}(0)}{n!} \int_0^\infty x^n\, e^{-t\,x}\, dx + R_m \\
&= \sum_{n=0}^{m} \frac{f^{(n)}(0)}{t^{n+1}} + R_m\,,
\end{aligned}$$

where when $t > \alpha$,

$$\begin{aligned}
|R_m| &< C \int_0^\infty x^{m+1}\, e^{\alpha\, x}\, e^{-t\,x}\, dx \\
&= C\, \frac{(m+1)!}{(t-\alpha)^{m+2}} \\
&= o\left(\frac{1}{t^{m+1}}\right), \qquad \text{as } t \to \infty\,.
\end{aligned}$$

Thus the series is an asymptotic expansion for the given integral. ∎

Some comments must be made at this point. The first is that there is nothing in the proof which requires $f(x)$ to be real-valued. So the lemma can be extended to complex-valued $f(x)$. A second point to be made is that the assumptions are stronger than necessary. It is sufficient to be able to decompose the integral as

$$\int_0^\infty e^{-t\,x} f(x)\, dx = \int_0^\delta e^{-t\,x} f(x)\, dx + \int_\delta^\infty e^{-t\,x} f(x)\, dx$$

so that the tail integral from δ to infinity contributes negligibly to the expansion. Specifically, we need

$$\int_\delta^\infty e^{-t\,x} f(x)\, dx = o(t^{-m})$$

for all $m \geq 1$. The second assumption in the lemma is sufficient, but not necessary for this result. Another point to be made is that Watson's lemma is particularly useful when combined with change of variables methods for integration. For example, using change of variables, we can obtain the useful asymptotic expansion

$$\sqrt{\frac{t}{2\pi}} \int_{-\delta}^{\delta} e^{-t\,x^2/2}\, h(x)\, dx \;\sim\; h(0) + \frac{h''(0)}{2\,t} + \frac{h^{(iv)}(0)}{2!\,(2\,t)^2} + \frac{h^{(vi)}(0)}{3!\,(2\,t)^3} + \cdots \tag{2.56}$$

as a corollary of Watson's lemma. A sufficient condition on $h(x)$ to ensure that the expansion is asymptotic is that $h(x)$ be analytic for all x such that $|x| < \delta + \epsilon$, where $\delta,\, \epsilon > 0$. The interval of integration can be extended to infinity provided the tail where $|x| > \delta$ is negligible. To prove this result using change of variables, we can partition the integral as

$$\int_{-\delta}^{\delta} e^{-t\,x^2/2}\, h(x)\, dx = \int_{-\delta}^{0} e^{-t\,x^2/2}\, h(x)\, dx + \int_{0}^{\delta} e^{-t\,x^2/2}\, h(x)\, dx$$

and then make the substitution $u = x^2/2$ on each component. The odd derivatives of $h(x)$ cancel in assembling the full integral.

2.4.6 Tail expansions for symmetric stable densities

Stable distributions can be regarded as generalisations of the normal distributions. A key property of each stable distribution is a parameter α, called its *exponent*, which controls the scaling law for the distribution when the distribution is convolved with itself. For example, the normal distributions are all symmetric stable laws with exponent $\alpha = 2$. On the other hand, the Cauchy distribution, which scales quite differently from the normal when convolved with itself, is a symmetric stable law with exponent $\alpha = 1$. This property can be made explicit as follows. Suppose that X, X_1 and X_2 are independent random variables with some common distribution F. The distribution F is said to be *(strictly) stable* if, there exists a number α such that for every pair of real numbers $a > 0$ and $b > 0$, we have

$$a^{1/\alpha}\, X_1 + b^{1/\alpha}\, X_2 \stackrel{d}{=} (a+b)^{1/\alpha}\, X\,. \tag{2.57}$$

Here, we use the symbol $\stackrel{d}{=}$ to denote the fact that two random variables have the same distribution.¶ It is possible to show that the exponent α satisfies $0 < \alpha \leq 2$. The random variable X is said to have a *symmetric stable distribution* if additionally $X \stackrel{d}{=} -X$. This can be generalised to a symmetric stable distribution about a center θ. For the purposes of our discussion here, we shall concentrate on stable distributions that

¶ Of course, the notation $U \sim V$ is more commonly used to denote the fact that U and V have the same distribution. However, in our context, this is too easily confused with our symbol for asymptotic equivalence.

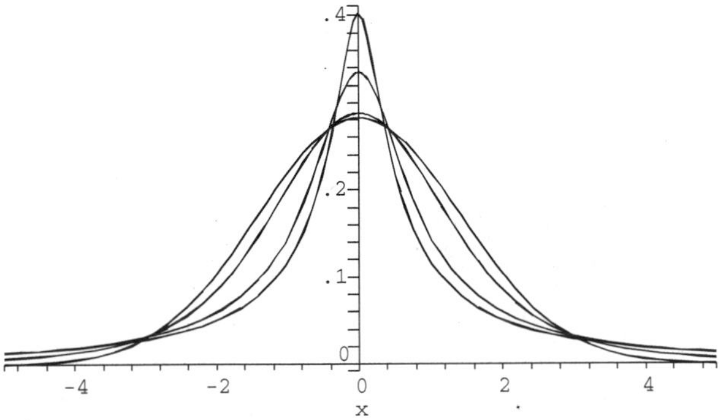

Figure 2.4 *Density functions for the symmetric stable distributions with exponent values* $\alpha = 0.7$, 0.9, 1.5 *and* 2.0

are symmetric about zero, so that $\theta = 0$. For any given choice of α between 0 and 2 there is a symmetric stable distribution centred at zero up to an arbitrary scale parameter. For the normal distribution (where $\alpha = 2$) this scale parameter determines the standard deviation. If we choose a canonical form for this scale parameter, then the particular symmetric stable distribution of given exponent α is unique. Among this family indexed by α, the only distributions which have densities that can be written in simple form are the normal, where $\alpha = 2$ and the Cauchy, where $\alpha = 1$. Nevertheless when calculating likelihoods or when modelling observations with stable laws, it is often necessary to be able to calculate the density function. While simple closed expressions for the densities are generally not available, there are both power series expansions and asymptotic expansions for the densities. In this section, we shall derive an asymptotic expansion for the symmetric stable density. The particular derivation of this asymptotic expansion, involving inverse Fourier transforms and contour integration, is of general interest as a technique for the approximation of densities, and will also be the basis for the saddle-point approximation in Chapter 7. Figure 2.4 displays the densities for symmetric stable distributions for $\alpha = 0.7$, 1.0, 1.5 and 2.0. Of the three densities plotted, the one with exponent $\alpha = 0.7$ has the heaviest tails and the narrowest "peak" at the mode. The density $\alpha = 1.5$ is closest in shape to a normal density ($\alpha = 2.0$), but still has heavier tails than the normal curve.

While most symmetric stable densities cannot be written in closed form, the characteristic functions of these distributions are very simple. We shall begin by noting that the characteristic function $\chi(t)$ of a random variable X having a stable distribution, symmetric about zero, with exponent α will be

$$\begin{aligned} \chi(t) &= E(e^{i\,t\,X}) \\ &= \exp(-c\,|t|^{\alpha}) \end{aligned} \tag{2.58}$$

for some constant $c > 0$. The constant c controls the scale of the distribution. Problem 25 asks the reader to check that random variables with this characteristic function satisfy the scaling law given in (2.57). For simplicity, we can standardise the distribution so that the constant $c = 1$. Note that for the normal distribution with $\alpha = 2$ this does not lead to the usual standardisation of the variance namely $\sigma = 1$, but rather to a version where $\sigma = \sqrt{2}$.

The density function $f(x)$ for a random variable X can be written quite generally as an integral of its characteristic function using the *inverse Fourier transform*. In the case of symmetric stable distributions, we can plug in the characteristic function $\chi(t)$ given above to obtain

$$f(x) = \frac{1}{2\,\pi} \int_{-\infty}^{\infty} e^{-i\,x\,t}\,\chi(t)\,dt \tag{2.59}$$

$$\begin{aligned} &= \frac{1}{2\,\pi} \int_{-\infty}^{\infty} \exp(-i\,x\,t - |t|^{\alpha})\,dt \\ &= \frac{1}{\pi} \int_{0}^{\infty} \exp(-i\,x\,t - t^{\alpha})\,dt\,. \end{aligned} \tag{2.60}$$

The last step in (2.60) follows from the fact that $f(x) = f(-x)$ for a symmetric law. Since $f(x)$ is real, the imaginary components of the integral will cancel each other to give zero.

One might be tempted to expand the integrand in (2.60) immediately using a power series for the exponential function. However, this direct approach does not work, because the resulting series cannot be integrated term-by-term. Instead, we use a trick that has become standard for dealing with such inverse Fourier transforms: we regard the given integral as a *contour integral*‖ in the complex plane, where the contour is the positive real axis. In this setting, we see that it is possible to perturb the contour of integration from zero to infinity along the real axis to some other contour in the complex plane with the same endpoints, provided that the integrand is a complex analytic function with no poles throughout a simply connected region which contains both contours. Since one of our endpoints is infinity, it is helpful to consider this as the

‖ Here I use the terms "contour" and "contour integral" to denote what some authors call a line integral or curvilinear integral in the complex plane. The term "closed contour" will be used for contours which start and end at the same point.

Joseph Fourier (1768–1830)

The great shock caused by his trigonometric expansions was due to his demonstration of a paradoxical property of equality over a finite interval between algebraic expressions of totally different form. ... For Fourier, every mathematical statement ... had to have physical meaning ...[T]his was expressed in his aphorism, "Profound study of nature is the most fertile source of mathematical discoveries."

Jerome R. Ravetz and I. Grattan-Guinness, *Dictionary of Scientific Biography*, Vol. 5, 1972, pp. 96–97.

"point at infinity" using a stereographic projection of the complex plane onto the Riemann sphere. See Figure 2.5. On the Riemann sphere, the positive real axis and the negative imaginary axis can be thought of as two contours with the same end points. It is obvious that they both start at the origin in the complex plane, represented here as the "south pole" of the Riemann sphere. They also end at the "north pole" or point at infinity. So the integral of our function from 0 to ∞ along the positive real axis will equal the integral of the principal branch of the function along the negative imaginary axis from 0 to $-i\,\infty$. Thus

$$f(x) = \frac{1}{\pi}\int_0^{-i\,\infty} \exp(-i\,x\,t - t^\alpha)\,dt\,. \tag{2.61}$$

Since $f(x) = f(-x)$, it is sufficient to restrict to the case $x \geq 0$ for computation of this integral. The next step is to change variables, by letting $u = i\,x\,t$, say. This restores the contour of integration to the real axis, and permits the following power series expansion. The integral of each term of the resulting power series reduces to the evaluation of the gamma function. We get

$$\begin{aligned} f(x) &= \frac{-i}{\pi\,x}\int_0^\infty e^{-u}\,\exp[-(-i\,u/x)^\alpha]\,du \\ &= \frac{-i}{\pi\,x}\int_0^\infty e^{-u}\left[1-(-i\,u/x)^\alpha + \frac{(-i\,u/x)^{2\,\alpha}}{2!} - \cdots\right]du \\ &\sim \frac{-i}{\pi\,x}\left[1-\left(\frac{-i}{x}\right)^\alpha \Gamma(\alpha+1) + \left(\frac{-i}{x}\right)^{2\,\alpha}\frac{\Gamma(2\,\alpha+1)}{2!} - \cdots\right]. \end{aligned}$$

Now as was stated above, the imaginary components of this series must vanish, so that the series equals its real part. But

$$\begin{aligned} \Re[(-i)^{k\alpha+1}] &= \Re\left[\left(e^{-i\,\pi/2}\right)^{k\alpha+1}\right] \\ &= \cos\left(-\frac{k\,\pi\,\alpha}{2} - \frac{\pi}{2}\right) \\ &= -\sin\left(\frac{k\,\pi\,\alpha}{2}\right). \end{aligned}$$

So taking the real part and using the well-known identity $\Gamma(u+1) = u\,\Gamma(u)$, we have

$$\begin{aligned} f(x) &\sim \frac{1}{\pi}\sum_{k=1}^{\infty}(-1)^{k-1}\sin\left(\frac{k\,\pi\,\alpha}{2}\right)\frac{\Gamma(k\alpha+1)}{k!}\,\frac{1}{x^{k\alpha+1}} \\ &\sim \frac{\alpha}{\pi}\sum_{k=1}^{\infty}(-1)^{k-1}\sin\left(\frac{k\,\pi\,\alpha}{2}\right)\frac{\Gamma(k\,\alpha)}{(k-1)!}\,\frac{1}{x^{k\alpha+1}} \end{aligned}$$

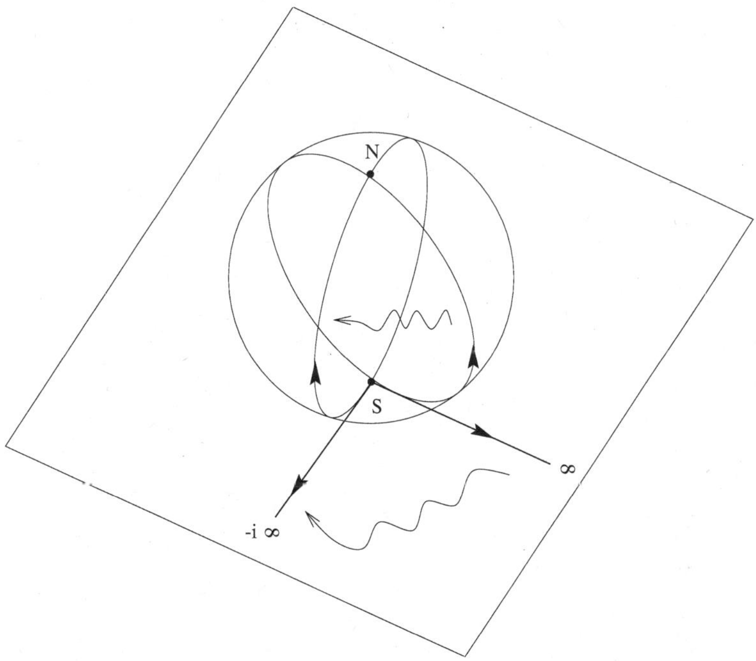

Figure 2.5 *Two contours of integration mapped into the Riemann sphere by a stereographic projection. The contours are both rays from the origin in the complex plane out to infinity in different directions, one along the positive real axis and the other along the negative imaginary axis. When mapped onto the Riemann sphere, these contours become arcs of great circles which start at the "south pole" labelled as S, and end at the "north pole" labelled as N. The value of a contour integral is unchanged if the contour over which it is evaluated is deformed into another contour with the same endpoints (S and N) provided that no singularity of the integrand is crossed by the deformation.*

This formula requires that x be positive. For more general $x \neq 0$ we can write

$$f(x) \sim \frac{\alpha}{\pi} \sum_{k=1}^{\infty} (-1)^{k-1} \sin\left(\frac{k\,\pi\,\alpha}{2}\right) \frac{\Gamma(k\,\alpha)}{(k-1)!} \frac{1}{|x|^{k\alpha+1}}. \tag{2.62}$$

When $\alpha < 1$, this series will converge because $\Gamma(k\alpha)$ goes to infinity more slowly than $(k-1)!$ does. On the other hand, when $\alpha > 1$, the series diverges. The case $\alpha = 1$ is the Cauchy density, for which the series converges for $|x| > 1$ and diverges elsewhere. However, in all cases regardless of the convergence properties, the series is asymptotic.

It is worth observing what happens to this series for $\alpha = 2$, which corresponds to the normal distribution. This case is an instructive example of what can and cannot be learned from such an asymptotic series. For $\alpha = 2$, all terms vanish, so that the asymptotic result reduces to the statement that $f(x) \sim 0$. This is quite correct, and informs us that the tail of the normal density goes to zero faster than any power law does. Unfortunately it is a weak and rather obvious result. For $\alpha < 2$, more information about tail behaviour of the densities can be gleaned. For these cases, we find that

$$f(x) \sim \frac{\alpha}{\pi} \sin\left(\frac{\pi\,\alpha}{2}\right) \frac{\Gamma(\alpha)}{|x|^{\alpha+1}} \qquad \text{as } x \to \pm\infty.$$

So all the other symmetric stable laws have tails that go relatively slowly to zero as power laws do, rather than as the normal does.

Note that the utility of series (2.62) for computing the stable density does *not* have a strong relationship to its convergence. In practice, if the series is divergent for given x, one would sum the terms for as long as the terms are decreasing in absolute value. Beyond that point, the series does not appear to have much utility. However, this can be quite satisfactory for computation when $|x|$ is large, because the terms go rapidly towards zero. However, when $|x|$ is small, the series is badly behaved even when it converges. In this case, the opposite phenomenon can appear, where the terms of the series increase in absolute value for many terms before finally decreasing rapidly to zero.

For completeness, it is worth mentioning that there is a power series for $f(x)$ which has complementary properties to the asymptotic series in (2.62). Returning to the integral given in (2.60), let us first make the substitution $u = t^\alpha$. We recall also that the imaginary part of the integral must vanish. So we may use the real part of the resulting integral. Therefore,

$$f(x) = \frac{1}{\pi} \Re \int_0^{\infty} \exp(-i\,x\,u^{1/\alpha})\, e^{-u}\, \alpha^{-1}\, u^{(1-\alpha)/\alpha}\, du.$$

If we now expand $\exp(-i\,x\,u^{1/\alpha})$ as a power series, and integrate term-by-term, we obtain a power series of the form

$$f(x) = \frac{1}{\alpha\,\pi}\sum_{k=0}^{\infty}(-1)^k\,\frac{\Gamma[(2k+1)/\alpha]}{(2k)!}\,x^{2k}. \tag{2.63}$$

In contrast to (2.62), the series in (2.63) is convergent for $\alpha > 1$ and divergent for $\alpha < 1$. Its properties mirror those in (2.62), with computational advantages for small $|x|$ and poor properties when $|x|$ is large.

2.5 Superasymptotic and hyperasymptotic series

The rather surprising thing about asymptotic series is that they work as well as they do. We have observed this phenomenon when comparing the power series expansions and the asymptotic expansions of the symmetric stable law densities. Interestingly, these series are not exceptional in their behaviour. These and other examples illustrate the principle that a divergent asymptotic series is often of greater practical value for computing approximations than is a slowly converging power series for the same function. Asymptotic series, while rapidly divergent eventually, are often in their initial partial sums rapidly "convergent." This situation is rather paradoxical, and deserves greater attention. Attempts to explain and exploit this bizarre behaviour lead us to the idea of a *superasymptotic series*.

This behaviour of asymptotic series is often summarised in an aphorism known as *Carrier's rule*, which states that divergent series converge faster than convergent series because they do not have to converge. At first glance this seems to be complete nonsense. However, the essence of the aphorism is that convergence and approximation are not the same, and that sometimes convergence is an obstacle to the speed of approximation of the partial sums of a series.**

Clearly as an explanation for the rapid initial "convergence" of asymptotic series, Carrier's rule is more of an observation than an explanation. Consider an asymptotic series

$$f(x) \sim a_0 + a_1 x^{-1} + a_2\,x^{-2} + \cdots$$

which diverges. Typically for given x, the terms of such a series will decrease in absolute value so that

$$\left|a_n\,x^{-n}\right| \geq \left|a_{n+1}\,x^{-(n+1)}\right|$$

** This aphorism is due to mathematician George F. Carrier of Harvard University.

for $n \le m$ and then increase after that point so that

$$\left|a_n\, x^{-n}\right| < \left|a_{n+1}\, x^{-(n+1)}\right|$$

for $n > m$. The value of m, where the switch occurs will depend on x, with m increasing as $x \to \infty$. For such a series, the natural, if not optimal, way to truncate the series to approximate $f(x)$ is at the value $m = m(x)$ where the terms are minimised in absolute value. A finite asymptotic series which is obtained by truncation in this way is said to be a *superasymptotic series.* The explanation for this terminology is that an asymptotic series, when truncated in this way, has an asymptotic error that is much smaller than an asymptotic series which is truncated after a fixed number of terms. We can write

$$f(x) = a_0 + a_1 x^{-1} + a_2\, x^{-2} + \cdots + a_{m(x)}\, x^{-m(x)} + \epsilon(x) \qquad (2.64)$$

where typically

$$\epsilon(x) = O\left(e^{-\alpha\, x^{\beta}}\right)$$

as $x \to \infty$, for some α, $\beta > 0$. This exponentially decreasing error at infinity is to be contrasted with the error associated with a fixed truncation: if m is chosen to be a fixed value independent of x, then the error is $O\left(x^{-(m+1)}\right)$.

Thus asymptotic series treat exponentially decreasing components as essentially negligible. We have already seen how this can occur when we noted that the asymptotic series does not uniquely determine a function. Recall that if $f(x) - g(x) = e^{-x}$, then the asymptotic series for $f(x)$ and $g(x)$ are the same.

To illustrate how superasymptotics can be used in practice, let us consider the asymptotic expansion for the *Stieltjes integral.* Let Y be a random variable distributed as $\mathcal{E}(\lambda)$. Then

$$\begin{aligned} E\left(\frac{1}{1+Y}\right) &= \int_0^\infty \frac{e^{-y}\, dy}{1+\lambda^{-1}\, y} \\ &\sim 1 - \frac{1!}{\lambda} + \frac{2!}{\lambda^2} - \frac{3!}{\lambda^3} + \cdots, \end{aligned} \qquad (2.65)$$

which can be obtained using a binomial expansion on $(1+\lambda^{-1}\, y)^{-1}$. The terms $(-1)^n\, n!\, \lambda^{-n}$ of this asymptotic series will decrease in absolute value for all n such that $\lambda > n$. So the superasymptotic series for this function is

$$\int_0^\infty \frac{e^{-y}\, dy}{1+\lambda^{-1}\, y} = \sum_{n=0}^{\lfloor\lambda\rfloor} (-1)^n\, \frac{n!}{\lambda^n} + \epsilon(\lambda)\,. \qquad (2.66)$$

Figure 2.6 shows the plots of the Stieltjes integral and the error function

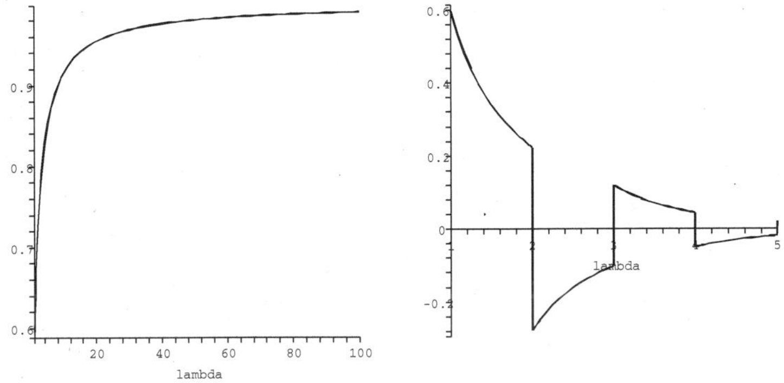

Figure 2.6 *A plot of the Stieltjes integral and its superasymptotic error function*

$\epsilon(\lambda)$ as functions of λ. The left-hand plot shows the integral plotted over a large range of values of λ so that the asymptotic nature of the integral is visible: it is seen to be monotonically increasing towards the upper bound of one. The right-hand plot of $\epsilon(x)$ for small values of λ clearly shows that the error is exponentially decreasing as $\lambda \to \infty$.

Additional attempts to construct asymptotic series for the remainder term of a superasymptotic series properly belong to the domain known as *hyperasymptotics.* By definition, a hyperasymptotic approximation is one which achieves an asymptotically better approximation than a superasymptotic series by adding additional terms that approximate the remainder term of the superasymptotic series. To illustrate this idea, let us consider how we can approximate the remainder term $\epsilon(\lambda)$ for the superasymptotic series of the Stieltjes integral. We have

$$\begin{aligned} \epsilon(\lambda) &= \int_0^\infty e^{-y} \sum_{n=\lfloor\lambda\rfloor+1}^{\infty} (-1)^n \frac{y^n}{\lambda^n}\, dy \\ &= \frac{(-1)^{\lfloor\lambda\rfloor+1}}{\lambda^{\lfloor\lambda\rfloor+1}} \int_0^\infty \frac{e^{-y}\, y^{\lfloor\lambda\rfloor+1}}{1+\lambda^{-1} y}\, dy \qquad (2.67) \end{aligned}$$

as an integral expression for the remainder term $\epsilon(\lambda)$ for large λ. It can be shown that

$$\epsilon(\lambda) \sim (-1)^{\lfloor\lambda\rfloor+1} \sqrt{\frac{\pi\,(\lfloor\lambda\rfloor+1)}{2}}\, e^{-(\lfloor\lambda\rfloor+1)} \qquad (2.68)$$

as $\lambda \to \infty$.

Ultimately, the desire for increased accuracy in the asymptotic approximation must give way. Increased complexity eventually overwhelms any analytic insight that can be gained from additional terms.

2.6 Asymptotic series for large samples

2.6.1 Asymptotic series in $\sqrt{n}$

In many cases in statistics, the expansion of some statistic T_n, based upon a random sample of size n, has the form

$$T_n = a_0 + \frac{a_1}{\sqrt{n}}\, Z + \frac{a_2}{n}\, Z^2 + \frac{a_3}{n\,\sqrt{n}}\, Z^3 + \ldots \tag{2.69}$$

Where Z is some random variable whose distribution does not depend upon the value of n (or approximately so). Such expansions are usually derived from a Taylor expansion of some function of a pivotal statistic. A corresponding expansion for the expectation of T_n is often obtained formally as

$$E(T_n) = a_0 + \frac{a_1\,\mu_1}{\sqrt{n}} + \frac{a_2\,\mu_2}{n} + \frac{a_3\,\mu_3}{n\,\sqrt{n}} + \cdots \tag{2.70}$$

where $\mu_j = E(Z^j)$. The expansion in (2.70) is formally derived by calculating the expectation of each term in (2.69). Of course, there is no guarantee that the resulting series converges to $E(T_n)$. Expansions such as (2.69) and (2.70) are not series in n^{-1}. However, the series is asymptotic in $\sqrt{n}$ instead provided that the series is analytic at infinity. Thus standard expansions in statistics are often asymptotic in this sense.

In some applications, the random variable Z in (2.69) will have a distribution that is symmetric (or approximately symmetric) about zero. Typically, this will arise when Z has a normal distribution centered at zero, or has a limiting normal distribution. Under these circumstances, the moments will often satisfy

$$\mu_{2k} = O(1)\,, \quad \mu_{2k+1} = O\left(\frac{1}{\sqrt{n}}\right)$$

so that the expansion of $E(T_n)$ will reduce to

$$E(T_n) = b_0 + \frac{b_1}{n} + \frac{b_2}{n^2} + \cdots$$

which has the form of an asymptotic expansion in n. Other types of asymptotics can also arise, particularly when an estimation or testing problem is nonstandard. The power of n, or more generally the function of n, for the asymptotics is usually governed by the rate at which the total information in the sample increases as a function of n. The following examples illustrate this.

2.6.2 Estimating a Poisson probability

Suppose $X_1, \ldots, X_n$ a random sample of $\mathcal{P}(\lambda)$ random variables. In order to estimate $P(X = 0) = e^{-\lambda}$ we can use the maximum likelihood estimator

$$T_n = e^{-\overline{X}_n}$$

where $\overline{X}_n$ is the sample mean. We stabilise the mean and variance of $\overline{X}_n$ by setting

$$Z = \sqrt{n}\,(\overline{X}_n - \lambda)$$

so that Z has mean zero and variance λ, which do not depend upon n. As $n \to \infty$, the limiting distribution of Z is well known to be $\mathcal{N}(0, \lambda)$. Performing a Taylor expansion on the exponential function, we find that

$$\begin{aligned} T_n &= e^{-\lambda}\, e^{-Z/\sqrt{n}} \\ &= e^{-\lambda}\left[1 - \frac{Z}{\sqrt{n}} + \frac{Z^2}{2\,n} - \frac{Z^3}{6\,n\,\sqrt{n}} + \cdots\right]. \end{aligned} \tag{2.71}$$

An expansion for the expectation of T_n can be obtained from the moments of Z. Problem 28 asks the reader to evaluate the moments of Z to show that

$$E(T_n) = e^{-\lambda}\left[1 + \frac{\lambda}{2\,n} - \frac{\lambda}{6\,n^2} + \frac{\lambda^2}{8\,n^2} + O\left(\frac{1}{n^3}\right)\right]. \tag{2.72}$$

2.6.3 Example: maximum of a sample of uniform random variables

Consider, for example, a random sample $X_1, \ldots, X_n$ of $\mathcal{U}(0, 1)$ random variables. Define

$$T_n = n\,[1 - \max(X_1, \ldots, X_n)]$$

Then for $0 < x < 1$,

$$P(T_n > x) = \left(1 - \frac{x}{n}\right)^n.$$

Applying the expansion obtained in Problem 16 of Chapter 1, we find that

$$P(T_n > x) = e^{-x}\left[1 - \frac{x^2}{2\,n} - \frac{x^3\,(8 - 3x)}{24\,n^2} - \cdots\right]. \tag{2.73}$$

2.6.4 Estimation of an integer-valued parameter

We should distinguish between the *asymptotic rate of convergence* of a sequence T_n of random variables as a function of n and the asymptotics of any series expansion of the distribution of T_n. Consider, for example, a random sample $X_1, X_2, \ldots, X_n$ of $\mathcal{N}(\theta,\, 1)$ random variables, where the mean θ is an unknown integer. Problems of this kind arise when taking continuous measurements of some physical quantity that is known to be an integer, the quantity in question usually being some integer multiple of some given unit. Estimation of the atomic number of an element is a case in point. Let $\overline{X}_n$ denote the sample average. An obvious estimate for θ is $[\overline{X}_n]$, where $[x]$ denotes the closest integer to x. Define

$$T_n = \begin{cases} 0 & \text{if } [\overline{X}_n] = \theta \\ 1 & \text{if } [\overline{X}_n] \neq \theta\,. \end{cases}$$

Then

$$\begin{aligned} E(T_n) &= P\left(|\overline{X}_n - \theta| > \frac{1}{2}\right) \\ &= 2\left[1 - \Phi\left(\frac{\sqrt{n}}{2}\right)\right]. \end{aligned}$$

Using the asymptotic expansion for Mills' ratio given in Section 2.4.2, we get

$$E(T_n) = \sqrt{\frac{8}{\pi\, n}}\, e^{-n/8}\left(1 - \frac{8}{n} + \frac{96}{n^2} - \cdots\right). \tag{2.74}$$

In this example, the asymptotic rate of convergence of T_n to zero is superexponential. However, the expansion is "n-asymptotic" in nature.

2.7 Generalised asymptotic expansions

The expansion of functions as asymptotic series is too restrictive to be useful for all purposes. For example, in the previous section, we saw that some expansions lead to asymptotics series in n, while others lead to asymptotic series in $\sqrt{n}$. Clearly, we can expand functions in other powers of n, as the context warrants.

Define a *generalised asymptotic series* of the *Poincaré type* about $x = x_0$ to be an expansion of the form

$$f(x) \;\sim\; a_0\, \varphi_0(x) + a_1\, \varphi_1(x) + a_2\, \varphi_2(x) + \cdots \tag{2.75}$$

where, for all k,

$$\varphi_{k+1}(x) = o[\varphi_k(x)] \quad \text{and} \quad f(x) = \sum_{j=0}^{k} a_j\, \varphi_j(x) + o[\varphi_k(x)]$$

as $x \to x_0$. Our main interest will be in the case $x_0 = \infty$, although finite values can easily arise.

Even more general definitions for asymptotic expansions can be given. Specifically, it is possible to define asymptotic expansions which are not of the Poincaré type. See Wong (2001). However, such generality can only be achieved by losing some of the results about asymptotic series that make the theory useful.

2.8 Notes

The classic textbook on series methods is Whittaker and Watson (1962). The first edition of this book appeared in 1902, but the many reprintings are a testimonial to the lasting value of its methods. The reader who would like to see rigorous proofs of basic theorems taught in most calculus courses will find that Spivak (1994) provides a readable account without sacrificing rigour. Whittaker and Watson (1962) give a brief description of the basic definitions and theorems for asymptotic series. More detail is to be found in Wong (2001). Although many asymptotic series are often enveloping series as well, the literature on enveloping series is much more limited. See Pólya and Szegö (1978, pp. 32–36) for some basic results. The reader is warned that the definition of an enveloping series given in this chapter is equivalent to the concept of a strictly enveloping series as defined in the reference cited. For an introduction to stable laws, see volume 2 of Feller's famous treatise (Feller, 1971, pp. 169–176, 574–583).

2.9 Problems

1. Suppose that the limit in formula (2.1) exists. Use the ratio test to prove that the interval of convergence of a power series is $(c-r, c+r)$ where r is given by formula (2.1). More generally, use the root test to prove that the interval of convergence of a power series is given by the Cauchy-Hadamard formula in equation (2.2).

2. Suppose that

$$\sum_{j=0}^{\infty} a_j\, x^j \cdot \sum_{k=0}^{\infty} b_k\, x^k = \sum_{n=0}^{\infty} c_n\, x^n$$

$$\sum_{j=0}^{\infty} \frac{\alpha_j}{j!}\, x^j \cdot \sum_{k=0}^{\infty} \frac{\beta_k}{k!}\, x^k = \sum_{n=0}^{\infty} \frac{\gamma_n}{n!}\, x^n .$$

Prove that

$$c_n = \sum_{j=0}^{n} a_j \, b_{n-j} \,, \quad \gamma_n = \sum_{j=0}^{n} \binom{n}{j} \alpha_j \, \beta_{n-j} \,.$$

3. Prove that the second cumulant κ_2 of a random variable equals its variance.

4. Let $\mu = \mu_1$ be the mean of a random variable X and σ^2 its variance. We define the *n-th central moment* of X to be $\nu_n = E(X - \mu)^n$. We define the *skewness* and the *kurtosis* of X to be $\gamma_1 = \nu_3/\sigma^3$ and $\gamma_2 = \nu_4/\sigma^4$, respectively. Find the skewness and kurtosis of a nonnegative random variable with gamma distribution $\mathcal{G}(\alpha, \lambda)$ and probability density function

$$f(x;\, \alpha,\, \lambda) = \frac{\lambda^\alpha \, x^{\alpha-1} \, \exp(-\lambda \, x)}{\Gamma(\alpha)}$$

where $x > 0$, $\alpha > 0$, and $\lambda > 0$.

5. Let $X^* = a\,X + b$. Show that the skewness and kurtosis of X^* are the same as the skewness and kurtosis of X. Show that the cumulants of the two random variables are related by the formula $\kappa_n^* = a^n \, \kappa_n$ for all $n > 1$, and that $\kappa_1^* = a\,\kappa_1 + b$.

6. Show that the cumulants of the Poisson distribution $\mathcal{P}(\mu)$ are $\kappa_n = \mu$ for all $n \geq 1$.

7. For a general distribution with finite moments, find formulas for the moments μ_r, where $1 \leq r \leq 5$ in terms of the cumulants κ_r, and vice versa.

8. The moments of a random variable may be recursively calculated from the cumulants and lower moments. To prove this, show that

$$\mu_n = \sum_{j=1}^{n} \binom{n-1}{j-1} \mu_{n-j} \, \kappa_j \,.$$

9. Using the results of Problems 6 and 8, find the first six moments for $\mathcal{P}(\mu)$.

10. Derive the recursion displayed in equation (2.26).

11. Let X be a nonnegative integer-valued random variable with probability generating function

$$p(\theta) = P(X = 0) + \theta\, P(X = 1) + \theta^2\, P(X = 2) + \theta^3\, P(X = 3) + \cdots.$$

(a) Prove that the domain of convergence of $p(\theta)$ includes the open set $-1 < \theta < 1$.

(b) Prove that $P(X = n) = p^{(n)}(0)/n!$.

(c) Let $(X)_k = X\,(X - 1)\,(X - 2)\,\cdots\,(X - k + 1)$. The value $\mu_{[k]} = E[(X)_k]$ is called the *k-th factorial moment of X*. Prove that

$$\mu_{[k]} = \lim_{\theta \to 1-} p^{(k)}(\theta)\,.$$

12. In equation (2.29) it was demonstrated that

$$M^*(t) = M(t) \exp[z\,(-t)^n]$$

However, one step of the argument was omitted. Fill in the details of this step. State explicitly any additional regularity assumptions necessary to derive the missing step.

13. Prove formula (2.30).

14. Let $X_1, X_2, X_3, \ldots$ be an independent sequence of random numbers that are uniformly distributed between 0 and 1. Let $0 \le x \le 1$. Define a positive integer-valued random variable by

$$N = \min(n \ge 1\ :\ X_1 + X_2 + \cdots + X_n > x)\,.$$

Prove that $P(N > n) = x^n/n!$. Use this to show that

$$E(N) = \sum_{n=0}^{\infty} P(N > n) = \exp(x)\,.$$

15. From the previous question, let $X_0 = x$, and

$$M = \min(n \ge 1\ :\ X_{n-1} > X_n)\,.$$

Prove that $P(M > m) = (1 - x)^m/m!$. Use this to show that

$$E(M) = \sum_{m=0}^{\infty} P(M > m) = \exp(1 - x)\,.$$

16. Prove Proposition 1 in Section 2.3, namely that if an alternating series is of the form $a_0 - a_1 + a_2 - a_3 + \cdots$, where $a_0 > a_1 > a_2 > \cdots > 0$, then the series is enveloping. Furthermore, if $a_n \to 0$ then the series envelops a unique value t to which it converges.

17. Prove Proposition 2 in Section 2.3.

18. Determine the values for x where the power series expansions for $\sin x$ and $\cos x$ expanded about zero are enveloping series.

19. Verify formula (2.52).

20. Stirling's approximation to the factorial function $n!$ is

$$n! = \sqrt{2\,\pi\,n}\,n^n\,e^{-n}\left[1 + O\left(\frac{1}{n}\right)\right].$$

This can also be thought of as an approximation for the gamma function using the identity $\Gamma(n+1) = n!$, when n is a nonnegative integer. Substituting, we have

$$\Gamma(1+x) = \sqrt{2\,\pi}\,x^{x+1/2}e^{-x}\left[1 + O\left(\frac{1}{x}\right)\right]$$

for real $x \to \infty$. An equivalent approximation is

$$\Gamma(x) = x^{-1}\,\Gamma(1+x) = \sqrt{2\,\pi}\,x^{x-1/2}e^{-x}\left[1 + O\left(\frac{1}{x}\right)\right]$$

The accuracy of Stirling's approximation to the gamma function was studied by J. P. M. Binet by examining the logarithmic difference

$$\varsigma(x) = \ln\,\Gamma(x) - \left[\frac{1}{2}\,\ln(2\,\pi) + \left(x - \frac{1}{2}\right)\,\ln\,x - x\right].$$

When $x > 0$, it can be shown that

$$\varsigma(x) = 2\int_0^\infty \frac{\arctan\,(t/x)}{e^{2\,\pi\,t} - 1}\,dt\,.$$

The reader can find this result in Whittaker and Watson (1962, pp. 251-252). Expand the arctangent function and set

$$\int_0^\infty \frac{t^{2\,n-1}}{e^{2\,\pi\,t} - 1}\,dt = \frac{B_n^*}{4\,n}$$

where B_n^* is the n-th *Bernoulli number*,[††] to show that

$$\frac{B_1^*}{1 \cdot 2 \cdot x} - \frac{B_2^*}{3 \cdot 4 \cdot x^3} + \frac{B_3^*}{5 \cdot 6 \cdot x^5} - \cdots \rightsquigarrow \zeta(x) .$$

[Note: the first few Bernoulli numbers can be evaluated to be $B_1^* = 1/6$, $B_2^* = 1/30$, $B_3^* = 1/42$, $B_4^* = 1/30$, $B_5^* = 5/66$. To prove this requires some contour integration.]

21. Let X have a distribution $\mathcal{G}(\alpha, 1)$ with probability density function

$$f(x) = \frac{x^{\alpha-1} \exp(-x)}{\Gamma(\alpha)}$$

when $x \geq 0$ and $f(x) = 0$ when $x < 0$. Use integration by parts to expand the required integral and thereby show that

$$P(X > t) \sim$$
$$f(t) \left[1 + \frac{\alpha - 1}{t} + \frac{(\alpha-1)(\alpha-2)}{t^2} + \frac{(\alpha-1)(\alpha-2)(\alpha-3)}{t^3} + \cdots \right]$$

for $t > 0$. Discuss the behaviour of the series for the case where α is, and the case where α is not, a positive integer.

22. Let X have distribution $\mathcal{E}(\lambda)$, that is, exponential with mean $\mu = \lambda^{-1}$. (For example, could be an interarrival time between events in a Poisson process of intensity λ.) Prove that

$$E\left(\frac{1}{1+X}\right) \sim 1 - \frac{1!}{\lambda} + \frac{2!}{\lambda^2} - \frac{3!}{\lambda^3} + \cdots .$$

Is the series enveloping? Justify your answer. Now suppose that conditionally on X, the random variable Y has a uniform distribution on the interval $(0, X)$. Show that the probability density function of Y has the form

$$f(y) = \int_y^\infty \frac{\lambda \exp(-\lambda x)}{x} \, dx$$
$$\sim \frac{\exp(-\lambda y)}{y} \left(1 - \frac{1!}{\lambda} + \frac{2!}{\lambda^2} - \frac{3!}{\lambda^3} + \cdots \right) .$$

In the special case where $\lambda = 1$, the integral above is called the *exponential integral*, and is denoted by $E_1(y)$.

[††] There are two distinct notations for the Bernoulli numbers, which is a source of some confusion. We shall label the order Whittaker and Watson notation as B_n^* and the more modern Ibramowitz and Stegun notation as B_n. The main difference is that in the modern notation $B_3 = B_5 = \cdots = 0$, and $B_n^* = \pm B_{2n}$, depending upon the value of n. There is an advantage to each notation in particular contexts.

23. Verify (2.55).

24. So far, our two main tools for expanding functions have been Taylor series and integration by parts. We shall show that these two methods are closely related to each other as follows. Using integration by parts, prove first that

$$\int_0^x f'(t)\,(x-t)^n\,dt = f'(0)\,\frac{x^{n+1}}{n+1} + \frac{1}{n+1}\int_0^x f''(t)\,(x-t)^{n+1}\,dt.$$

Apply the formula for consecutive values of $n \geq 0$ to obtain the Taylor expansion for $f(x)$ about $x = 0$.

25. Prove that a random variable whose characteristic function is of the form

$$\chi(t) = \exp(-c\,|t|^\alpha)$$

will satisfy the scaling law given in (2.57).

26. The characteristic function of

$$f(x) = \frac{1}{\pi\,\sqrt{1-x^2}} \qquad -1 < x < 1$$

is

$$\sum_{k=0}^{\infty} \frac{(-1)^k}{k!\,k!}\left(\frac{t}{2}\right)^{2k} = J_0(t)$$

where $J_0(t) = I_0(i\,t)$ and I_0 is the Bessel function.

27. The *arc sine law* on the interval (0, 1) is associated with random walks and the limiting distribution of the zero crossings of symmetric random walks. It has density

$$f(x) = \frac{1}{\pi\,\sqrt{x\,(1-x)}} \qquad 0 < x < 1\,.$$

Modify the previous example to show that this distribution has characteristic function $exp(i\,t/2)\,J_0(t/2)$.

28. Let $\overline{X}_n$ denote the sample average of n independent $\mathcal{P}(\lambda)$ random variables.

    (a) Determine the first six cumulants of $\overline{X}_n$, as well as the first six cumulants of $Z = \sqrt{n}\,(\overline{X}_n - \lambda)$.
    (b) Determine the first six moments of $\overline{X}_n$ and Z.
    (c) From these results, prove (2.72).

CHAPTER 3

Padé approximants and continued fractions

3.1 The Padé table

A *rational function* is a function which can be represented in the form $p(x)/q(x)$ where $p(x)$ and $q(x)$ are polynomials. When $p(x)$ has degree m and $q(x)$ has degree n, we shall say that the ratio has degree (m, n).

As was discussed earlier in the last chapter, both power series and asymptotic series share the common property that their partial sums are rational functions. So a natural extension of our investigations is a theory by which general functions may be locally approximated by rational functions. Taylor's theorem and Taylor polynomials provide the theoretical basis for the use of polynomial approximations to analytic functions. The corresponding theory for rational functions leads to the class of Padé approximants.

Although named after Henri Padé, these rational approximations to power series were known long before his time. However, it was Padé who provided the first systematic study of their properties. In particular, it was Padé who arranged these approximations in a doubly indexed table and studied the types of continued fractions which are also Padé approximants.

Let $f(x)$ be a function that can be represented by a power series. For simplicity, we shall assume that the power series is about zero. So

$$f(x) = \sum_{r=0}^{\infty} c_r\, x^r\,.$$

By the *Padé approximant* of degree (m, n) for $f(x)$ we shall mean the rational function

$$\begin{aligned} f_{[m,\,n]}(x) &= \sum_{j=0}^{m} a_j\, x^j \bigg/ \sum_{k=0}^{n} b_k\, x^k \\ &= p(x)/q(x) \end{aligned}$$

of degree (m, n) which satisfies the condition

$$p(x) - f(x)\,q(x) = o(x^{m+n}) \tag{3.1}$$

as $x \to 0$. This is the same as the limiting condition

$$\lim_{x\to 0} x^{m+n}\,[p(x) - f(x)\,q(x)] = 0\,.$$

To see how this order condition in (3.1) works, let us write it out as

$$\left(\sum_{k=0}^{n} b_k\,x^k\right)\cdot\left(\sum_{r=0}^{\infty} c_r x^r\right) = \sum_{j=0}^{m} a_j\,x^j + o(x^{m+n})\,. \tag{3.2}$$

We are only interested in the terms of this equation up to order $m+n$. So we can write $f(x) = \sum_{r=0}^{m+n} c_r\,x^r + o(x^{m+n})$, and reformulate equation (3.2) as

$$\left(\sum_{k=0}^{n} b_k\,x^k\right)\cdot\left(\sum_{r=0}^{m+n} c_r x^r\right) = \sum_{j=0}^{m} a_j\,x^j + o(x^{m+n})\,. \tag{3.3}$$

Next, we expand the left-hand side of (3.3), and collect terms with common powers of x, to get

$$\sum_{j=0}^{m+n}\left(\sum_{k=0}^{\min(j,n)} b_k\,c_{j-k}\right)x^j = \sum_{j=0}^{m} a_j\,x^j + o(x^{m+n})\,. \tag{3.4}$$

The coefficients on the left can be equated with corresponding coefficients on the right for each value of j. This gives us the following set of simultaneous equations.

$$\sum_{k=0}^{\min(j,n)} b_k\,c_{j-k} = a_j \quad \text{for } 0 \le j \le m \tag{3.5}$$

$$\sum_{k=0}^{\min(j,n)} b_k\,c_{j-k} = 0 \quad \text{for } m < j \le m+n\,. \tag{3.6}$$

This is seen to be a set of $m+n+1$ simultaneous equations on the $m+1$ coefficients of $p(x)$ and the $n+1$ coefficients of $q(x)$. This is just the right number of equations, because we can choose to set any nonzero term in $q(x)$ equal to one by multiplying numerator and denominator by an appropriate constant.*

* The definition we have given here oversimplifies some of the regularity assumptions for Padé approximants. However, it is helpful not to get stuck on these issues at this point. See Problem 1 at the end of the chapter for an exploration of an example where the regularity fails.

The full set of Padé approximants of a given function $f(x)$ can be arranged in a table (known as the Padé table) as follows.

$$\begin{array}{ccccc} f_{[0,0]}(x) & f_{[0,1]}(x) & f_{[0,2]}(x) & f_{[0,3]}(x) & \cdots \\ f_{[1,0]}(x) & f_{[1,1]}(x) & f_{[1,2]}(x) & f_{[1,3]}(x) & \cdots \\ f_{[2,0]}(x) & f_{[2,1]}(x) & f_{[2,2]}(x) & f_{[2,3]}(x) & \cdots \\ \vdots & \vdots & \vdots & \vdots & \ddots \end{array}$$

Down the left-hand column of the table, the functions are all polynomials, which can easily be recognised as the Taylor polynomials from the power series expansion about $x = 0$. Other elements of the table are *bona fide* rational functions whose denominators are of positive degree. The functions $f_{[n,n]}(x)$ down the main diagonal are of particular interest for certain approximations of functions with singularities as shall be seen below. Functions above and below this main diagonal are related by the formula

$$\left[\frac{1}{f(x)}\right]_{[m,\,n]} = \frac{1}{f_{[n,\,m]}(x)}\,. \tag{3.7}$$

See Problem 2. So for example, the Padé approximants along the top row are the reciprocals of Taylor polynomials–but polynomials approximating $1/f(x)$ and not $f(x)$ itself.

What advantages can be obtained by using Padé approximants rather than Taylor polynomials? To answer this question, let us compare the three functions

$$f_{[2n,\,0]}(x), \quad f_{[n,\,n]}(x), \quad f_{[0,\,2n]}(x),$$

the first being a standard Taylor polynomial fit to $f(x)$ and the other two alternative rational approximations to $f(x)$. These three functions lie on the same small diagonal of the Padé table running from lower left to upper right. So all three fit $f(x)$ to the same order about $x = 0$, and there is no obvious reason to prefer one over the others from any error analysis close to zero. However, the global properties of $f(x)$ may suggest that one of these approximations should work better than the others when x is not close to zero. Consider, for example the following three cases. (See Figure 3.1.)

1. The function $f(x)$ is bounded on finite intervals, and also satisfies

$$\lim_{x\to\pm\infty} f(x) = \pm\infty.$$

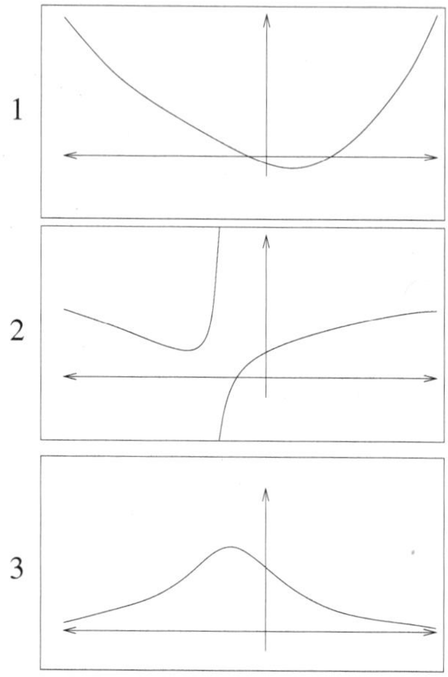

Figure 3.1 *Three cases of functions with different types of global behaviour*

2. The function $f(x)$ satisfies

$$\lim_{x \to a} f(x) = \pm\infty.$$

3. The function $f(x)$ is bounded away from zero on finite intervals and also satisfies

$$\lim_{x \to \pm\infty} f(x) = 0.$$

Obviously, this list is not exhaustive. Note that the first property is shared by all nonconstant polynomials. Therefore, a function satisfying this is, to a certain extent, rather "polynomial-like" in its properties. On the other hand, polynomials never satisfy the second and third properties. So functions satisfying the second and third conditions are quite "unpolynomial-like" so to speak. On the other hand, rational functions can easily satisfy the second and third conditions. The reader will observe that the third condition is satisfied by many density functions such as the density of the normal distribution. If we take the reciprocal of any function which satisfies property three, we have a function which is "polynomial-like" in its global behaviour. When approximating functions whose global properties are quite different from the behaviour of a polynomial, we may reasonably hope that a rational Padé approximant can perform better than any Taylor polynomial.

Alongside the reasons given above, there are also computational advantages to Padé approximants. For example, when solving an equation of the form $f(x) = y_0$ for given y_0, there are computational advantages to using a diagonal Padé approximant $f_{[n,\,n]}(x)$ rather than the polynomial $f_{[2n,0]}(x)$. This is because the approximating equation

$$\frac{\sum_{j=0}^{n} a_j\, x^j}{\sum_{k=0}^{n} b_k\, x^k} = y_0$$

reduces to

$$\sum_{j=0}^{n} (a_j - y_0\, b_j)\, x^j = 0$$

which requires the solution of a polynomial of degree n rather than of degree $2\,n$ in the case of $f_{[2n,\,0]}(x) = y_0$.

Because Padé approximants require the solution of a number of simultaneous equations, it is often difficult to find an expression for the approximants of a given function. This task is made substantially easier by symbolic computation using Maple. The *pade* command is to be found in the *numapprox* package, and is used as follows. The command

$> \; pade(f(x),\, x,\, [m, n])$

returns the Padé approximation to $f(x)$ to order (m, n) about $x = 0$. The more general command

$> \; pade(f(x),\, x = x0,\, [m, n])$

performs the same calculation with the approximation centred about $x = x0$, rather than $x = 0$ as we have used above.

3.2 Padé approximations for the exponential function

To illustrate the Padé functions, let us find the Padé approximant of degree $(m,\, n)$ for the exponential function locally about zero. Let $p(x) = a_0 + a_1 x + \cdots + a_m x^m$ and $q(x) = b_0 + b_1 x + \cdots + b_n x^n$ be such that

$$\exp_{[m,\,n]}(x) = p(x)/q(x)\,.$$

Performing a Taylor expansion on the exponential function to order $m+n$, we get

$$\exp(x) = 1 + x + \frac{x^2}{2!} + \cdots + \frac{x^{m+n}}{(m+n)!} + O(x^{m+n+1}) \quad \text{as } x \to 0. \quad (3.8)$$

Henri Eugène Padé (1863–1953)

In his 1892 thesis, Henri Eugène Padé systematically studied rational approximants to power series. He arranged these approximants in a table now known as the Padé table and showed the relationship between the elements of this table and continued fraction expansions.

"The fundamental motive for studying Padé approximants is the extraction of information about a function from the first few terms of its power series."

Baker & Graves-Morris, 1996, p. 372.

Inserting this into conditions (3.5) and (3.6), with $c_r = 1/r!$, we get

$$\begin{aligned}
a_0 - b_0 &= 0 \\
a_1 - (b_1 + b_0) &= 0 \\
a_2 - (b_2 + b_1 + b_0/2) &= 0 \\
\vdots &\quad \vdots \\
\frac{b_0}{(m+n)!} + \frac{b_1}{(m+n-1)!} + \cdots + \frac{b_n}{m!} &= 0
\end{aligned}$$

to which we can append the additional equation $b_0 = 1$. For example, when $m,\ n = 1$, these equations reduce to $b_0 = 1$, $a_0 = b_0$, $a_1 = b_1 + b_0$, and $b_1 = -b_0/2$. So the Padé approximant of degree $(1,\ 1)$ is

$$\exp_{[1,\,1]}(x) = \frac{1 + x/2}{1 - x/2}.$$

More generally, we have

$$\exp_{[m,\,n]}(x) = p_{m\,n}(x)/p_{n\,m}(-x), \tag{3.9}$$

where

$$p_{m\,n}(x) = \sum_{k=0}^{m} \frac{(m+n-k)!\,m!}{(m+n)!\,k!\,(m-k)!}\,x^k. \tag{3.10}$$

3.3 Two applications

3.3.1 Padé approximants to stable densities

Among the symmetric stable distributions, only the Cauchy distribution has a density which is a rational function. Nevertheless, the symmetric stable distributions with exponent $\alpha < 2$ have tails which obey a power law much as rational functions do. Therefore, it is of interest to see how well Padé approximants behave. While no simple expression is available for most stable densities, we do have representations via power series such as formula (2.63). As Padé approximants are calculated via the power series of a function, we are no worse off for failing to have a simple formula.

As a case in point, let us consider the approximation of a symmetric stable density with $\alpha = 1.5$. The power series in (2.63) is formally convergent. However, as noted earlier, this is of little use. To compute the density over a reasonable range, we need a power series of at least one hundred terms. Unfortunately, most simple naive calculation of the terms this far out produces serious overflow and underflow problems.

The alternative is to find a Padé approximant of much lower order which is competitive with a polynomial of degree 100. To calculate the Padé approximant does not require that we compute the tail of the density accurately.

Only the appropriate number of derivatives at zero need to be found. To compute the Taylor polynomial of degree 20, say, which approximates the density at zero, we can use the Maple code

$$> f := (x, \alpha) \to \frac{1}{\pi \cdot \alpha} \cdot \sum_{k=0}^{10} \frac{(-1)^k \cdot \Gamma\left(\frac{(2 \cdot k + 1)}{\alpha}\right) \cdot x^{2 \cdot k}}{(2 \cdot k)!}$$

We call on the Maple command

$$> pade(f(x, 1.5),\ x = 0,\ [8, 11])$$

which can be found in the *numapprox* package. From the Maple output we find to three significant figures that

$$f_{[8,\,11]}(x,\, 1.5) =$$

$$\frac{0.287 + 0.0^2557x^2 + 0.0^2197x^4 + 0.0^4131x^6 + 0.0^5114x^8}{1 + 0.389x^2 + 0.0649x^4 + 0.0^2587x^6 + 0.0^3291x^8 + 0.0^5644x^{10}}.$$

While the order of this fit is only $m + n = 19$, the quality of the approximation holds up surprisingly well when compared with a Taylor polynomial of order 180 as shown in Figure 3.2. As can be seen, the two approximations are close in the centre of the distribution. In the tails, however, the "polynomial nature" of the latter function starts to dominate. Despite the fact that the order of the Padé approximant is much lower, it continues to be well behaved beyond the region where the polynomial works. Although this Padé approximant has greater stability than the Taylor polynomial, it does not obey the correct power law in the tails. A more accurate approximation for extreme tail values can be obtained from the asymptotic series in (2.62).

3.3.2 Padé approximants for likelihood functions

In statistical inference, it is commonplace to use a likelihood function to construct an approximate confidence interval for a parameter. Suppose $x_1, \ldots, x_n$ are observations based on random sampling from a distribution whose probability density function is $f(x; \theta)$, where $\theta \in \Theta$. For

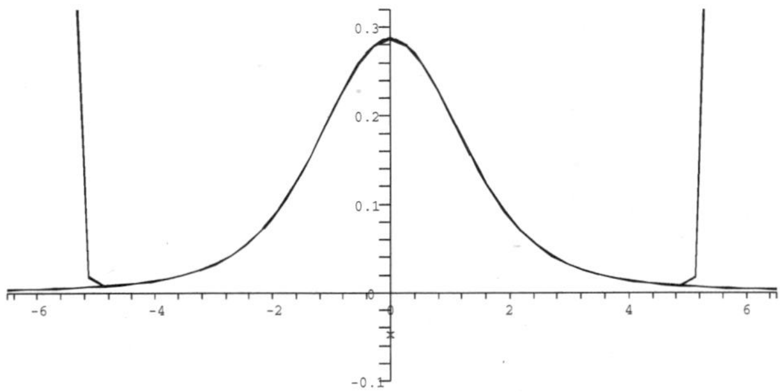

Figure 3.2 *Comparison between a Padé approximant of order* $(8, 11)$ *and a Taylor polynomial of order* 180 *for the centred and scale symmetric stable law with exponent* $\alpha = 1.5$

simplicity, we shall assume that Θ is an open subset of $\mathbb{R}$. The *likelihood function* for θ is defined as

$$L_n(\theta) = c(x_1, \ldots, x_n)\, f(x_1;\, \theta)\, f(x_2;\, \theta) \cdots f(x_n;\, \theta)\,.$$

In this definition, $c(x_1, \ldots, x_n)$ is an arbitrary function that does not depend on θ. It is often chosen to standardise the likelihood at its maximum or to simplify the expression. The *maximum likelihood estimator* for θ is that value $\widehat{\theta}_n$ which maximises the likelihood, namely

$$\widehat{\theta}_n = \arg\max_{\theta \in \Theta} L_n(\theta)\,,$$

so that $L_n(\widehat{\theta}_n) \geq L_n(\theta)$ for all $\theta \in \Theta$. In practice, the maximisation of the likelihood is often achieved by maximising the *log-likelihood* function, defined as

$$\ell_n(\theta) = \ln L_n(\theta)\,.$$

A $100\,p\,\%$ *likelihood region* for θ is defined as

$$\left\{\theta \in \Theta\ :\ L_n(\theta)\,/\,L(\widehat{\theta}_n) \geq p\right\} = \left\{\theta \in \Theta\ :\ \ell_n(\theta) - \ell_n(\widehat{\theta}_n) \geq \ln p\right\}.$$

In many examples, this set will be an interval. So we typically refer to this as a $100\,p\,\%$ *likelihood interval.* Unfortunately, even in simple models, it is often impossible to find explicit formulas for the endpoints of such a likelihood interval. When this is the case, we have to rely on an iterative numerical method such as Newton-Raphson to approximate the endpoints. However, as we shall see below, highly accurate approximations to likelihood intervals can be obtained with explicit formulas using Padé approximants.

For example, let us consider a set of observations $x_1, \ldots, x_n$ drawn from

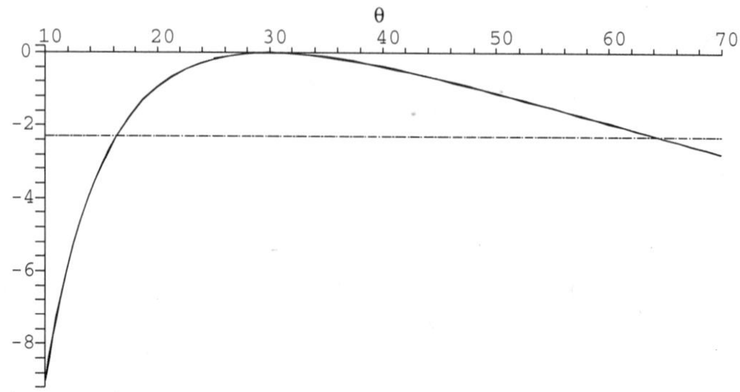

Figure 3.3 *The function* $\ell_n(\theta) - \ell_n(\widehat{\theta}_n)$ *for an exponential model with* $n = 10$ *observations, showing also the 10 % likelihood interval cutoff.*

an exponential distribution with mean θ. The maximum likelihood estimator for θ in this model is $\widehat{\theta}_n = \overline{x}_n$. Figure 3.3 shows the *log-relative likelihood*

$$\begin{aligned} r_n(\theta) &= \ell_n(\theta) - \ell_n(\widehat{\theta}_n) \\ &= n\left(\ln\frac{\overline{x}_n}{\theta} + 1 - \frac{\overline{x}_n}{\theta}\right) \end{aligned} \tag{3.11}$$

for $n = 10$ observations with $\overline{x}_n = 30$. Also plotted is the horizontal line $y = \ln 0.1$ which marks the cutoff points for the 10% likelihood interval. In this example, the endpoints of a $100\,p\,\%$ likelihood interval can be found numerically as a function of p, but not analytically. A common solution to this problem is to use a quadratic approximation to $r_n(\theta)$ via the Taylor expansion to $r_n(\theta)$ about $\theta = \widehat{\theta}$. This does not seem appropriate here because of the strong asymmetry of the log-relative likelihood about its maximum. Using a Taylor expansion of order three or four leads to very complicated approximations for the endpoints of the likelihood interval.

An alternative approach is to use a Padé approximant of order (2, 2), $r_{[2,\,2]}(\theta)$ say, which fits the curve to fourth order, but requires only the solution to quadratic equations when solving for the endpoints. Upon calculating the coefficients of the rational function, we find that

$$r_{[2,\,2]}(\theta) = \frac{9\,n\,(\,\theta - \widehat{\theta}_n\,)^2}{\widehat{\theta}_n^{\,2} - 14\,\widehat{\theta}_n\,\theta - 5\,\theta^2}\,. \tag{3.12}$$

Solving the inequality $r_{[2,\,2]}(\theta) \geq \ln p$ yields an interval with endpoints

$$\widehat{\theta}\left[1 - \frac{12 \ln p \pm 3\sqrt{6\,(\ln p)^2 - 18\,n \ln p}}{9\,n + 5 \ln p}\right] \qquad (3.13)$$

For $\widehat{\theta} = 30$, $n = 10$ and $p = 0.1$, the Padé approximate $100\,p\,\%$ likelihood interval is

$$[\,\underline{\theta}_{2,2},\, \overline{\theta}_{2,2}\,] = [16.337,\, 64.785]$$

which can be compared with the more precise numerical solution

$$[\,\underline{\theta},\, \overline{\theta}\,] = [16.304,\, 64.461]\,.$$

The formula for the (2, 2) Padé approximant for a general log-likelihood can also be written down fairly easily. We find that

$$r_{[2,\,2]}(\theta) = \frac{3\,(\widehat{\ell_n''})^3\,(\,\theta - \widehat{\theta}_n\,)^2/2}{3\,(\widehat{\ell_n''})^2 - \widehat{\ell_n'''}\,\widehat{\ell_n''}(\,\theta - \widehat{\theta}_n\,) + [4\,(\widehat{\ell_n'''})^2 - 3\,\widehat{\ell_n''}\,\widehat{\ell_n^{(iv)}}]\,(\,\theta - \widehat{\theta}_n\,)^2/12}$$

where $\widehat{\ell_n^{(k)}} = \ell_n^{(k)}(\widehat{\theta}_n)$ for $k = 1, 2, \ldots$.

3.4 Continued fraction expansions

By a continued fraction expansion of a function $f(x)$ we shall mean a representation of the function which typically has the form

$$f(x) = b_0 + \cfrac{a_1\,x}{b_1 + \cfrac{a_2\,x}{b_2 + \cfrac{a_3\,x}{b_3 + \cfrac{a_4\,x}{b_4 + \cdots}}}} \qquad (3.14)$$

which represents the limit of a sequence of rational function approximations, namely,

$$b_0\,, \qquad b_0 + \frac{a_1\,x}{b_1}\,, \qquad b_0 + \cfrac{a_1\,x}{b_1 + \cfrac{a_2\,x}{b_2}}\,, \qquad \text{and so on.}$$

For those who are used to the tidy organisation of Taylor series, the amount of freedom available in constructing continued fraction expansions of a given function may be surprising. For example, we can standardise the denominators of this sequence of functions so that the expansion has the form

$$b_0 + \cfrac{a_1\,x}{1 + \cfrac{a_2\,x}{1 + \cfrac{a_3\,x}{1 + \cfrac{a_4\,x}{1 + \cdots}}}} \qquad (3.15)$$

which is called a *regular C-fraction* when $a_j \neq 0$ for all j. An alternative form of the regular C-fraction expansion is

$$b_0 + \cfrac{x}{b_1 + \cfrac{x}{b_2 + \cfrac{x}{b_3 + \cfrac{x}{b_4 + \cdots}}}} \tag{3.16}$$

Here, the letter C stands for the word "corresponding" because there is a one-to-one correspondence between C-fractions and power series which do not represent rational functions. There is also a one-to-one correspondence between finite (*i.e.*, terminating) C-fractions and the power series of rational functions. The C-fraction expansions arise in the conversion of Maclaurin power series into continued fraction form, or through Viskovatov's method for converting power series into continued fractions, as we shall discuss below. The sequence of rational approximations that arise as C-fraction expansions in this way can be shown to be equivalent to staircase sequences of Padé approximants such as

$$\begin{array}{ccccc}
\boxed{f_{[0,0]}(x)} & f_{[0,1]}(x) & f_{[0,2]}(x) & f_{[0,3]}(x) & \cdots \\
\boxed{f_{[1,0]}(x)} & \boxed{f_{[1,1]}(x)} & f_{[1,2]}(x) & f_{[1,3]}(x) & \cdots \\
f_{[2,0]}(x) & \boxed{f_{[2,1]}(x)} & \boxed{f_{[2,2]}(x)} & f_{[2,3]}(x) & \cdots \\
f_{[3,0]}(x) & f_{[3,1]}(x) & \boxed{f_{[3,2]}(x)} & \boxed{f_{[3,3]}(x)} & \cdots \\
\vdots & \vdots & \vdots & \vdots & \ddots
\end{array}$$

in the case of expansions (3.15) or the mirror image of this about its diagonal in the case of (3.16).

At this stage, it is useful to introduce a more compact notation for continued fractions, as the notation used above takes up a large amount of space. We can write (3.15) and (3.16) as

$$b_0 + \frac{a_1\,x}{1\ +}\,\frac{a_2\,x}{1\ +}\,\frac{a_3\,x}{1\ +}\cdots \tag{3.17}$$

and

$$\frac{a_1}{1\ +}\,\frac{a_1\,x}{1\ +}\,\frac{a_2\,x}{1\ +}\,\frac{a_3\,x}{1\ +}\cdots. \tag{3.18}$$

We now turn to the problem of deriving a continued fraction expansion from a power series. Suppose we can write

$$f(x) = c_0 + c_1\,x + c_2\,x^2 + c_3\,x^3 + \cdots$$

We reorganise this series as follows

$$
\begin{aligned}
f(x) &= c_0 + c_1\,x\left[1 + \frac{c_2}{c_1}\,x + \frac{c_3}{c_1}\,x^2 + \cdots\right] \\
&= c_0 + c_1\,x\left[1 + c_1^*\,x + c_2^*\,x^2 + \cdots\right]^{-1} \\
&= c_0 + c_1\,x\left[1 + c_1^*\,x\left(1 + \frac{c_2^*}{c_1^*}\,x + \cdots\right)\right]^{-1} \\
&= c_0 + c_1\,x\left[1 + c_1^*\,x\,(1 + c_1^{**}\,x + \cdots)^{-1}\right]^{-1}
\end{aligned}
$$

and so on. Continuing in this way, we obtain a C-fraction expansion of the form

$$
f(x) = c_0 + \frac{c_1\,x}{1\ \ +}\,\frac{c_1^*\,x}{1\ \ +}\,\frac{c_1^{**}\,x}{1\ \ +}\cdots. \tag{3.19}
$$

In practice, the following algorithm, known as Viskovatov's method, can be used. This method works on ratios of power series—a generalisation of the situation above—and repeatedly uses the following step for the basic algorithm:

$$
\begin{aligned}
\frac{\sum_{j=0}^{\infty} a_j\,x^j}{\sum_{j=0}^{\infty} b_j\,x^j} &= \frac{a_0}{b_0} + \cfrac{x}{\cfrac{\sum_{j=0}^{\infty} b_j\,x^j}{\sum_{j=0}^{\infty}\left(a_{j+1} - a_0 b_{j+1}/b_0\right)x^j}} \qquad (3.20) \\
&= \frac{a_0}{b_0} + \cfrac{x}{\cfrac{\sum_{j=0}^{\infty} a'_j\,x^j}{\sum_{j=0}^{\infty} b'_j\,x^j}}\,. \qquad (3.21)
\end{aligned}
$$

The algorithm then proceeds by applying the same procedure to the quotient

$$
\frac{\sum_{j=0}^{\infty} a'_j\,x^j}{\sum_{j=0}^{\infty} b'_j\,x^j} = \frac{a'_0}{b'_0} + \cdots,
$$

and so on. For example, Problem 10 asks the reader to verify that the exponential function has continued fraction expansion

$$
\begin{aligned}
\exp(x) &= 1 + \frac{x}{1\ \ +}\,\frac{x}{-2\ \ +}\,\frac{x}{-3\ \ +}\,\frac{x}{2\ \ +} \\
&\qquad \frac{x}{5\ \ +}\,\frac{x}{-2\ \ +}\,\frac{x}{-7\ \ +}\cdots. \qquad (3.22)
\end{aligned}
$$

Viskovatov's method can also be used to turn an asymptotic series into a continued fraction. This is because an asymptotic series can be formally written as a power series in x^{-1}. So Viskovatov's method on an asymptotic series proceeds in the same way as (3.21) above, with x^{-j} replacing x^j.

In Maple, a rational function or function represented by a power series can be converted into a continued fraction equivalent to a staircase sequence of the Padé table using the *convert* command. The basic form of this command for a continued fraction is

$> convert(f(x), confrac, x)$

When $f(x)$ is a rational function, the output from this is the exact function in continued fraction form. When $f(x)$ is not rational, then the output is a finite continued fraction expansion with a default number of partial quotients (*i.e.*, the number of steps of the expansion). To control the number of partial quotients, an optional positive integer may be appended to the list such as

$> convert(f(x), confrac, x, n)$

which carries the expansion out to n steps for positive integer n. For example, the command

$> convert(x \cdot \exp(-x), confrac, x, 5)$

gives

$$\cfrac{x}{1 + \cfrac{x}{1 + \cfrac{x}{-2 - \frac{1}{3}x}}}$$

as output.

3.5 A continued fraction for the normal distribution

Let us now return to consider the approximation of Mills' ratio

$$\Psi(x) = \frac{1 - \Phi(x)}{\phi(x)}$$

for the normal distribution, which we discussed previously in Chapter 2. We remind the reader that $\phi(x)$ and $\Phi(x)$ are the standard normal density function and distribution function, respectively. In that section, we found an asymptotic series for $\Psi(x)$ whose partial sums are rational functions. The rational approximations for Mills' ratio which converge reliably and quickly are often those provided by continued fractions. One such approximation, due to Laplace, can be obtained as follows. Let

$$\Omega(t) = x\,\Psi[x\,(1-t)]. \tag{3.23}$$

It is not hard to check that $\Omega(t)$ satisfies the differential equation

$$\frac{d\Omega(t)}{dt} = -x^2\left[(1-t)\,\Omega(t) - 1\right]. \tag{3.24}$$

Now, let us write out $\Omega(t)$ as a power series. Set

$$\Omega(t) = \sum_{j=0}^{\infty} \omega_j\, t^j \tag{3.25}$$

so that

$$\Omega'(t) = \sum_{j=0}^{\infty} (j+1)\,\omega_{j+1}\, t^j.$$

Plugging these two power series into (3.24), we get

$$\omega_1 + \sum_{j=1}^{\infty}(j+1)\,\omega_{j+1}\,t^j = x^2\,(1-\omega_0) + \sum_{j=1}^{\infty} x^2\,(\omega_{j-1} - \omega_j)\,t^j. \tag{3.26}$$

Equating coefficients, we see that

$$\omega_1 = x^2\,(1-\omega_0) \tag{3.27}$$

and

$$(j+1)\,\omega_{j+1} = x^2\,\omega_{j-1} - x^2\,\omega_j \quad j = 1,\, 2,\, \ldots\,. \tag{3.28}$$

Equation (3.28) can be interpreted recursively, if it is rewritten as

$$\frac{\omega_j}{\omega_{j-1}} = \frac{x}{x + \frac{j+1}{x}\,\frac{\omega_{j+1}}{\omega_j}}\,. \tag{3.29}$$

We can recognise this as the kind of recursion used in Viskovatov's method in equation (3.21). Both recursions lead to a continued fraction expansion. Starting with $j = 1$, and applying this formula recursively, we see that

$$\begin{aligned} \frac{\omega_1}{\omega_0} &= \frac{x}{x + \frac{2}{x}\,\frac{\omega_2}{\omega_1}} \\ &\;\vdots \qquad\qquad \ddots \\ &= \frac{x}{x\,+}\,\frac{2}{x\,+}\,\frac{3}{x\,+}\,\frac{4}{x\,+}\,\cdots \end{aligned} \tag{3.30}$$

Now from (3.23) and (3.25) we have

$$\Psi(x) = \frac{\Omega(0)}{x} = \frac{\omega_0}{x} \tag{3.31}$$

Additionally, from (3.27) we get

$$\omega_0 = 1 - \frac{\omega_1}{x^2}. \tag{3.32}$$

Combining equations (3.30–3.32), we get

$$\begin{aligned}\Psi(x) &= \cfrac{1}{x + \frac{1}{x}\left(\frac{\omega_1}{\omega_0}\right)} \\ &= \frac{1}{x\ +}\,\frac{1}{x\ +}\,\frac{2}{x\ +}\,\frac{3}{x\ +}\cdots.\end{aligned} \tag{3.33}$$

This continued fraction converges for all $x > 0$, and quite rapidly when x is large. Using the fact that $1 - \Phi(x) = \phi(x)\,\Psi(x)$, we may write the normal distribution function as

$$\Phi(x) = 1 - \frac{\exp(-x^2/2)}{\sqrt{2\,\pi}}\left[\frac{1}{x\ +}\,\frac{1}{x\ +}\,\frac{2}{x\ +}\,\frac{3}{x\ +}\cdots\right]. \tag{3.34}$$

These expansions were used by Sheppard (1939) to compute tables of the standard normal distribution. The well-known Biometrika Tables For Statisticians used Sheppard's calculations to compute the standard normal distribution function. The reader will find more examples of continued fraction expansions in the problems at the end of the chapter.

3.6 Approximating transforms and other integrals

A rather useful consequence of the Padé formulation is in the approximation of integrals. Techniques for expanding integrals were introduced in Chapter 2, and will be a central topic of the next chapter. In this section, we shall consider a way to approximate an integral using a rational approximation directly on the integrand itself.

Suppose we wish to find an approximation for the characteristic function of a random variable whose probability density function is $f(x)$. We will need to evaluate

$$\chi(t) = \int_{-\infty}^{\infty} e^{i\,t\,x}\, f(x)\, dx \tag{3.35}$$

for all real t. While numerical methods can be used to approximate this integral for any desired t, it would be preferable to have an explicit function of t that is a good approximation when t is any value in some appropriate range. One approach is to replace the density $f(x)$ by a rational approximant such as $f_{[m,\,n]}(x)$, where $n > m$. The next step is to perform a partial fraction decomposition on $f_{[m,\,n]}(x)$. The following proposition is given without detailed proof.

Proposition 1. Suppose $f_{[m,\,n]}(x) = p_m(x)/q_n(x)$ where $p_m(x)$ and $q_n(x)$ are polynomials of degree m and n, respectively, with $n > m$. Let $q_n(x)$ have n distinct complex roots denoted by $\rho_1, \ldots, \rho_n$. In addition, we suppose that $p_m(x)$ and $q_n(x)$ have no common root. Then there exist complex constants $a_1, \ldots, a_n$ such that

$$f_{[m,\,n]}(x) = \sum_{j=1}^{n} \frac{a_j}{x - \rho_j}. \tag{3.36}$$

The right-hand side is the partial fraction decomposition of $f_{[m,\,n]}(x)$.

Proof. We omit the details of the proof. The coefficient a_j can be shown to be the complex residue of the function $f_{[m,\,n]}(x)$ at the point ρ_j, denoted by

$$\begin{aligned} a_j &= \operatorname{Res}\left[f_{[m,\,n]}(x),\, \rho_j\right] \\ &= \oint_{C(\rho_j)} f_{[m,\,n]}(z)\, dz, \end{aligned}$$

where $C(\rho_j)$ is a circle taken in the counterclockwise sense around ρ_j such that $f_{[m,\,n]}(z)$ is analytic on and in the interior of $C(\rho_j)$ with the exception of the point ρ_j itself. For more on the theory of residues, the reader is referred to Rudin (1987). ■

In practice, we do not evaluate the contour integral directly, but solve for the coefficients a_j by clearing the denominators in (3.36) and equating coefficients on left- and right-hand sides with common powers of x. This is the standard procedure for partial fractions taught in elementary calculus, adapted here for this context.

Let us now assume that $q_n(x)$ has no zeros on the real axis. This requires that n be even if $p_m(x)/q_n(x)$ is in lowest form. Let $n = 2k$. As there are no roots on the real axis, the roots are paired with k of them being above the real axis, and the remaining k being their complex conjugates below the real axis. It will be useful to order $\rho_1, \ldots, \rho_n$ so that the roots are of the form $\rho_1, \ldots \rho_k, \overline{\rho_1}, \ldots \overline{\rho_k}$, where $\Im(\rho_j) > 0$ for all $j = 1, \ldots, k$. Here, the bar notation denotes the complex conjugate.

Now

$$\int_{-\infty}^{\infty} e^{itx} f_{[m,\,n]}(x)\, dx = \sum_{j=1}^{n} a_j \int_{-\infty}^{\infty} \frac{e^{itx}\, dx}{x - \rho_j}. \tag{3.37}$$

We can evaluate each integral in formula (3.37) by means of contour integration in the complex plane so that

$$\int_{-\infty}^{\infty} \frac{e^{itx}\, dx}{x - \rho_j} = \begin{cases} 2\pi i\, e^{it\rho_j} & \text{when } \Im(t\,\rho_j) > 0 \\ 0 & \text{when } \Im(t\,\rho_j) < 0 \\ \text{undefined} & \text{when } t = 0. \end{cases} \tag{3.38}$$

Putting equations (3.37) and (3.38) together gives

$$\int_{-\infty}^{\infty} e^{i\,t\,x}\, f_{[m,\,n]}(x)\, dx = \begin{cases} 2\,\pi\,i\sum_{j=1}^{k} a_j\, e^{i\,t\,\rho_j} & \text{when } t > 0 \\ \\ 2\,\pi i\sum_{j=1}^{k} a_j\, e^{i\,t\,\overline{\rho_j}} & \text{when } t < 0\,. \end{cases} \tag{3.39}$$

For an application of (3.39), the reader can check that, if $n \geq 2$, formula (3.39) gives the exact solution for the Cauchy distribution, which has a rational density function.

In pursuing the approximation in (3.39), we have made a number of assumptions that may not be satisfied in particular examples. Most notably, we have assumed that the zeros of $q_n(x)$ are all distinct. When the zeros are not distinct, the method must obviously be modified. However, the spirit of the argument remains much the same, with a more general partial fraction decomposition formula.

In the method above, we applied the Padé approximation before integrating. An obvious alternative is to expand the integrand as a power series or asymptotic series and then formally integrate term by term. The resulting series can then be approximated by a rational function in the hope that the rational approximation has better properties than the formally integrated series itself. We have already encountered such an approach to approximation in Section 3.3.1, where we obtained the Padé approximant for a power series expansion of a symmetric stable law. The reader should recall that this power series was formally constructed from an integrated expansion itself.

3.7 Multivariate extensions

In principle, the methods of this chapter can be freely extended to real-valued functions of several variables. Rational functions of several variables are ratios of polynomials; their coefficients can be matched to the coefficients of a power series as in the univariate case. Unlike the univariate case, however, there is so much freedom that it is a challenge to choose an appropriate set of equations. Recall that in the univariate case, we match a rational function to a power series by requiring the equality of the first $n + m$ derivatives about some point, which we have usually taken to be zero. In higher dimensions, this can also be done. However, a power series of several variables does not have a canonical (total) ordering of its terms. Thus, it is much less clear what should be matched to what.

In the discussion which follows, I shall restrict to the case of rational functions of two variables. The extension of these ideas to higher dimensions is not difficult, but does involve the use of additional indices that impede the exposition by emphasising notation over insight. Let $Z^+ \times Z^+$ denote the lattice of all points (j, k) in the plane where j and k are nonnegative integers.[†] By a rational approximation to a function $f(x, y)$ we shall mean a function

$$f_{[M,N]}(x, y) = \frac{\sum\sum_{(j,k)\in M} a_{jk}\, x^j\, y^k}{\sum\sum_{(j,k)\in N} b_{jk}\, x^j\, y^k}, \tag{3.40}$$

where M and N are finite subsets of $Z^+ \times Z^+$, and the coefficients a_{jk} and b_{jk} are chosen so that

$$\sum_{(j,k)\in M}\sum a_{jk}\, x^j\, y^k - f(x, y) \sum_{(j,k)\in N}\sum b_{jk}\, x^j\, y^k \tag{3.41}$$

is vanishingly small to high order close to $x, y = 0$. But this, of course, begs the question of what we mean by vanishingly small to high order in the case of several variables.

A reasonable interpretation is the following: The coefficient on the term involving $x^r\, y^s$ in the power series expansion of

$$\sum_{(j,k)\in M}\sum a_{jk}\, x^j\, y^k - f(x, y) \sum_{(j,k)\in N}\sum b_{jk}\, x^j\, y^k = o(x^r\, y^s) \tag{3.42}$$

equals zero for all (r, s) in some appropriately chosen finite subset L of $Z^+ \times Z^+$.

How should subsets M, N and L be chosen? There is much freedom. In order to ensure that we are solving as many equations as there are unknown coefficients, there must be a constraint on the cardinality of L in terms of the cardinalities of M and N. Let us use the symbol $\#(B)$ to denote the cardinality (*i.e.*, number of elements) of a finite set B. Since we can set one coefficient in the denominator equal to one, the number of coefficients in the rational function is $\#(M) + \#(N) - 1$. The number of equations to solve is $\#(L)$. Therefore

$$\#(L) = \#(M) + \#(N) - 1. \tag{3.43}$$

3.8 Notes

The theory of Padé approximants receives an excellent treatment by Baker and Graves-Morris (1996). In this chapter, we have only had room

[†] Note that 0 is included in Z^+.

to touch briefly upon the basic ideas. For more information about the invariance properties of Padé approximants, methods for calculating them, their connection to continued fractions, and results on convergence, the reader is referred to the book mentioned. Space limitations prevent us from considering the numerous algorithms that are available for calculating Padé approximants. Fortunately, the Maple algorithms spare the reader from having to spend much time on solving the simultaneous equations necessary to calculating the coefficients of Padé approximants. Maple uses two different algorithms. When there are exact rational coefficients, Maple uses a fast variant of the extended Euclidean algorithm. When there are exact coefficients which involve parameters, then the algorithm used is a variation of fraction-free Gaussian elimination. The linear system has a coefficient matrix having Hankel structure and the variation takes into consideration that this matrix is symmetric.

The reader who wishes to study the theoretical properties of continued fractions would do well to read Wall (1973). The theory of convergence of continued fractions is a rich area for study, and, unlike the theory of convergence of series, cannot be said to be a "closed book." Wall (1973) treats the subject at an advanced level, and will require a solid background by the reader in analysis.

3.9 Problems

1. Our first problem examines some (much delayed) issues of regularity in the definition of Padé approximants.

   (a) Equation (3.1) is the defining equation for a Padé approximant. Dividing both sides by $q(x)$, and assuming that $q(x) \neq 0$ in a neighbourhood of $x = 0$, we obtain the order condition

   $$f_{[m,\,n]}(x) = f(x) + o(x^{m+n})\,. \tag{3.44}$$

   It is natural to ask whether this order condition provides a satisfactory definition in its own right. This definition is known as the *Baker definition*. Let us consider the function $f_{[1,\,1]}(x)$ where $f(x) = 1 + x^2$. In this case the new order condition would be

   $$f_{[1,\,1]}(x) = f(x) + o(x^2).$$

   Show that this does not work by solving for the coefficients a_n, b_n where $n = 0, 1$ for

   $$\frac{a_0 + a_1\,x}{b_0 + b_1\,x} = 1 + x^2 + o(x^2)$$

and checking that

$$a_0 = b_0, \quad a_1 = b_1, \quad b_0 = 0, \tag{3.45}$$

so that

$$\frac{a_0 + a_1 x}{b_0 + b_1 x} = 1 \neq 1 + x^2 + o(x^2).$$

(b) In the calculations above, we observed that the function $f(x) = 1 + x^2$ failed to have a Padé approximant of order (1, 1) according to the Baker definition. Show that the values given in (3.45) do satisfy the definition given in (3.1) early in this chapter. Can we write $f_{[1,\,1]}(x) = 1$ when $f(x) = 1 + x^2$? Yes and no! If we write

$$f_{[1,\,1]}(x) = \frac{0 + x}{0 + x}$$

then $f_{[1,\,1]}(x)$ satisfies (3.1). However, verify that

$$f_{[1,\,1]}(x) = \frac{1 + 0\,x}{1 + 0\,x}$$

does not!

(c) Our calculations lead us to an understanding that (3.1) and (3.44) are not equivalent, although they may lead to the same results in many examples.

Note that a given rational function can be represented as the ratio of two polynomials in many different ways. Our original definition of a Padé approximant $f_{[m,\,n]}(x)$ is dependent upon the particular polynomials used, whereas the Baker definition is not. Henri Padé defined the *deficiency index* of $f_{[m,\,n]}(x)$ to be the smallest integer ω such that

$$f(x) = f_{[m,\,n]}(x) + o(x^{m+n-\omega}).$$

Comment upon the following quote from Baker and Graves-Morris (1996):

> "Because the accuracy-through-order requirement [the requirement $\omega = 0$] is fundamental, a definition which preserves it [the Baker definition] is essential ...."

Your comments should consider both the strengths and the weaknesses of each definition.

2. Prove formula (3.7).

3. Let $f(x)$ be the probability density function for a random variable having a $\mathcal{N}(0,\,1)$ distribution. Find $f_{[8,\,0]}(x)$ and $f_{[0,\,8]}(x)$.

4. Let $f(x)$ and $g(x)$ be two probability density functions related by the equation

$$g(x) = \frac{1}{\sigma} f\left(\frac{x}{\sigma}\right).$$

Prove that

$$g_{[m,\,n]}(x) = \frac{1}{\sigma} f_{[m,\,n]}\left(\frac{x}{\sigma}\right).$$

5. Prove that every rational function $f(x)$ can be written as $g(x)+h(x)$, where $g(x)$ is a polynomial–it may be zero–and $h(x)$ is a rational function with the property that the degree of the numerator is less than the degree of the denominator.

6. Suppose $p(x)/q(x)$ is a (real) rational function such that the degree of the numerator is strictly less than the degree of the denominator. Furthermore, suppose that

$$q(x) = (x - \rho_1)\,(x - \rho_2)\,\cdots\,(x - \rho_k)$$

where $\rho_1, \ldots, \rho_k$ are distinct real numbers. Prove that $p(x)/q(x)$ can be written in the form

$$\frac{p(x)}{q(x)} = \sum_{j=1}^{k} \frac{c_j}{x - \rho_j}$$

where $c_j = p(\rho_j)/q'(\rho_j)$.

7. The general formula for the Padé approximants for $exp(x)$ given in equations (3.9) and (3.10) is due to Padé from his thesis work. Derive it from the equations given in (3.5) and (3.6).

8. Let U be a random number that is uniformly distributed on the interval (0, 1). Define

$$V = U + \frac{1}{U\ \ +}\,\frac{1}{U\ \ +}\,\frac{1}{U\ \ +}\cdots.$$

Find the probability density function for V.

9. In Problem 20 of Chapter 2, we considered the logarithmic difference

$$\zeta(x) = \ln\,\Gamma(1+x) - \left[\frac{1}{2}\,\ln(2\,\pi) + \left(x + \frac{1}{2}\right)\,\ln\,x - x\right].$$

which expressed the error in using Stirling's approximation for $\Gamma(1+x)$. A continued fraction expansion for $\zeta(x)$ is

$$\zeta(x) = \frac{1}{12\,x\ \ +}\,\frac{2}{5\,x\ \ +}\,\frac{53}{42\,x\ \ +}\,\frac{1170}{53\,x\ \ +}\,\frac{22999}{429\,x\ \ +}\cdots.$$

Use this continued fraction expansion to construct the first four rational approximations to $\zeta(x)$ which are convergents of this continued fraction.

10. Use Viskovatov's method or otherwise to derive the continued fraction

$$\exp(x) = 1 + \frac{x}{1\ +}\ \frac{x}{-2\ +}\ \frac{x}{-3\ +} \cdots .$$

11. Asymptotic series can be changed formally into power series using the inversion $y = 1/x$. Thus, Viskovatov's method can be applied to asymptotic series as well as power series. Using the asymptotic series for $\zeta(x)$ given in Chapter 2, show that

$$\zeta(x) = \frac{1/12}{x\ +}\ \frac{1/30}{x\ +}\ \frac{53/210}{x\ +} \cdots .$$

Note that Viskovatov's method derives a continued fraction in C-fraction form. You will need to transform it into the S-fraction given above. An operation which multiplies numerator and denominator of a continued fraction by some nonzero quantity is called an *equivalence transformation*. It can be seen that such an operation "carries forward" throughout all the convergents of the continued fraction. Compare this with the series in Problem 9 above. Is the difference in form explained by an equivalence transformation?

12. Let X have a distribution $\mathcal{G}(\alpha, 1)$ with probability density function

$$f(x) = \frac{x^{\alpha-1}\exp(-x)}{\Gamma(\alpha)}$$

when $x \geq 0$ and $f(x) = 0$ when $x < 0$. In Problem 21 of Chapter 2, we saw that the asymptotic series

$$P(X > t) \sim f(t)\left[1 + \frac{\alpha-1}{t} + \frac{(\alpha-1)(\alpha-2)}{t^2} + \frac{(\alpha-1)(\alpha-2)(\alpha-3)}{t^3} + \cdots\right]$$

could be obtained for $t > 0$ using integration by parts. In the manner similar to the previous question, use Viskovatov's method or otherwise on the asymptotic series to show that

$$P(X > t) \sim f(t)\left[\frac{1}{1\ +}\ \frac{1-\alpha}{t\ +}\ \frac{1}{1\ +}\ \frac{2-\alpha}{t\ +}\ \frac{2}{1\ +}\ \frac{3-\alpha}{t\ +}\ \frac{3}{1\ +} \cdots\right].$$

CHAPTER 4

The delta method and its extensions

4.1 Introduction to the delta method

Suppose $X_1, X_2, \ldots, X_n$ are independent, identically distributed random variables with unknown mean μ. If we wish to estimate some parameter $\theta = h(\mu)$ of the distribution, then a natural estimator is $h(\overline{X}_n)$, where $\overline{X}_n$ is the sample mean. To evaluate the asymptotic performance of $h(\overline{X}_n)$ we might find the limiting moments of $h(\overline{X}_n)$ as the sample size $n \to \infty$.

For example, we could look for an expansion of the form

$$E[h(\overline{X}_n)] = a_0 + \frac{a_1}{n} + \frac{a_2}{n^2} + \cdots + \frac{a_m}{n^m} + o\left(n^{-m}\right) \tag{4.1}$$

as $n \to \infty$. Such an expansion can be used to study convergence of $h(\overline{X}_n)$ to $h(\mu)$ through the moments of the distribution. We would also want to have similar sorts of expansions for the variance of $h(\overline{X}_n)$ such as

$$\mathrm{Var}[h(\overline{X}_n)] = \frac{b_1}{n} + \frac{b_2}{n^2} + \cdots + \frac{b_m}{n^m} + o\left(n^{-m}\right) . \tag{4.2}$$

The constant term will be zero under mild regularity because the variance goes to zero as $n \to \infty$.

Another way to evaluate the asymptotic performance of $h(\overline{X}_n)$ is to consider its limiting distribution as $n \to \infty$. Since $\overline{X}_n$ converges almost surely to μ, the limit distribution of $h(\overline{X}_n)$ is degenerate with mass one on $h(\mu)$. However, we can try to centre and scale $h(\overline{X}_n)$ so that a nondegenerate distribution is the result. We might centre and scale so that

$$\sqrt{n}\,[h(\overline{X}_n) - h(\mu)] \stackrel{d}{\Longrightarrow} \mathcal{N}(0, \sigma^2) \tag{4.3}$$

for some $\sigma^2 > 0$ as $n \to \infty$. The normality of the limit distribution will follow from the local linearity of $h(x)$ close to $x = \mu$ and the central limit approximation to $\overline{X}_n$.

At first glance, the two methods in (4.1) and (4.3) seem to be roughly equivalent—and indeed they often are. The observation that the limit normal distribution in (4.3) has mean zero and variance σ^2 corresponds to $a_0 = h(\mu)$ in (4.1), and $b_1 = \sigma^2$ in (4.2). However, the correspondence between the limit moment formulas in (4.1) and (4.2) and the limit distribution formula in (4.3) is not exact. For example, in general

$$Y_n \stackrel{d}{\Longrightarrow} Y \qquad \text{does not imply} \qquad E(Y_n) \to E(Y)\,.$$

However, convergence in distribution will imply convergence of means for uniformly bounded random variables.*

The two methods leading to the formulas above are called the delta method for moments and the delta method for distributions, respectively.

4.2 Preliminary results

The following proposition will be useful for the derivation of the delta method for moments.

Proposition 1. *Let $X_1, \ldots, X_n$ be independent and identically distributed with mean μ. Let $\overline{X}_n$ be the sample mean of $X_1, \ldots, X_n$. Suppose that for a given integer $k \geq 1$, X_1 has finite $(2\,k-1)^{\text{st}}$ moment. Then*

$$E\left(\overline{X}_n - \mu\right)^{2\,k-1} = O(n^{-k})\,.$$

as $n \to \infty$. Furthermore, if X_1 has finite $(2\,k)^{\text{th}}$ moment, then

$$E\left(\overline{X}_n - \mu\right)^{2\,k} = O(n^{-k})\,,$$

also as $n \to \infty$.

Proof. Note that we can combine these two statements together as

$$\begin{aligned} E\left(\overline{X}_n - \mu\right)^m &= O\left(n^{-\lfloor (m+1)/2 \rfloor}\right) \\ &= O\left(n^{-\lceil m/2 \rceil}\right). \end{aligned}$$

(The equality $\lfloor (m+1)/2 \rfloor = \lceil m/2 \rceil$ holds for integer m but not for general real numbers.) It suffices to prove this result for the case where $\mu = 0$. Let μ_j be the jth moment of X_1. Expanding $\overline{X}_n^m$ we have

$$E\,\overline{X}_n^m = n^{-m} \sum_{1 \leq j_1\; j_2\; \cdots\; j_m \leq n} E\left(X_{j_1} X_{j_2} \cdots X_{j_m}\right). \tag{4.4}$$

* What is needed here is uniform integrability. Random variables which are uniformly bounded are uniformly integrable. However, boundedness is a very strong condition. A weaker sufficient condition for uniform integrability is that $\sup_n E\,|Y_n|^{1+\epsilon} < \infty$ for some $\epsilon > 0$.

Since $E(X_j) = 0$, the only nonzero terms in this sum are those where every j appears at least twice. Let us drop the zero terms and assume henceforth that each index j which is present in a term must appear two or more times. For each term of the sum, let p denote the total number of distinct indices j that appear in that term. For example, if $m = 5$ the term $E(X_{j_1}^2\, X_{j_2}^3)$ is assigned the value $p = 2$. Since each j appears two or more times, it follows that $p \le \lfloor m/2 \rfloor$. Furthermore, for each possible value of p, there are

$$O\left[\binom{n}{p}\right] = O(n^p)$$

nonzero terms in (4.4) which are assigned that particular value of p. Here $\binom{n}{p}$ is the number of ways of picking p random variables out of n. This binomial coefficient needs to be multiplied by the number of ways of ordering m random variables selected with replacement from a set of p given distinct random variables so that each of the p variables appears at least twice. This combinatorial factor is

$$\sum_{\substack{s_1 + \cdots + s_p = m \\ s_k \ge 2 \text{ for all } k}} \binom{m}{s_1\, s_2\, \cdots\, s_p}.$$

However, the precise evaluation of this combinatorial factor need not concern us, because for fixed p this factor does not depend on the value of n. Therefore, the total number of nonzero terms in (4.4) is of order

$$\sum_{p=1}^{\lfloor m/2 \rfloor} O(n^p) = O(n^{\lfloor m/2 \rfloor}).$$

Next, we note that the terms in (4.4) are uniformly bounded. This follows from the fact that

$$|\, E\,(X_{j_1}\, X_{j_2} \cdots X_{j_m})\, | \le E\,|X_1|^m. \tag{4.5}$$

See Problem 1. So

$$\begin{aligned} E\,\overline{X}_n^m = n^{-m}\, O(n^{\lfloor m/2 \rfloor}) &= O(n^{\lfloor m/2 \rfloor - m}) \\ &= O(n^{-\lceil m/2 \rceil}), \end{aligned}$$

as required. ∎

For small values of k it is easy to check these order statements by direct calculation of the moments. For example, if the rth cumulant† of X_1 is

† See Section 2.2.5. In particular $\mu = \kappa_1$ and $\sigma^2 = \kappa_2$.

κ_r, then

$$
\begin{aligned}
E\left(\overline{X}_n - \mu\right)^2 &= \frac{\kappa_2}{n} = O(n^{-1}), \\
E\left(\overline{X}_n - \mu\right)^3 &= \frac{\kappa_3}{n^2} = O(n^{-2}), \\
E\left(\overline{X}_n - \mu\right)^4 &= \frac{\kappa_4 + 3\,n\,\kappa_2^2}{n^3} = O(n^{-2}), \\
E\left(\overline{X}_n - \mu\right)^5 &= \frac{\kappa_5 + 10\,n\,\kappa_3\,\kappa_2}{n^4} = O(n^{-3}), \\
E\left(\overline{X}_n - \mu\right)^6 &= \frac{\kappa_6 + 15\,n\,\kappa_4\,\kappa_2 + 10\,n^2\,\kappa_3^2}{n^5} = O(n^{-3}), \qquad (4.6)
\end{aligned}
$$

and so on.

A few observations about these moments are in order. First, we should note that Proposition 1 is closely related to the Marcinkiewicz-Zygmund inequality for sums of centred independent random variables. As the use of this inequality is not our main intention here, we shall state the result without proof. The reader is referred to Chow and Teicher (1988, 367—368) for a complete proof.

Proposition 2 (Marcinkiewicz–Zygmund Inequality). *Let $Y_1, Y_2, \cdots$ be independent random variables with mean zero. For every $m \geq 1$ there exist positive constants α_m and β_m which are universal in the sense that they depend only on m, such that*

$$
\alpha_m\, E\left[\left(\sum_{j=1}^{n} Y_j^2\right)^{m/2}\right] \leq E\left[\left|\sum_{j=1}^{n} Y_j\right|^m\right] \leq \beta_m\, E\left[\left(\sum_{j=1}^{n} Y_j^2\right)^{m/2}\right].
$$

for all n.

The right-hand side of this inequality is particularly useful. Using Hölder's inequality with $p = m/2$ and $q = m/(m-2)$ we see that

$$
\begin{aligned}
\sum_{j=1}^{n} Y_j^2 &= \sum_{j=1}^{n} 1 \cdot Y_j^2 \\
&\leq \left(\sum_{j=1}^{n} 1^q\right)^{1/q} \left(\sum_{j=1}^{n} |Y_j|^{2p}\right)^{1/p} \\
&\leq n^{(m-2)/m} \left(\sum_{j=1}^{n} |\, Y_j\,|^m\right)^{2/m}.
\end{aligned}
$$

Plugging this inequality into the right-hand inequality of Proposition 2 gives us

$$E\left[\left|\sum_{j=1}^{n} Y_j\right|^m\right] \le \beta_m\, n^{(m-2)/2}\, E\left(\sum_{j=1}^{n} |\, Y_j\,|^m\right). \tag{4.7}$$

Now suppose that $X_1, X_2, \ldots$ are independent and identically distributed with mean μ and finite m^{th} moment. Set $Y_j = X_j - \mu$. Then the expectation on the right-hand side of (4.7) is $O(n)$. So

$$\begin{aligned} E\left(|\overline{X}_n - \mu|^m\right) &= n^{-m}\, E\left[\left|\sum_{j=1}^{n} Y_j\right|^m\right] \\ &= n^{-m}\, n^{(m-2)/2}\, O(n) \\ &= O(n^{-m/2}) \end{aligned} \tag{4.8}$$

for all positive integers m. This inequality should be compared with those in Proposition 1. Note that (4.8) provides a similar sort of bound to those in Proposition 1, but with an absolute value sign within the expectation. For even values of m the introduction of an absolute value has no effect, and in these cases our latest order statement in (4.8) agrees with Proposition 1.

The second observation that should be made on the results of Proposition 1, is that we may write the moments more precisely as

$$E\left(\overline{X}_n - \mu\right)^m = \sum_{j=\lfloor (m+1)/2 \rfloor}^{m-1} c_{jm}\, n^{-j} \tag{4.9}$$

where the coefficients c_{jm} do not depend upon n. The values of these coefficients for $m \le 6$ can be read from the table of values in (4.6) above.

4.3 The delta method for moments

We are now in a position to state and prove the basic results for the delta method for moments.

Proposition 3. *Suppose $X_1, X_2, \ldots, X_n$ are independent identically distributed random variables. Let $k \ge 0$ be an integer.*

1. *Assume that $E\,|X_1|^{2\,k+1} < \infty$. Let κ_j denote the jth cumulant of X_1, for $1 \le j \le 2\,k+1$, so that $\mu = \kappa_1$ and $\sigma^2 = \kappa_2$, and so on.*

2. *Let $h(x)$ be a real-valued function which is $2\,k+1$ times differentiable for all x. We shall suppose that there exists some constant $B > 0$ such that $|h^{(2\,k+1)}(x)| \le B$ for all x.*

Then

$$E\,h(\overline{X}_n) = a_0 + \frac{a_1}{n} + \frac{a_2}{n^2} + \cdots + \frac{a_k}{n^k} + O\left[\frac{1}{n^{(2\,k+1)/2}}\right],$$

where

$$a_0 = h(\mu)\,, \qquad a_1 = h''(\mu)\,\frac{\sigma^2}{2}\,, \qquad a_2 = h'''(\mu)\,\frac{\kappa_3}{6} + h^{(iv)}(\mu)\,\frac{\sigma^4}{8}\,.$$

Higher order coefficients can be calculated using the method outlined in the proof below.

Proof. Expanding $h(\overline{X}_n)$ in a Taylor series about μ we get

$$h(\overline{X}_n) = h(\mu) + \sum_{j=1}^{2\,k} \frac{h^{(j)}(\mu)}{j!}\,(\overline{X}_n - \mu)^j + \frac{h^{(2\,k+1)}(\xi)}{(2\,k+1)!}\,(\overline{X}_n - \mu)^{2\,k+1}$$

where ξ lies between $\overline{X}_n$ and μ. Taking expectations of both sides yields

$$E\,h(\overline{X}_n) = h(\mu) + \sum_{j=1}^{2\,k} \frac{h^{(j)}(\mu)}{j!}\,E(\overline{X}_n - \mu)^j + R_{2\,k+1}\,, \tag{4.10}$$

where $R_{2\,k+1}$ is the expectation of the remainder term involving ξ. Since $|h^{(2\,k+1)}(\xi)| \le B$, we conclude that

$$\begin{aligned}
|R_{2\,k+1}| &\le \left| E\,\frac{h^{(2\,k+1)}(\xi)}{(2\,k+1)!}\,(\overline{X}_n - \mu)^{2\,k+1}\right| \\
&\le E\left|\frac{h^{(2\,k+1)}(\xi)}{(2\,k+1)!}\,(\overline{X}_n - \mu)^{2\,k+1}\right| \\
&\le \frac{B}{(2\,k+1)!}\,E\,|\,\overline{X}_n - \mu\,|^{2\,k+1} \\
&= O\left[\frac{1}{n^{(2\,k+1)/2}}\right],
\end{aligned}$$

using the order result given in (4.8). Using (4.9), we can express the moments in (4.10) as sums, so that

$$E\,h(\overline{X}_n) \;=\; h(\mu) + \sum_{j=1}^{2\,k}\left[\sum_{r=\lfloor (j+1)/2\rfloor}^{j-1} \frac{c_{rj}}{n^r}\right]\frac{h^{(j)}(\mu)}{j!}$$

$$+O\left[\frac{1}{n^{(2k+1)/2}}\right]$$

$$= h(\mu) + \sum_{r=1}^{k} n^{-r}\left[\sum_{j=r+1}^{2r} \frac{c_{rj}\, h^{(j)}(\mu)}{j!}\right] + O\left[\frac{1}{n^{k+1}}\right]$$

$$+O\left[\frac{1}{n^{(2k+1)/2}}\right]$$

$$= h(\mu) + \sum_{r=1}^{k} n^{-r}\left[\sum_{j=r+1}^{2r} \frac{c_{rj}\, h^{(j)}(\mu)}{j!}\right] + O\left[\frac{1}{n^{(2k+1)/2}}\right].$$

We set $a_0 = h(\mu)$ and $a_r = \sum_{r+1}^{2r} c_{rj}\, h^{(j)}(\mu)/j!$ for $r \geq 1$.

From this derivation, we see that

$$a_1 = \frac{c_{12}\, h^{(2)}(\mu)}{2!} = \frac{\sigma^2\, h^{(2)}(\mu)}{2!}$$

and

$$a_2 = \frac{c_{23}\, h'''(\mu)}{3!} + \frac{c_{24}\, h^{(iv)}(\mu)}{4!} = \frac{4\, h'''(\mu)\, \kappa_3 + 3\, h^{(iv)}(\mu)\, \sigma^4}{24},$$

as required. ■

With extra regularity on $h(x)$ and X_1, the order of the remainder term can be improved, as shown in the next proposition.

Proposition 4. *Suppose that Assumptions 1 and 2 in Proposition 3 are replaced by the following conditions.*

1. *Assume that* $E\,|X_1|^{2k+2} < \infty$.
2. *Let* $h(x)$ *be* $2k+2$ *times differentiable for all* x*, and suppose that there exists some constant* $B > 0$ *such that* $|h^{(2k+2)}(x)| \leq B$ *for all* x.

Then

$$E\, h(\overline{X}_n) = a_0 + \frac{a_1}{n} + \frac{a_2}{n^2} + \cdots + \frac{a_k}{n^k} + O\left(\frac{1}{n^{k+1}}\right),$$

where the coefficients are determined as in Proposition 3.

Proof. The proof is similar to that of Proposition 3 except that an extra term is carried in the Taylor expansion. So $E\,h(\overline{X}_n)$ equals

$$h(\mu) + \sum_{j=1}^{2k} \frac{h^{(j)}(\mu)}{j!}\,E\,(\overline{X}_n - \mu)^j + \frac{h^{(2k+1)}(\mu)}{(2k+1)!}\,E\,(\overline{X}_n - \mu)^{2k+1} + R_{2k+2}\,.$$

Then $R_{2k+2} = O[n^{-(k+1)}]$ for reasons similar to the argument in Proposition 3. In addition, the penultimate term is also of order $O[n^{-(k+1)}]$ from the assumptions and Proposition 1. ∎

A widely used special case of Propositions 3 and 4 arises when $k = 1$ so that $h(x)$ is three (respectively, four) times differentiable with bounded third (respectively, fourth) derivative.

Proposition 5. *Under the conditions of Proposition 3 with $k = 1$, we have*

$$E\,h(\overline{X}_n) = h(\mu) + \frac{h''(\mu)\,\sigma^2}{2\,n} + O\left(\frac{1}{n\sqrt{n}}\right).$$

Similarly, under the conditions of Proposition 4 with $k = 1$, we have

$$E\,h(\overline{X}_n) = h(\mu) + \frac{h''(\mu)\,\sigma^2}{2\,n} + O\left(\frac{1}{n^2}\right),$$

in both cases, as $n \to \infty$.

Several observations should be made on Propositions 3 and 4. First, the assumption of a bounded higher derivative of $h(x)$ is stronger than necessary for the conclusions.‡ We can weaken the assumptions to some extent by noticing that the conclusions in Propositions 3 and 4 follow for a given value of k if the assumptions hold for some higher value $k^\star$, where $k^\star \geq k$. This follows from the fact that the order properties of the series "propagate backwards" to all lower orders. However, functions $h(x)$ with bounded derivatives at some order are polynomial-like in their behaviour at infinity, and there are many functions for which we would like to apply the delta method where the assumptions fail. We may try to solve this problem by attacking the remainder term with more powerful analytical methods. If these difficulties prove too much to handle, the delta method for distributions, discussed below, is an alternative approach.

‡ Nevertheless, the assumption cannot simply be removed, as counterexamples can be constructed if $h(x)$ is differentiable to high order but does not have a bounded derivative of some order.

A perusal of the proofs of Propositions 3 and 4 shows that the terms of the Taylor expansion of $h(\overline{X}_n)$ must be grouped in pairs to isolate them in orders of n. The reason for this can be seen in Proposition 1, where the $(2\,k-1)^{\text{st}}$ moment and the $(2\,k)^{\text{th}}$ moment of $\overline{X}_n - \mu$ are seen to have the same order in n. This can be a trap for the unwary: when working through the delta method from first principles in an example, we may accidentally terminate the expansion with a remainder term that is the same order as the last term—the term of smallest order—in the expansion. There is nothing wrong with such a "do it yourself" approach to the delta method provided the order of the remainder term is checked carefully to avoid omitting an important term from the expansion.

Propositions 3 and 4 also provide a mechanism for calculating the higher moments of $h(\overline{X}_n)$. No new propositions are needed for this purpose. For example, to calculate the *second* moment of $h(\overline{X}_n)$ it is sufficient to replace the function $h(x)$ by $g(x)$, where $g(x) = [h(x)]^2$. To calculate the coefficients in the expansion, it is necessary to write the derivatives of $g(x)$ in terms of the derivatives of $h(x)$. Note that the jth derivative of $g(x)$ depends upon all lower derivatives of $h(x)$. So, to impose a bound on the absolute value of $g^{(2\,k+1)}(x)$ or $g^{(2\,k+2)}(x)$ some regularity must hold on all lower order derivatives of $h(x)$. For example, if $|h^{(r)}(x)|$ is bounded for all $0 \le r \le j$, then $|g^{(r)}(x)|$ is bounded. However, this condition is much too strong to be practical for many applications. It is usually better to check the assumptions on $g(x)$ directly.

Proposition 6. *Suppose that the assumptions of Proposition 4 hold with $k = 1$, and that $|g^{(iv)}(x)|$ is bounded where $g(x) = [h(x)]^2$. Then*

$$\operatorname{Var} h(\overline{X}_n) = \frac{[h'(\mu)]^2\,\sigma^2}{n} + O\left(\frac{1}{n^2}\right),$$

as $n \to \infty$.

Proof. Applying Proposition 4 to $g(x) = [h(x)]^2$, we find that

$$\begin{aligned} E\,\left[h(\overline{X}_n)\right]^2 &= g(\mu) + \frac{g''(\mu)\,\sigma^2}{2\,n} + O\left(\frac{1}{n^2}\right) \\ &= [h(\mu)]^2 + \frac{h''(\mu)\,h(\mu)\,\sigma^2}{n} + \frac{[h'(\mu)]^2\,\sigma^2}{n} + O\left(\frac{1}{n^2}\right). \end{aligned}$$

From Proposition 5, we obtain

$$\begin{aligned} \left[E\,h(\overline{X}_n)\right]^2 &= \left[h(\mu) + \frac{h''(\mu)\,\sigma^2}{2\,n} + O\left(\frac{1}{n^2}\right)\right]^2 \\ &= [h(\mu)]^2 + \frac{h''(\mu)\,h(\mu)\,\sigma^2}{n} + O\left(\frac{1}{n^2}\right). \end{aligned}$$

Subtracting the second expansion from the first gives us the required result. ∎

It is worth noting that the delta method is applicable even when the function $h(x)$ is defined implicitly through an equation or the optimisation of an objective function. For example, when $\overline{X}_n$ is a sufficient statistic for the estimation of a parameter θ, then the principle of sufficiency requires that any estimator $\widehat{\theta}$ be a function of $\overline{X}_n$. This function need not be explicit, or have any closed form representation. However, to apply the delta method for moments, we must be able to evaluate or approximate the function and its derivatives close to the expected value of $\overline{X}_n$.

4.4 Using the delta method in Maple

The delta method is particularly easy to implement in Maple when coupled with the *powseries* package. This package was briefly described in Section 2.2.6. To illustrate the method here, let us consider the task of computing $E\left(\overline{X}_n - \mu\right)^7$ based upon a random sample of size n from the $\mathcal{P}(\lambda)$ distribution. The Poisson distribution has the property that its cumulants κ_n are all equal to λ for $n \geq 1$. So, in particular, $\mu = \lambda$ and $\sigma^2 = \lambda$. The Maple commands

$> \mathit{with}(\mathit{powseries})$

$> \mathit{powcreate}\left(\mathit{cgf}(j) = \frac{\lambda}{n^{j-1} \cdot j!}, \mathit{cgf}(0) = 0, \mathit{cgf}(1) = 0\right)$

formally generate a recursive sequence *cgf* whose elements are the cumulants of $\overline{X}_n - \mu$. To formally generate the sequence *mgf* of coefficients of the moment generating function for $\overline{X}_n - \mu$, we use

$> \mathit{mgf} := \mathit{powexp}(\mathit{cgf})$

and the job is done. To obtain $E\left(\overline{X}_n - \mu\right)^m$, for $m = 7$ say, we use

$> m := 7$

$> \mathit{collect}(m! \cdot \mathit{mgf}(m), n)$

which generates the output

$$\frac{105\,\lambda^3}{n^4} + \frac{56\,\lambda^2}{n^5} + \frac{\lambda}{n^6}\,.$$

To illustrate the use of these moments in the delta method, let us consider an example with a function such as $h(x) = \sqrt{x}$. First we use the command

$> powcreate\,(\,h(j) = subs\,(\,x = \lambda\,,\, diff\,(\sqrt{x},\, x\$j)\,)\,)$

to generate the sequence of derivatives of $h(x)$. The elements of this sequence need to be multiplied element–by–element by the sequence of moments of $\overline{X}_n - \mu$. This is accomplished by

$> powcreate(\,\delta(j) = mgf(j) \cdot h(j)\,)$

The sequence of coefficients in δ needs to be summed. To do this, we can use the command

$> result := subs(\,t = 1\,,\, mtaylor(tpsform(\delta,\, t,\, 6),\, t,\, 6)\,)$

The displayed command sums the first five terms of δ by using *tpsform* to generate a polynomial in t from δ. The *mtaylor* command simply strips off the order term from this polynomial. The coefficients are then summed by setting $t = 1$. The final step is to organise the expression obtained by the last command, by grouping common powers of n together. This is accomplished using

$> collect(result,\, n)$

The first three terms of the Maple output give us

$$E\sqrt{\overline{X}_n} = \sqrt{\lambda} - \frac{1}{8\sqrt{\lambda}}\frac{1}{n} - \frac{7}{128\,\lambda\sqrt{\lambda}}\frac{1}{n^2} + O\left(\frac{1}{n^3}\right). \tag{4.11}$$

To apply the delta method to any other function $h(x)$, we need only replace $\sqrt{x}$ above by the appropriate choice. For example, in the next section we shall consider $h(x) = e^{-x}$.

4.5 Asymptotic bias

In this section, we shall consider the asymptotic bias of maximum likelihood estimators using the Poisson example of the previous section. Let $X_1, \ldots, X_n$ be a random sample from a Poisson distribution with parameter λ. The maximum likelihood estimator for λ based upon this random sample is $\widehat{\lambda} = \overline{X}_n$. Let $\pi_0 = P(X_1 = 0)$. Then π_0 provides an alternative parametrisation of the Poisson model and is related to λ by the smoothly invertible transformation $\pi_0 = e^{-\lambda}$.

The maximum likelihood estimator for π_0 is

$$\begin{aligned}\widehat{\pi}_0 &= \exp(-\widehat{\lambda}) \\ &= \exp(-\overline{X}_n)\,,\end{aligned}$$

a result which follows immediately from the functional equivariance of the method of maximum likelihood estimation. Using the delta method for moments, it can be checked that

$$\begin{aligned} E\,\widehat{\pi}_0 &= E\,\exp(-\overline{X}_n) \\ &= \pi_0\left[1+\frac{\lambda}{2\,n}-\frac{\lambda}{6\,n^2}+\frac{\lambda^2}{8\,n^2}+O\left(\frac{1}{n^3}\right)\right]. \end{aligned} \quad (4.12)$$

This result can be obtained by calculating the coefficients a_0, a_1 and a_2 in Proposition 4. For the $\mathcal{P}(\lambda)$ distribution, the first three cumulants are

$$\mu = \lambda\,,\ \sigma^2 = \lambda\,,\ \text{and } \kappa_3 = \lambda\,.$$

Note that the assumptions of Proposition 4 hold in this example: the Poisson distribution has moments of all orders, and all derivatives of $h(x) = e^{-x}$ are bounded for positive x.

Such an expansion can be used to obtain an asymptotic correction for the bias of $\widehat{\pi}_0$. For example, we can write the bias $B(\widehat{\pi}_0) = E(\widehat{\pi}_0) - \pi_0$ as

$$B(\widehat{\pi}_0) = \frac{\lambda\,e^{-\lambda}}{2\,n} + O(n^{-2})$$

so that the bias in $\widehat{\pi}_0$ is $\lambda\,e^{-\lambda}/(2\,n)$ to lowest order in n. Since λ is unknown, we cannot simply subtract this asymptotic bias to produce a new estimator. However, we can obtain a "plug–in" bias correction by replacing λ by $\overline{X}_n$ in this bias formula. This leads to the estimator

$$\widehat{\pi}_0^\star = \widehat{\pi}_0 - \frac{\overline{X}_n\,\exp(-\overline{X}_n)}{2\,n} \quad (4.13)$$

which has asymptotic bias

$$\begin{aligned} B(\widehat{\pi}_0^\star) &= E\left[\exp(-\overline{X}_n) - e^{-\lambda}\right] - E\left[\frac{\overline{X}_n\,\exp(-\overline{X}_n)}{2\,n}\right] \\ &= \left[\frac{\lambda\,e^{-\lambda}}{2\,n} + O(n^{-2})\right] - E\left[\frac{\overline{X}_n\,\exp(-\overline{X}_n)}{2\,n}\right] \ \text{(delta method)} \\ &= \left[\frac{\lambda\,e^{-\lambda}}{2\,n} + O(n^{-2})\right] - \left[\frac{\lambda\,e^{-\lambda}}{2\,n} + \frac{O(n^{-1})}{n}\right] \ \text{(delta m. again!)} \\ &= O(n^{-2})\,. \end{aligned}$$

The second application of the delta method in the penultimate step is applied to the function $h(x) = x\,\exp(-x)$.

The principle illustrated through this Poisson example is a special case of a more general property of maximum likelihood estimators for models involving random samples from a distribution. The estimator $\widehat{\pi}_0 = \exp(\overline{X}_n)$ is the maximum likelihood estimator for the parameter $e^{-\lambda}$, and maximum likelihood estimators typically have a bias which is asymptotically $O(n^{-1})$. Thus if $\widehat{\theta}$ is the maximum likelihood estimator for some real parameter θ, we can usually write

$$E(\widehat{\theta}) = \theta + \frac{b(\theta)}{n} + O(n^{-2}) \qquad \text{so that} \qquad B(\widehat{\theta}) \sim \frac{b(\theta)}{n} \tag{4.14}$$

as $n \to \infty$, where $b(\theta)$ does not depend upon n. This bias does not appear in the calculations of the first order asymptotic properties of the maximum likelihood estimator. It enters into second order calculations when calculating the mean square error of $\widehat{\theta}$ to order $O(n^{-2})$.

4.6 Variance stabilising transformations

4.6.1 General principles

The asymptotic normality of $\overline{X}_n$ is often used to construct an approximate confidence interval for the mean of a distribution. As is well known, the interval

$$\overline{X}_n \pm 1.96\sqrt{\text{Var}(\overline{X}_n)} \qquad \text{that is,} \qquad \overline{X}_n \pm 1.96\,\frac{\sigma}{\sqrt{n}}$$

will contain the mean with probability close to 95% when n is large. However, this formula can only be implemented directly when σ is known. When this is not the case, we often use the sample standard deviation instead of σ to produce a confidence interval for the mean based on the t-statistic. An alternative is to transform $\overline{X}_n$ to some new statistic $U_n = h(\overline{X}_n)$ whose asymptotic variance is fixed and not dependent upon any unknown parameter.

Consider a random sample from one-parameter model where we may write both μ and σ as functions of an unknown real-valued parameter θ, namely

$$\mu = \mu(\theta)\,, \qquad \sigma^2 = \sigma^2(\theta)\,. \tag{4.15}$$

Under the assumptions of Proposition 6, we have

$$\text{Var}\, h(\overline{X}_n) = \frac{\{h'[\mu(\theta)]\}^2\,\sigma^2(\theta)}{n} + O\left(\frac{1}{n^2}\right).$$

So if we wish to make $\text{Var}\, h(\overline{X}_n) = n^{-1}$, say, to order $O(n^{-1})$, then we want to have

$$h(x) = \int^x \frac{dt}{\sigma[\mu^{-1}(t)]}\,. \tag{4.16}$$

See Problem 6. We might also seek a transformation $h(x)$ which stabilises other types of statistics, which are not simple averages or sums. Such an example is the variance stabilisation of the sample correlation coefficient, which we shall consider in Section 4.6.5 below.

4.6.2 The linearised Poisson distribution

When $X_1, \ldots, X_n$ are $\mathcal{P}(\lambda)$, then the uniformly minimum variance unbiased estimator (UMVUE) for λ is $\overline{X}$. In this case,

$$\mu(\lambda) = \lambda\,, \qquad \sigma(\lambda) = \sqrt{\lambda}$$

in (4.16), and

$$\begin{aligned} h\left(\overline{X}_n\right) &= \int^{\overline{X}_n} t^{-1/2}\,dt \\ &= 2\sqrt{\overline{X}_n}\,, \end{aligned}$$

with the constant of integration set to zero. Therefore

$$\mathrm{Var}\left(\sqrt{\overline{X}_n}\right) = \frac{1}{4n} + O(n^{-2})\,. \tag{4.17}$$

A variant of this formula is the following result.

Proposition 7. *If $X \stackrel{d}{=} \mathcal{P}(\lambda)$, then $\sqrt{X}$ is asymptotically normal with mean*

$$E\left(\sqrt{X}\right) = \sqrt{\lambda} + O\left(\frac{1}{\sqrt{\lambda}}\right)$$

and variance

$$\mathrm{Var}\left(\sqrt{X}\right) = \frac{1}{4} + O\left(\frac{1}{\lambda}\right)$$

as $\lambda \to \infty$.

See Problem 7. Noting this property of the square root transformation, Fraser (1968) observed[§] that the parameter $\sqrt{\lambda}$ for the *linearised Poisson* $\sqrt{X}$ behaves like a location parameter. He proposed the quantity

$$V = 2\left(\sqrt{X} - \sqrt{\lambda}\right) \tag{4.18}$$

as an approximate normal pivotal for inference about λ as $\lambda \to \infty$.

[§] G. Van Belle obtained an asymptotic expression for the linearising transformation in a 1967 Ph. D. dissertation at the University of Toronto, entitled *Location analysis and the Poisson distribution.*

Asymptotically, this quantity is $\mathcal{N}(0, 1)$. This approximate normal pivotal can be compared with the usual approximate normal pivotal found by standardisation of the mean and variance, namely

$$W = \frac{X - \lambda}{\sqrt{\lambda}} . \tag{4.19}$$

Neither W nor V converges uniformly faster than the other to $\mathcal{N}(0, 1)$. However, the density of V is closer than W to $\mathcal{N}(0, 1)$ to lowest order in n except for the interval from 1.4 to 2.4 standard deviations away from zero.

Before leaving this example, we should note that the function $h(x) = \sqrt{x}$ does not satisfy the assumptions of Propositions 4 and 6, because the derivatives of $h(x)$ are infinite at $x = 0$. However, the asymptotic contribution of $x = 0$ is negligible, and the conclusion of Proposition 6 follows even though the assumptions fail. This can be shown by *ad hoc* methods.

4.6.3 Averaging over Bernoulli trials

A discrete random variable taking the value one with probability θ and zero with probability $1 - \theta$ is often called a Bernoulli random variables. A sequence $X_1, \ldots, X_n$ of independent, identically distributed Bernoulli random variables is called a sequence of Bernoulli trials. In a sequence of Bernoulli trials, $\overline{X}$ is the best unbiased estimator for the parameter θ.

Using formula (4.16) with

$$\mu(\theta) = \theta\,, \qquad \sigma(\theta) = \sqrt{\theta\,(1-\theta)}$$

gives us

$$\begin{aligned} h(\overline{X}_n) &= \int^{\overline{X}_n} \frac{dt}{\sqrt{t\,(1-t)}} \\ &= 2\,\sin^{-1}\sqrt{\overline{X}_n}\,. \end{aligned}$$

Again, we set the constant of integration to zero for simplicity.

4.6.4 Gamma distributions

Consider the two-parameter model where $X_1, \ldots, X_n$ are $\mathcal{G}(\alpha, \lambda)$. Let us consider α as fixed so that we can stabilise the variance of $h\left(\overline{X}_n\right)$

with λ varying and α constant. The mean and variance in this restricted model are $\alpha\,\lambda^{-1}$ and $\alpha\,\lambda^{-2}$, respectively. Therefore

$$\begin{aligned} h(\overline{X}_n) &= \int^{\overline{X}_n} \frac{\sqrt{\alpha}\,dt}{t} \\ &= \sqrt{\alpha}\,\ln \overline{X}_n\,. \end{aligned}$$

4.6.5 The correlation coefficient

Suppose $(X_1, Y_1), \ldots, (X_n, Y_n)$ are n independent pairs. The coefficient of linear correlation between X and Y is determined by

$$\rho = \frac{\text{Cov}\,(X_j, Y_j)}{\sqrt{\text{Var}(X_j)\,\text{Var}(Y_j)}}$$

The sample correlation coefficient

$$R = \frac{\sum_{j=1}^n (X_j - \overline{X}_n)\,(Y_j - \overline{Y}_n)}{\sqrt{\sum_{j=1}^n (X_j - \overline{X}_n)^2 \sum_{j=1}^n (Y_j - \overline{Y}_n)^2}}$$

has mean $\mu(\rho)$ and standard deviation $\sigma(\rho)/\sqrt{n}$, where

$$\mu(\rho) = \rho\,, \qquad \sigma(\rho) = 1 - \rho^2\,. \tag{4.20}$$

The statistic R differs from previous examples, because it is not a simple average over a number of independent replications. However, the basic methodology works. The stabilising transformation is

$$\begin{aligned} h(R) &= \int^R \frac{dt}{1 - t^2} \\ &= \frac{1}{2} \ln \left(\frac{1 - R}{1 + R} \right) \\ &= \tanh^{-1} R\,. \end{aligned}$$

4.7 Normalising transformations

4.7.1 General theory

The variance stabilisation method of the previous section allowed us to derive approximate pivotal quantities that could be used to construct approximate confidence intervals for a real-valued parameter. The success of this method depends upon the asymptotic normality of $\overline{X}_n$, and *a fortiori* the transformed statistic $h(\overline{X}_n)$. However, the statistic $h(\overline{X}_n)$

may not be closer to normal than $\overline{X}_n$. So, transformations which try to "normalise" $\overline{X}_n$ are also of interest.

How can we do this? Once again, let us assume that the moments and cumulants are functions of a real parameter θ. In particular, we should have

$$\mu = \mu(\theta)\,, \qquad \sigma^2 = \sigma^2(\theta)\,, \qquad \text{and } \kappa_3 = \kappa_3(\theta)\,.$$

One measure of non-normality is the third cumulant or third central moment of $\overline{X}_n$. From Proposition 1, we have

$$E\left(\overline{X}_n - \mu\right)^3 = O(n^{-2})\,.$$

The transformation $h(x)$ has effect of normalising $\overline{X}_n$ if

$$E\left[h(\overline{X}_n) - E\,h(\overline{X}_n)\right]^3 = o(n^{-2})\,. \tag{4.21}$$

To find a transformation which satisfies this, we expand the left-hand side as

$$E\,h^3(\overline{X}_n) - 3\,E\,h^2(\overline{X}_n)\,E\,h(\overline{X}_n) + 2\left[E\,h(\overline{X}_n)\right]^3\,.$$

In turn, each of the expectations in this expression can be expanded using Proposition 3 applied to the functions h, h^2 and h^3. Terms which involve either n^{-1} or n^0 will cancel. So, it is sufficient to find the coefficient of the term involving n^{-2} and to set this coefficient to zero. This leads to the equation

$$\left[h'(\mu)\right]^3\,\kappa_3 + 3\left[h'(\mu)\right]^2\left[h''(\mu)\right]\sigma^4 = 0\,. \tag{4.22}$$

See Problem 8. This differential equation turns out to be simpler than it looks. We can assume that $h(x)$ is a non-constant function, and therefore that $h'(\mu)$ does not vanish. We cancel $[h'(\mu)]^2$, and set $g(\mu) = h'(\mu)$. Then we get the differential equation

$$g(\mu)\,\kappa_3 + 3\,g'(\mu)\,\sigma^4 = 0\,. \tag{4.23}$$

At this stage, we should recall that μ, σ^2, and κ_3 are all functions of the parameter θ. Let us consider some examples.

4.7.2 Poisson distribution

Suppose $\overline{X}_n$ is the average of n Poisson random variables with mean λ. Then

$$\mu(\lambda) = \lambda\,, \qquad \sigma^2(\lambda) = \lambda\,, \qquad \text{and } \kappa_3(\lambda) = \lambda\,.$$

Then (4.23) reduces to

$$g(\lambda) + 3\,\lambda\,g'(\lambda) = 0\,.$$

This differential equation is solved by $g(\lambda) = C\,\lambda^{-1/3}$. In turn, $g(\lambda) = h'(\lambda)$. This implies that the normalising transformation for the Poisson is

$$h\left(\overline{X}_n\right) = C_1\,\overline{X}_n^{\,2/3} + C_2 \tag{4.24}$$

where C_1 and C_2 are arbitrary constants with $C_1 \neq 0$.

4.7.3 Exponential distribution

Next we consider a random sample from $\mathcal{E}(\lambda)$. In this case, the first three cumulants are

$$\mu(\lambda) = \lambda^{-1}\,, \qquad \sigma^2(\lambda) = \lambda^{-2}\,, \qquad \text{and } \kappa_3(\lambda) = 2\,\lambda^{-3}\,.$$

In this case, it is more convenient to reparametrise the model in terms of $\mu = \lambda^{-1}$ rather than λ. With this transformation, (4.23) becomes

$$2\,\mu^3\,g(\mu) + 3\,\mu^4\,g'(\mu) = 0\,.$$

We can cancel μ^3 here. Again, letting $g(\mu) = h'(\mu)$, it is easily seen that the normalising family of transformations for $\overline{X}_n$ is

$$h\left(\overline{X}_n\right) = C_1\,\overline{X}_n^{\,1/3} + C_2 \tag{4.25}$$

where, once again, the constants are arbitrary and $C_1 \neq 0$.

4.8 Parameter transformations

In several simple exponential family models, the statistic $\overline{X}_n$ is the maximum likelihood estimator $\widehat{\mu}_n$ for the mean μ of the common distribution. If the parameter space undergoes a reparametrisation, say to $\nu = h(\mu)$, then the maximum likelihood estimator for ν is $\widehat{\nu}_n = h(\overline{X}_n)$. Thus the variance stabilising transformations and the normalising transformations of the previous sections can be treated as special cases of the problem of stabilising and normalising the maximum likelihood estimator for more general models.

Hougaard (1982) noted that Robert Wedderburn considered this for one-dimensional exponential families. Unfortunately, Wedderburn's paper was not published. Suppose that $X_1, X_2, \ldots$ have a common density that can be written in the form

$$f(x;\,\theta) = \exp[\theta\,t(x) - K(\theta)]\,f_0(x)$$

for some $K(\theta)$. Then $\widehat{\theta}_n$ is a solution to the equation

$$\overline{T}_n = K'(\widehat{\theta}_n)$$

where $\overline{T}_n = n^{-1} \sum_{j=1}^{n} t(X_j)$.

Proposition 8. *Consider a transformation of θ to*

$$\nu = \int^{\theta} \left[K''(u) \right]^{\delta} du \,. \tag{4.26}$$

When $\delta = 0$, we obtain the identity transformation. Also, the following hold.

- *When $\delta = 1/3$, the likelihood is approximately normal in the sense that $E_\nu \left[\ell_n'''(\nu) \right] = 0$ where $\ell_n(\nu)$ is the reparametrised log-likelihood function.*
- *When $\delta = 1/2$, the variance of the likelihood is stabilised in the sense that the information function $-E_\nu \left[\ell_n''(\nu) \right]$ is constant.*
- *When $\delta = 2/3$, then $\widehat{\nu}_n = h(\widehat{\theta}_n)$ is normalised in the sense that $E_\nu \left[\widehat{\nu}_n - \nu \right]^3 = o(n^{-2})$, which is asymptotically negligible.*
- *Finally, when $\delta = 1$, then $\widehat{\nu}_n$ is bias adjusted in the sense that $E_\nu(\widehat{\nu}_n) = \nu + o(n^{-1})$.*

Proof. This follows as a special case of Hougaard's result, which we state below. ■

Hougaard (1982) extended this result to curved exponential families, showing that for a transformation satisfying a particular differential equation in h involving a parameter δ. Suppose the random variables $X_1, X_2, \ldots$ have joint density

$$f(x;\, \theta) = \exp\left\{ \theta(\beta)\, t(x) - K[\,\theta(\beta)\,] \right\} f_0(x) \tag{4.27}$$

where, in this case, β is a real parameter, while $\theta(\beta)$ and $t(x)$ are row and column vectors of the same dimension, respectively. Hougaard proved the next result under mild regularity as given.

Theorem 1. *Suppose that the model is a curved exponential family as given in (4.27), where $\theta(\beta)$ is twice continuously differentiable with $\partial\theta/\partial\beta \neq 0$. Suppose also that*

$$\nu(\beta) = h[\,\theta(\beta)\,]$$

is a twice differentiable real-valued function of β with $\partial\nu/\partial\beta \neq 0$ satisfying the differential equation

$$\frac{\partial^2\nu/\partial\beta^2}{\partial\nu/\partial\beta} = \left[\delta\, \frac{\partial^3 K}{\partial\theta^3} \left(\frac{\partial\theta}{\partial\beta} \right) + \frac{\partial^2 K}{\partial\theta^2} \left(\frac{\partial\theta}{\partial\beta} \right) \right] \Big/ \left[\frac{\partial^2 K}{\partial\theta^2} \left(\frac{\partial\theta}{\partial\beta} \right) \right] \tag{4.28}$$

for some choice of δ.¶ *Then the following hold.*

- *When* $\delta = 1/3$, *the likelihood is approximately normal in the sense that*

$$E_\nu \left[\frac{\partial^3}{\partial \nu^3} \ell_n(\nu) \right] = 0$$

where, again, $\ell_n(\nu)$ *is the reparametrised likelihood function.*

- *When* $\delta = 1/2$, *the variance of the likelihood is stabilised in the sense that the information function*

$$-E_\nu \left[\frac{\partial^2}{\partial \nu^2} \ell_n(\nu) \right]$$

is constant.

- *When* $\delta = 2/3$, *then* $\widehat{\nu} = h(\widehat{\theta}_n)$ *is normalised in the sense that*

$$E_\nu \left[\widehat{\nu}_n - \nu\right]^3 = o(n^{-2})$$

which is asymptotically negligible.

- *Finally, when* $\delta = 1$, *then* $\widehat{\nu}_n$ *is bias adjusted in the sense that*

$$E_\nu(\widehat{\nu}_n) = \nu + o(n^{-1}) .$$

Proof. See Hougaard (1982). ∎

This result of Hougaard was very influential to Shun-ichi Amari in developing his family of α-connections. See Amari (1985). Hougaard's δ and Amari's α both define a concept of "flatness" on the parameter space. In Hougaard's work, this flatness is interpreted as a measure of how close the likelihood and maximum likelihood estimate are to being pivotal as in the normal location model. In Amari's work the concept of flatness is formalised for general models by an affine connection on the parameter space. Unfortunately, in both cases there is no single transformation (or connection) which works successfully for all criteria.‖

¶ In this expression, $\partial^2 K/\partial\theta^2$ is the matrix of second partials acting as a bilinear form on vectors so that

$$\frac{\partial^2 K}{\partial \theta^2}\left(\frac{\partial \theta}{\partial \beta}\right) = \sum_j \sum_k \frac{\partial^2 K}{\partial\theta_j \partial\theta_k} \frac{\partial \theta_j}{\partial \beta} \frac{\partial \theta_k}{\partial \beta} .$$

Also, $\partial^3 K/\partial\theta^3$ is an array of mixed third partials acting as a trilinear form on vectors so that

$$\frac{\partial^3 K}{\partial \theta^3}\left(\frac{\partial \theta}{\partial \beta}\right) = \sum_j \sum_k \sum_m \frac{\partial^3 K(\theta)}{\partial\theta_j \partial\theta_k \partial\theta_m} \frac{\partial \theta_j}{\partial \beta} \frac{\partial \theta_k}{\partial \beta} \frac{\partial \theta_m}{\partial \beta} .$$

‖ I am grateful to Paul Marriott for pointing this out.

4.9 Functions of several variables

The delta method extends to functions $h(\overline{X}_n, \overline{Y}_n)$ of two or more random variables. In these expansions, it is necessary to consider both product moments of X and Y as well as mixed derivatives.

Suppose (X_j, Y_j), $j \geq 1$ are independent and identically distributed bivariate vectors, and that we wish to calculate

$$E\, h(\overline{Y}_n, \overline{X}_n) \qquad \textit{and} \qquad \mathrm{Var}\; h(\overline{X}_n, \overline{Y}_n)$$

where

$$\overline{X}_n = \frac{\sum_{j=1}^n X_j}{n} \qquad \text{and} \qquad \overline{Y}_n = \frac{\sum_{j=1}^n Y_j}{n}.$$

We assume the first moments are given by

$$E(X_j) = \mu_x, \qquad E(Y_j) = \mu_y$$

with second moment assumptions

$$\mathrm{Var}(X_j) = \sigma_x^2, \qquad \mathrm{Var}(Y_j) = \sigma_y^2, \qquad \text{and} \qquad \mathrm{Cov}(X_j, Y_j) = \sigma_{xy}^2 .$$

As $n \to \infty$, the marginal means $\overline{X}_n$ and $\overline{Y}_n$ are asymptotically normal. For simplicity, let $\mu = (\mu_x, \mu_y)$. Then the two-variable versions of Propositions 5 and 6 are

$$E\, h(\overline{X}_n, \overline{Y}_n) = h(\mu) + \frac{\partial^2 h}{\partial x^2}\frac{\sigma_x^2}{2\,n} + \frac{\partial^2 h}{\partial y^2}\frac{\sigma_y^2}{2\,n} + \frac{\partial^2 h}{\partial x\,\partial y}\frac{\sigma_{xy}^2}{n} + O(n^{-2}), \quad (4.29)$$

and

$$\mathrm{Var}\, h(\overline{X}_n, \overline{Y}_n) = \left[\frac{\partial h}{\partial x}\right]^2 \frac{\sigma_x^2}{n} + \left[\frac{\partial h}{\partial y}\right]^2 \frac{\sigma_y^2}{n} + \left[\frac{\partial h}{\partial x}\frac{\partial h}{\partial y}\right] \frac{2\,\sigma_{xy}^2}{n} + O(n^{-2}), \quad (4.30)$$

where the partial derivatives are all evaluated at μ. Higher order terms can be calculated by pushing this further in the usual way.

4.10 Ratios of averages

One particular case of a function of two variables occurs so frequently that it is important to mention by itself. Suppose (X_j, Y_j), $j \geq 1$ are independent and identically distributed bivariate vectors, and that we wish to calculate $E\left(\overline{Y}_n / \overline{X}_n\right)$ where

$$\overline{X}_n = \frac{\sum_{j=1}^n X_j}{n} \qquad \text{and} \qquad \overline{Y}_n = \frac{\sum_{j=1}^n Y_j}{n}.$$

We assume the first moments are given by

$$E(X_j) = \mu_x, \qquad E(Y_j) = \mu_y$$

with second moment assumptions

$$\text{Var}(X_j) = \sigma_x^2, \qquad \text{Var}(Y_j) = \sigma_y^2, \qquad \text{and} \qquad \text{Cov}(X_j, Y_j) = \sigma_{xy}^2 .$$

As $n \to \infty$, the marginal means $\overline{X}_n$ and $\overline{Y}_n$ are asymptotically normal.

Now $\overline{X}_n - \mu_x$ and $\overline{Y}_n - \mu_y$ are both $O_p(1/\sqrt{n})$. Suppose $\mu_x \neq 0$. Then a binomial expansion gives

$$\begin{aligned}
\overline{X}_n^{-1} &= \mu_x^{-1}\left[1 + \frac{\overline{X}_n - \mu_x}{\mu_x}\right]^{-1} \\
&= \mu_x^{-1}\left[1 - \frac{\overline{X}_n - \mu_x}{\mu_x} + \left(\frac{\overline{X}_n - \mu_x}{\mu_x}\right)^2 + O_p\left(\frac{1}{n\sqrt{n}}\right)\right].
\end{aligned}$$

So

$$\begin{aligned}
\frac{\overline{Y}_n}{\overline{X}_n} &= \frac{\overline{Y}_n}{\mu_x} - \frac{(\overline{X}_n - \mu_x)\,\overline{Y}_n}{\mu_x^2} + \frac{(\overline{X}_n - \mu_x)^2\,\overline{Y}_n}{\mu_x^3} + O_p\left(\frac{1}{n\sqrt{n}}\right) \\
&= \frac{\overline{Y}_n}{\mu_x} - \frac{(\overline{X}_n - \mu_x)\,\overline{Y}_n}{\mu_x^2} + \frac{(\overline{X}_n - \mu_x)^2\,\mu_y}{\mu_x^3} \\
&\qquad + \frac{(\overline{X}_n - \mu_x)^2\,(\overline{Y}_n - \mu_y)}{\mu_x^3} + O_p\left(\frac{1}{n\sqrt{n}}\right) \\
&= \frac{\overline{Y}_n}{\mu_x} - \frac{(\overline{X}_n - \mu_x)\,\overline{Y}_n}{\mu_x^2} + \frac{(\overline{X}_n - \mu_x)^2\,\mu_y}{\mu_x^3} + O_p\left(\frac{1}{n\sqrt{n}}\right)
\end{aligned}$$

In regular cases, this order term will have expectation $O(n^{-2})$, or at least will be $o(n^{-1})$. So by taking expectations of the terms, the reader may check that

$$E\left(\frac{\overline{Y}_n}{\overline{X}_n}\right) = \frac{\mu_y}{\mu_x} + \frac{1}{n}\left(\frac{\sigma_x^2\,\mu_y}{\mu_x^3} - \frac{\sigma_{xy}^2}{\mu_x^2}\right) + o(n^{-1}). \tag{4.31}$$

See Problem 12. When X_j and Y_j are independent for all j, then $\sigma_{xy}^2 = 0$. In this case we can write

$$E\left(\frac{\overline{Y}_n}{\overline{X}_n}\right) = \frac{\mu_y}{\mu_x}\left(1 + \frac{C_x^2}{n}\right) + o(n^{-1}) \tag{4.32}$$

where $C_x = \sigma_x/\mu_x$ is the coefficient of variation for X_j. See Problem 13.

It is left to the reader to prove similarly that

$$\text{Var}\left(\frac{\overline{Y}_n}{\overline{X}_n}\right) = \frac{1}{n}\left(\frac{\mu_y^2\,\sigma_x^2}{\mu_x^4} + \frac{\sigma_y^2}{\mu_x^2} - \frac{2\,\sigma_{xy}^2\,\mu_y}{\mu_x^3}\right) + o(n^{-1}). \tag{4.33}$$

See Problem 14.

4.11 The delta method for distributions

A key difficulty in implementing the delta method for moments is that its usual regularity conditions are very restrictive. When the assumptions of Propositions 3 and 4 fail, the conclusions may still hold but this is often hard to verify. The delta method for distributions offers an alternative approach that is often more easily applicable with weaker assumptions for large sample theory.

Henceforth, we shall write $U_n \stackrel{d}{\Longrightarrow} F$ to denote the fact that the sequence of random variables $U_n\,, n = 1, 2, \ldots$ converges in distribution to F, where F may be a distribution function or the name of some common distribution such as $\mathcal{N}(0, \sigma^2)$, say. We shall also abuse notation and write $U_n \stackrel{d}{\Longrightarrow} U$, where U is a random variable. This latter expression shall be understood to mean that U_n converges in distribution to F where $U \stackrel{d}{=} F$. Note in particular, that when μ is a constant, then

$$U_n \stackrel{d}{\Longrightarrow} \mu \qquad \text{and} \qquad U_n = \mu + o_p(1)$$

are equivalent, and both are equivalent to convergence in probability to μ.

The next result that we shall state is Slutsky's Theorem, which is useful for proving results on convergence in distribution.

Proposition 9. *Suppose that U_n, $n \geq 1$ is a sequence of random variables such that $U_n \stackrel{d}{\Longrightarrow} U$. Let V_n, $n \geq 1$ and W_n, $n \geq 1$ be sequences of random variables.*

1. *If $V_n = U_n + o_p(1)$ as $n \to \infty$, then $V_n \stackrel{d}{\Longrightarrow} U$.*
2. *If $W_n = U_n\,[1 + o_p(1)]$ as $n \to \infty$, then $W_n \stackrel{d}{\Longrightarrow} U$.*

Proof. This is essentially a special case of Proposition 5 in Chapter 1, setting $c = 0$ in the first part of that proposition and $c = 1$ in the second part. ∎

The following proposition may be thought of as the delta method for general distributions.

Proposition 10. *Suppose that T_n, $n \geq 1$ is a sequence of random variables such that*

$$T_n = \mu + o_p(1)$$

as $n \to \infty$. *Let* $\{c_n, n \geq 1\}$ *be a sequence of real constants such that*

$$c_n (T_n - \mu) \stackrel{d}{\Longrightarrow} X$$

as $n \to \infty$, *for some random variable* X. *Let* $h(x)$ *be a real-valued function of a real variable which is* k *times differentiable at* $x = \mu$, *such that* $h^{(k)}(\mu) \neq 0$ *and such that* $h^{(j)}(\mu) = 0$ *for all* $j < k$.

Then

$$(c_n)^k [h(T_n) - h(\mu)] \stackrel{d}{\Longrightarrow} \frac{h^{(k)}(\mu)}{k!} X^k$$

as $n \to \infty$.

Proof. Expanding $h(T_n)$ using Taylor's Theorem gives us

$$h(T_n) = h(\mu) + \frac{h^{(k)}(\mu)}{k!} (T_n - \mu)^k [1 + o_p(1)] .$$

Therefore,

$$(c_n)^k [h(T_n) - h(\mu)] = \frac{h^{(k)}(\mu)}{k!} [c_n (T_n - \mu)]^k [1 + o_p(1)] . \qquad (4.34)$$

By assumption, $c_n (T_n - \mu)$ converges in distribution to X. Therefore

$$\frac{h^{(k)}(\mu)}{k!} [c_n (T_n - \mu)]^k \stackrel{d}{\Longrightarrow} \frac{h^{(k)}(\mu)}{k!} X^k .$$

We can now apply Slutsky's Theorem as stated in Proposition 9, part 2, with

$$\begin{aligned} U_n &= \frac{h^{(k)}(\mu)}{k!} [c_n (T_n - \mu)]^k \\ W_n &= (c_n)^k [h(T_n) - h(\mu)] , \text{ and} \\ U &= \frac{h^{(k)}(\mu)}{k!} X^k . \end{aligned}$$

We conclude that the left-hand side converges as required. ∎

Using Proposition 10, we obtain the following two results, which are commonly known as the delta method for distributions.

Proposition 11. *Let* $X_1, X_2, \ldots$ *be independent, identically distributed random variables with mean* μ *and variance* σ^2. *Then the following hold as* $n \to \infty$.

1. *Suppose that* $h(x)$ *is differentiable at* $x = \mu$, *where* $h'(\mu) \neq 0$. *Then*

$$\sqrt{n} \left[h(\overline{X}_n) - h(\mu) \right] \stackrel{d}{\Longrightarrow} \mathcal{N}\left(0, [h'(\mu)]^2 \sigma^2\right) .$$

2. *Suppose $h(x)$ is twice differentiable at $x = \mu$ with $h'(\mu) = 0$ and $h''(\mu) \neq 0$. Then*

$$\frac{2\,n\,\left[h(\overline{X}_n) - h(\mu)\right]}{h''(\mu)\,\sigma^2} \stackrel{d}{\Longrightarrow} \mathcal{X}(1)\,,$$

where $\mathcal{X}(1)$ is the chi-square distribution with one degree of freedom.

Proof. The statements of this proposition follow as a special case of Proposition 10. Set $c_n = \sqrt{n}$ and $T_n = \overline{X}_n$. By the central limit theorem,

$$\sqrt{n}\,(\overline{X}_n - \mu) \stackrel{d}{\Longrightarrow} \mathcal{N}(0,\,\sigma^2)\,.$$

Statement 1 now follows using $k = 1$ in Proposition 10, and statement 2 follows using $k = 2$. ∎

4.12 The von Mises calculus

4.12.1 Statistical functionals

One of the main difficulties with the delta method is its restricted domain of application. There are many other statistics of a random sample $X_1,\,\ldots,\,X_n$ which are not of the form $h(\overline{X}_n)$ for which an asymptotic expansion would be useful. Although these statistics are not functions of $\overline{X}_n$, a common feature of many of them is that they are functions of the empirical distribution function $\widehat{F}_n$, defined as

$$\widehat{F}_n(t) = \frac{\#\{j\,:\,X_j \le t,\ \text{where } 1 \le j \le n\}}{n} \tag{4.35}$$

for all $-\infty < t < \infty$. These include the sample median

$$M_n = \widehat{F}_n^{-1}\left(\frac{1}{2}\right)\,,$$

the sample variance

$$S^2 = \int_{-\infty}^{\infty} t^2\,d\widehat{F}_n(t) - \left[\int_{-\infty}^{\infty} t\,d\widehat{F}_n(t)\right]^2\,,$$

and the Cramér-von Mises statistic

$$\Omega^2 = \int_{-\infty}^{\infty}\left[\widehat{F}_n(t) - F(t)\right]^2\,dF(t)$$

where $F(t)$ is the theoretical distribution function of the sample. It would be helpful to extend the delta method to statistics of this sort. In such an extension, the difference $\widehat{F}_n - F$ should replace the sample average $\overline{X}_n -$

Richard von Mises (1883–1953)

Richard von Mises was an applied mathematician who made diverse contributions to the theories of probability and statistics. In addition to the von Mises calculus, he contributed to the philosophy of probability. His best known contribution to probability is the statement of the "birthday problem," which has been popular in introductory probability courses ever since. This particular problem, posed in 1939, was to determine the number of people who must be gathered together to ensure that the probability that two share a birthday is greater than fifty percent. Outside probability and statistics, von Mises worked on airfoil design and positivist philosophy.

μ. One such extension of the delta method to a wider class of statistics is due to Richard von Mises, and is called the *von Mises calculus.*

To approach the von Mises calculus, it is helpful to reconsider the concept of a parameter in a statistical model. We usually think of a parameter θ as an index for a family of distributions $\{F_\theta : \theta \in \Theta\}$. When we write a model in this form, the parameter may simply serve as a label for the distribution of the data. A different interpretation is that a parameter is a *statistical functional***, namely $\theta = h(F)$, where h is defined on a space of distribution functions F, and takes values among the real numbers. More precisely, a statistical functional is a function defined on a space of distribution functions including the set of all empirical distribution functions $\widehat{F}_n$.

We often estimate a parameter by its sample version. Such estimates are called "plug-in" estimates. So if θ is defined by a statistical functional $\theta = h(F)$, then the plug-in estimate for θ based upon a random sample of size n is $\widehat{\theta} = h(\widehat{F}_n)$. One such example is the distribution mean, whose plug-in estimate is the sample mean, to wit

$$\mu(F) = \int_{-\infty}^{\infty} t\, dF(t)\,, \qquad \overline{X}_n = \int_{-\infty}^{\infty} t\, d\widehat{F}_n(t)\,.$$

Not every plug-in estimate will be a successful estimator. For example, the statistical functional

$$h(F) = \sum_t [F(t) - F(t-)]$$

measures the probability weight of the discrete component of the distribution. (In this expression, the sum is over all real t. The infinite sum is well defined because a distribution function can have at most countably many jumps, and at these points the summands are positive.) However, its plug-in estimate is

$$\begin{aligned} h(\widehat{F}_n) &= \sum_{j=1}^{n} \widehat{F}_n(X_j) - \widehat{F}_n(X_j-) \\ &= 1\,, \end{aligned}$$

for all data sets, which is not a useful estimator.

** A *functional* is a function whose domain is a set of functions.

4.12.2 Algebraic functionals

To extend the delta method to $h(\widehat{F}_n)$ we will need to construct extensions of the polynomial approximations that are the basis for the delta method for $h(\overline{X}_n)$.

Definition 12. *We say that $h(F)$ is an* algebraic functional *of order k provided that there exists a real-valued symmetric function $\zeta_k(t_1, \ldots, t_k)$ such that*

$$h(F) = \int^{[k]} \zeta_k \, d^k F \,,$$

where $\int^{[k]} = \int \cdots \int$ represents a k-fold integral over all t_j, and

$$d^k F = \prod_{j=1}^{k} dF\,(t_j)\,,$$

is the k-dimensional Stieltjes integrating function. The integrand ζ_k for the algebraic functional is called the representation function.

Algebraic functionals play much the same role within the family of statistical functionals that symmetric homogeneous polynomials play in algebra.[††] Many statistical functionals are not algebraic, but they can often be locally approximated by algebraic functionals.

Another way to represent algebraic functionals is as expectations. If $h(F)$ is an algebraic functional of order k as in Definition 12, then we may also write

$$h(F) = E_F\, \zeta_k(X_1, X_2, \cdots, X_k)$$

where $X_1, \ldots, X_k$ is a random sample of size k from F. When evaluated at $\widehat{F}_n$, algebraic functionals become symmetric functions of the random variables that are closely related to U-statistics. For example, the kth order functional can be written as

$$\begin{aligned} h(\widehat{F}_n) &= \int^{[k]} \zeta_k \, d^k \widehat{F}_n \\ &= \frac{1}{n^k} \sum_{j_1=1}^{n} \cdots \sum_{j_k=1}^{n} \zeta_k(X_{j_1}, \ldots, X_{j_k})\,. \end{aligned} \tag{4.36}$$

[††] Removing the assumption of symmetry in ζ_k produces no greater generality. For example when $k = 2$,

$$\int\!\!\int \zeta_2(t_1, t_2)\, dF(t_1)\, dF(t_2) = \int\!\!\int \frac{\zeta_2(t_1, t_2) + \zeta_2(t_2, t_1)}{2}\, dF(t_1)\, dF(t_2)\,.$$

This differs from a U-statistic only in the fact that $j_1, \ldots, j_k$ are not necessarily distinct. When $k = 1$ this reduces to the average of $\zeta_1(X_1), \ldots, \zeta_1(X_n)$.

A functional of order one is called a *linear statistical functional.* It has representation

$$\begin{aligned} F &\mapsto E_F\, \zeta_1(X) \\ &= \int^{[1]} \zeta_1\, dF\,. \end{aligned}$$

For example, the mth moment functional $\mu_m(F)$ is an example of a linear statistical functional. It has representation

$$\begin{aligned} \mu_m(F) &= E_F\,(X^m) \\ &= \int_{-\infty}^{\infty} t^m\, dF(t)\,. \end{aligned} \tag{4.37}$$

As we noted previously, when $m = 1$, we simply write $\mu(F)$, which is the mean functional. The evaluation of the mean functional at the empirical distribution yields the sample mean $\mu(\widehat{F}_n) = \overline{X}_n$.

Similarly, a functional of order two is called a *quadratic statistical functional.* Its integral representation is

$$F \mapsto \int^{[2]} \zeta_2\, d^2F(t_1)\,.$$

where $\zeta_2(t_1, t_2)$ is a symmetric function. Perhaps the best known quadratic statistical functional is the variance functional defined by

$$\sigma^2(F) = \int_{-\infty}^{\infty}\int_{-\infty}^{\infty} \frac{(t_1 - t_2)^2}{2}\, dF(t_1)\, dF(t_2)\,. \tag{4.38}$$

Evaluating the variance functional at $\widehat{F}_n$ yields the sample variance $\sigma^2(\widehat{F}_n) = S_n^2$, where the sample variance has n and not $n - 1$ in the denominator.

Functionals of general order k can easily be expanded locally around a given distribution. Suppose we wish to evaluate $h(G)$ where G is in some neighbourhood of F. Write

$$\begin{aligned} d^kG &= d^k[\,F + (G - F)\,] \\ &= \sum_{j=0}^{k} \binom{k}{j} d^{k-j}F\, d^j(G - F)\,. \end{aligned}$$

Then

$$h(G) = \int^{[k]} \zeta_k\, d^kG$$

$$= \int^{[k]} \zeta_k \, d^k [\, F + (G - F)\,]$$

$$= \sum_{j=0}^{k} \int^{[j]} \left[\int^{[k-j]} \binom{k}{j} \zeta_k \, d^{k-j} F \right] d^j (G - F)\,.$$

The $j = 0$ term in this sum is simply $h(F)$. Define

$$\xi_j = k\,(k-1) \cdots (k-j+1) \int^{[k-j]} \zeta_k \, d^{k-j} F\,. \tag{4.39}$$

The function ξ_j is a symmetric function of j real variables (which have been suppressed in the notation). Also for any integrating function V, define

$$D^j h_F(V) = \int^{[j]} \xi_j \, d^j V\,. \tag{4.40}$$

which is a functional of order j. With these definitions we note that the expansion of $h(G)$ above can be written in terms of $D^j h_F(G - F)$. We obtain the following expansion.

Proposition 13. *If h is an algebraic functional of order k, then*

$$\begin{aligned} h(G) &= h(F) + \sum_{j=1}^{k} \int^{[j]} \xi_j \, d^j (G - F) \\ &= h(F) + \sum_{j=1}^{k} \frac{1}{j!} D^j h_F(G - F)\,, \end{aligned}$$

where $D^j h_F(G - F)$ is a functional of order j in the difference $G - F$.

This expansion of $h(G)$ about F is analogous to a Taylor expansion about some general point. Since $h(G)$ is algebraic—that is, "polynomial-like" in its properties—the expansion is finite. More generally, we might wish to expand a functional $h(G)$ which is not algebraic in an infinite expansion of the form

$$h(G) = h(F) + Dh_F(G - F) + \frac{1}{2} D^2 h_F(G - F) + \cdots\,,$$

where $D^j h_F$ is an algebraic functional of order j. When truncated, such a series will be an approximation of $h(G)$ about $h(F)$ by an algebraic functional. As with Taylor expansions of real-valued functions of real variables the existence of such a representation cannot be assumed in general.

4.12.3 Derivatives of statistical functionals

Up to this point, we have been rather informal about the space in which distributions F, G and differences $G - F$ lie. We may suppose a certain set of distribution functions including F and G, all relevant distributions for a statistical model, and all empirical distributions are in some space $\mathcal{V}$. We suppose that $\mathcal{V}$ is vector space closed under all finite linear combinations of the functions. Thus

$$F, G \in \mathcal{V} \qquad \text{implies} \qquad \alpha F + \beta G \in \mathcal{V}$$

for all real values α and β. (Here, addition and scalar multiplication are performed pointwise.) So in particular $G - F$ is in $\mathcal{V}$.

Within $\mathcal{V}$ lie some subsets and subspaces of interest to us. One of these is $\mathcal{K}$ which is the set of all distribution functions within $\mathcal{V}$. This subset is convex in the sense that

$$F, G \in \mathcal{K} \qquad \text{implies} \qquad \alpha F + (1 - \alpha) G \in \mathcal{K},$$

for all $0 \le \alpha \le 1$.

Another subset of interest to us is the set $\mathcal{K}_0$ of all differences $G - F$ where F, $G \in \mathcal{K}$. This subset is also convex like $\mathcal{K}$, but it has no elements in common with $\mathcal{K}$. The set of all finite linear combinations of elements of $\mathcal{K}_0$ is a proper subspace of $\mathcal{V}$ that we shall call $\mathcal{V}_0$. To define a linear functional on $\mathcal{V}_0$ it is sufficient to define it on $\mathcal{K}_0$, as we do in the next definition.

Definition 14. *Let $h : F \mapsto h(F)$ be a statistical functional defined on $\mathcal{K}$. Then h is said to be* Gâteaux differentiable *at $F \in \mathcal{K}$ if there exists a statistical functional Dh_F defined on $\mathcal{V}_0$, satisfying the following two conditions.*

1. *The functional Dh_F is a linear statistical functional on $\mathcal{V}_0$ in the sense that it has integral representation*

$$Dh_F(V) = \int_{-\infty}^{\infty} \xi_F(t)\, dV(t) \tag{4.41}$$

for all $V \in \mathcal{V}_0$.

2. *When $V = G - F \in \mathcal{K}_0$, the functional Dh_F satisfies the identity*

$$Dh_F(G - F) = \left[\frac{d}{d\alpha}\, h\{(1 - \alpha) F + \alpha G\}\right]_{\alpha=0} \tag{4.42}$$

for all $G \in \mathcal{K}$. On the right-hand side, the derivative is a one-sided derivative at $\alpha = 0$ for $0 \le \alpha \le 1$.

The linear functional Dh_F is called the Gâteaux derivative *of h at F. The representation function $\xi_F(t)$ displayed in (4.41) above is called the* influence function *of h at F.*

The Gâteaux derivative of a functional on $\mathcal{K}$ is uniquely defined on $\mathcal{V}_0$ but does not have a unique extension to $\mathcal{V}$. Also, the influence function $\xi_F(t)$ is also not unique, because any function $\xi_F(t)$ which works in Definition 14 can be replaced by $\xi_F(t) + c$ for any constant c. However, $\xi_F(t)$ will be uniquely defined up to an arbitrary additive constant. Because an influence function is "almost" unique in this sense, it is commonplace in the literature to refer to "the" influence function as if it were unique. Below, we shall consider a version of the influence function called the influence curve, which is uniquely defined.

Now suppose h is Gâteaux differentiable at F, so that we can write

$$h(G) = h(F) + Dh_F(G - F) + \text{Rem}_F(G - F)\,, \tag{4.43}$$

where $\text{Rem}_F(G - F)$ is a remainder term.

Is $\text{Rem}_F(G - F)$ negligible as G gets close to F? In particular, we are interested in the case where $G = \widehat{F}_n$, with $\widehat{F}_n$ converging empirically to F. Loosely speaking, we expect that $\widehat{F}_n - F = O_p(n^{-1/2})$ by the central limit theorem. So for the remainder to be negligible we need $\text{Rem}_F(\widehat{F}_n - F) = o_p(n^{-1/2})$.

Before we consider the empirical convergence of $\widehat{F}_n$ to F, let us consider a non-random sequence converging to F. For any distribution G, let

$$F_\alpha = (1 - \alpha)\, F + \alpha\, G\,,$$

so that $Dh_F(F_\alpha - F) = \alpha\, Dh_F(G - F)$ is of order $O(\alpha)$ as $\alpha \to 0$. Applying Taylor's Theorem to $h(F_\alpha)$ and Definition 14, we can write

$$h(F_\alpha) = h(F) + \underbrace{Dh_F(F_\alpha - F)}_{O(\alpha)} + \underbrace{\text{Rem}_F(F_\alpha - F)}_{o(\alpha)}\,.$$

This says that as we approach the distribution F along any line in $\mathcal{K}$ passing through F and G, the value of the functional $h(F_\alpha)$ is well approximated by $h(F) + Dh_F(F_\alpha - F)$ for small α, with a negligible remainder.

At first glance, this seems to prove that $\text{Rem}_F(\widehat{F}_n - F)$ is negligible when the empirical distribution $\widehat{F}_n$ converges to F. However, here our finite dimensional intuition can be misleading. The non-random sequence F_α converges to F along a straight line. Although the empirical distribution $\widehat{F}_n$ gets close to F as $n \to \infty$, it does not converge along a straight line passing through F. So there is still some work to do to show that the remainder is negligible, *viz.*,

$$h(\widehat{F}_n) = h(F) + \underbrace{Dh_{_F}(\widehat{F}_n - F)}_{O_p(n^{-1/2})} + \underbrace{\mathrm{Rem}_{_F}(\widehat{F}_n - F)}_{o_p(n^{-1/2})} .$$

Indeed, this result may fail, even if h is Gâteaux differentiable. The problem is that Gâteaux differentiability of h is not strong enough to prove that the remainder term is *uniformly negligle* in small neighbourhoods of F.

Although Gâteaux differentiability is not strong enough, many statistical functionals are differentiable in a stronger sense. For example, a functional which is Fréchet differentiable is automatically Gâteaux differentiable, although the reverse is not true in general. Fréchet differentiability forces the kind of uniform convergence that we need to show that $\mathrm{Rem}_{_F}(\widehat{F}_n - F) = o(n^{-1/2})$. In order to define Fréchet differentiability, it is necessary to impose more structure on the space $\mathcal{V}$ by making it into a normed vector space. For a vector space of bounded functions, a commonly used norm is the supremum norm defined by

$$||\, V\, || = \sup\{|\, V(t)\,| \ : \ -\infty < t < \infty\} . \qquad (4.44)$$

Other norms are possible, and the type of Fréchet differentiability will vary depending on the choice of norm. Suppose a norm $||V||$ is defined for each $V \in \mathcal{V}$.

Definition 15. *The statistical functional h is said to be Fréchet differentiable at F if there exists a linear functional*

$$Dh_{_F}(V) = \int_{-\infty}^{\infty} \xi(t)\, dV(t)$$

such that

$$\frac{|\, h(G) - h(F) - Dh_{_F}(G-F)\,|}{||\, G - F\,||} \to 0 ,$$

whenever G converges to F in the sense that $||G - F|| \to 0$.

Fréchet differentiable functionals satisfy the order conditions

$$|\,\mathrm{Rem}_{_F}(V)\,| = o(\,||\, V\,||\,) \qquad \text{and} \qquad |\, Dh_{_F}(V)\,| = O(\,||\, V\,||\,)$$

as $||V||$ goes to zero. In these cases, the remainder term is asymptotically negligible, and $h(G) - h(F)$ can be approximated by $Dh_{_F}(G - F)$. Fréchet differentiability is stronger than Gâteaux differentiability. When the Fréchet derivative exists, it will be identical to the Gâteaux derivative. However, the latter has the advantage that it exists in greater generality and is often easier to calculate.‡‡

When $G = \widehat{F}_n$, it is possible to express the order of each term in (4.43) as a function of the sample size n. For many norms on $\mathcal{V}$ we find that

$$|| \widehat{F}_n - F || = O_p\left(\frac{1}{\sqrt{n}}\right) .$$

For example, the supremum norm defined in (4.44) satisfies this property, a fact that may be deduced from the asymptotic properties of the Kolmogorov-Smirnov statistic. Therefore, a Fréchet differentiable function will satisfy

$$Dh_F(\widehat{F}_n - F) = O_p\left(\frac{1}{\sqrt{n}}\right) \qquad \text{and} \qquad \text{Rem}_F(\widehat{F}_n - F) = o_p\left(\frac{1}{\sqrt{n}}\right) .$$

4.12.4 Two examples

To illustrate the role of the Gâteaux derivative, let us consider the derivative of the mean functional $\mu(F)$ and the variance functional $\sigma^2(F)$. The mean function is itself a linear functional. As we might expect, the derivative is idempotent on the class of linear functionals. This is easy to check from the definition.

$$\begin{aligned} D\mu_F(G - F) &= \left[\frac{d}{d\alpha}\,\mu\{\,(1-\alpha)\,F + \alpha\,G\,\}\right]_{\alpha=0} \\ &= \left[\frac{d}{d\alpha}\,\{\,(1-\alpha)\,\mu(F) + \alpha\,\mu(G)\,\}\right]_{\alpha=0} \qquad \text{(linearity of } \mu) \\ &= \mu(G) - \mu(F) \\ &= \mu(G - F) \quad \text{(linearity again)} . \end{aligned}$$

Therefore $D\mu_F = \mu$ for all values of F. The linearity of $\mu(F)$ also provides a quick proof that μ is Fréchet differentiable. The linearity also implies that for the functional μ, the remainder term $\text{Rem}_F(G - F)$ vanishes.

Let us now consider the variance functional,

$$\sigma^2(F) = \mu_2(F) - \mu^2(F)$$

‡‡ L. Turrin Fernholz has shown that although Gâteaux differentiability is too weak, Fréchet differentiability is often too strong. Hadamard differentiability is offered as a compromise. Whatever form of differentiability is used, it must be checked in a given example. Often the most direct thing to check is the remainder term itself, by evaluating and determining its order using analytic methods.

which is a quadratic statistical functional as noted above. The reader can check that

$$\begin{aligned} D\sigma^2_F(G-F) &= \left[\frac{d}{d\alpha}\,\sigma^2\{\,(1-\alpha)\,F+\alpha\,G\,\}\right]_{\alpha=0} \\ &= \mu_2(G-F)-2\,\mu(F)\,\mu(G-F)\,. \qquad (4.45) \end{aligned}$$

Substituting $G=\widehat{F}_n$, equation (4.45) becomes

$$D\sigma_F{}^2(\widehat{F}_n-F)=\mu_2(\widehat{F}_n-F)-2\,\mu(F)\,\mu(\widehat{F}_n-F)\,.$$

It is easily shown that this can be rewritten as

$$\sigma^2(\widehat{F}_n)=\sigma^2(F)+D\sigma^2_F(\widehat{F}_n-F)+\mathrm{Rem}_F(\widehat{F}_n-F)\,,$$

where $\mathrm{Rem}_F(\widehat{F}_n-F)=-\mu^2(\widehat{F}_n-F)$. In this case, we are provided with an explicit formula for the remainder. We see directly that

$$\begin{aligned} \mu(\widehat{F}_n-F) &= \mu(\widehat{F}_n)-\mu(F) \\ &= \overline{X}_n-\mu(F) \\ &= O_p(n^{-1/2})\,. \end{aligned}$$

which implies that for the functional $\sigma^2(\widehat{F}_n)$ the remainder term is

$$\begin{aligned} \mathrm{Rem}_F(\widehat{F}_n-F) &= -\mu^2(\widehat{F}_n-F) \\ &= O_p(n^{-1})\,, \end{aligned}$$

as $n\to\infty$.

4.12.5 The delta method for functionals

As required in Definitions 14 and 15, the derivative

$$Dh_F(V)=\int_{-\infty}^{\infty}\xi_F(t)\,dV(t)$$

is a linear functional of V, where $\xi_F(t)$ is the influence function of h at F. In addition, as we noted in (4.36), when evaluated at $V=\widehat{F}_n$ such a linear functional becomes an average over the data. When $V=\widehat{F}_n-F$, the result is an average over the data, centred with expectation zero, namely

$$Dh_F(\widehat{F}_n-F)=\frac{1}{n}\sum_{j=1}^{n}[\,\xi_F(X_j)-E_F\,\xi_F(X_j)\,]\,.$$

So we can apply the central limit theorem to get

$$\sqrt{n}\,Dh_F(\widehat{F}_n-F) \;=\; \frac{1}{\sqrt{n}}\sum_{j=1}^{n}[\,\xi_F(X_j)-E_F\xi_F(X_j)]$$

$$\stackrel{d}{\Longrightarrow} \mathcal{N}(0, v)$$

where

$$\begin{aligned} v &= \operatorname{Var}_F \xi_F(X) \\ &= \int_{-\infty}^{\infty} [\xi_F(t)]^2 \, dF - \left[\int_{-\infty}^{\infty} \xi_F(t) \, dF(t) \right]^2 . \end{aligned}$$

is such that $0 < v < \infty$.

Suppose that h is Gâteaux differentiable at F and that

$$\operatorname{Rem}_F(\widehat{F}_n - F) = o_p(n^{-1/2})$$

or equivalently

$$\sqrt{n} \operatorname{Rem}_F(\widehat{F}_n - F) = o_p(1) .$$

Provided that $v > 0$, the remainder term will be asymptotically negligible, so that

$$\begin{aligned} \sqrt{n}[\, h(\widehat{F}_n) - h(F) \,] &= \sqrt{n} \, Dh_F(\widehat{F}_n - F) + o_p(1) \\ &\stackrel{d}{\Longrightarrow} \mathcal{N}(0, v) . \end{aligned}$$

by Slutsky's Theorem. This central limit theorem for statistical functionals extends the delta method that was developed earlier in the chapter.

4.13 Obstacles and opportunities: robustness

We have seen that the von Mises calculus can be used to prove the asymptotic normality of a large class of statistics of a random sample. This class includes M-estimators, which are generalisations of maximum likelihood estimators, as the next definition explains.

Definition 16. *A functional h is said to be an M-functional if there exists a real-valued function $u(\theta, t)$ such that*

$$\int_{-\infty}^{\infty} u(\theta, t) \, dF(t) = 0 \qquad \text{whenever} \qquad \theta = h(F) .$$

When h is an M-functional, then $\widehat{\theta}_n = h(\widehat{F}_n)$ is said to be an M-estimator for θ.

For example, maximum likelihood estimators are M-estimators. In such cases, the function $u(\theta, x)$ is the score function defined by

$$u(\theta, x) = \frac{\partial}{\partial \theta} \ln f(x; \theta)$$

where $f(x;\,\theta)$ is the probability density function or probability mass function of the distribution $F(t)$. Although the von Mises calculus is applicable to likelihood estimation, it is rare to find the asymptotic normality of a maximum likelihood estimator proved in this way. In most course textbooks asymptotic normality is proved from a Taylor expansion of the likelihood equation and the use of regularity conditions on the model due to H. Cramér. With some modest modification, the Cramér argument for asymptotic normality can be extended to a general class of M-estimators. See Bickel and Doksum (2001).

Why is the von Mises calculus not routinely used to prove asymptotic normality of M-estimator such as the maximum likelihood estimator? Both von Mises' and Cramér's methods involve regularity assumptions. However, von Mises' assumptions, whether or not modified for Fréchet or Hadamard differentability, are more difficult to check than the Cramér assumptions. Concepts of statistical functionals and linear functionals on normed vector spaces are mathematically mature, and an elementary argument must be considered superior to any advanced argument.

For this reason, the von Mises calculus has not become as popular as Cramér's methodology. However, it received a new lease on life when its connection to the theory of robust statistics was noticed. Suppose we plug in a specific choice for G in Definition 14, namely that corresponding to a distribution which places all its probability mass on a point t. This is the Dirac delta distribution δ_t, and the mixture $F + \alpha(\delta_t - F)$ represents a perturbation of F obtained by placing a probability mass of size α at t and correspondingly downweighting F by a factor of $1-\alpha$. Now suppose that h is Gâteaux differentiable with influence function $\xi_F(t)$, as in Definition 14. Recall that the influence function of Definition 14 is defined up to an arbitrary additive constant, which can be determined in a variety of ways. Suppose we choose the additive constant so that $E_F\,\xi_F(X) = 0$, when the random variable X has distribution F. Substituting $G = \delta_t$ in (4.42) yields

$$\int_{-\infty}^{\infty} \xi_F(s)\, d(\delta_t - F)(s) = \left[\frac{d}{d\alpha}\, h\{\, F + \alpha\,(\delta_t - F)\,\}\right]_{\alpha=0} .$$

On the left-hand side of this equation, we see that

$$\int_{-\infty}^{\infty} \xi_F(s)\, d\delta_t(s) = \xi_F(t) \qquad \text{and} \qquad \int_{-\infty}^{\infty} \xi_F(s)\, dF(s) = 0\,.$$

So it follows that

$$\xi_F(t) = \left[\frac{d}{d\alpha}\, h\{\, F + \alpha\,(\delta_t - F)\,\}\right]_{\alpha=0} . \tag{4.46}$$

for all $-\infty < t < \infty$. This particular version of the influence function is often called the *influence curve*.

The influence curve was introduced into the study of robustness in Hampel (1968). In that context, it is a tool for measuring the sensitivity of a parameter or estimator—both in our terminology here are defined by a statistical functional—to the contamination of a distribution F by a datum positioned at point t. The behaviour of the influence function as $t \to \pm\infty$ is particularly important, because it can be used to measure how robust an estimator is to outlying data points. For example, the mean functional has influence function $\xi_F(t) = t - \mu(F)$ which is unbounded at $t = \pm\infty$. The influence curve for the sample mean is obtained by plugging in $F = \widehat{F}_n$, to obtain $\xi(t) = t - \overline{X}_n$. Calculating the influence curve for the variance functional $\sigma^2(F)$ is left as Problem 17.

Let us compare the influence curve of the distribution mean $\mu(F)$ with the influence curve for the distribution median $m(F)$. Suppose that F is a continuous distribution having a continuous density f with the property that $f[\,m(F)\,] > 0$. With this assumption, the median is unique and

$$m(F) = F^{-1}\,(1/2)\,.$$

Let $m = m(F)$ and suppose that $t > m$. Define

$$m^\star = m[\,(1-\alpha)\,F + \alpha\,\delta_t\,]\,.$$

It is left to the reader to check that as $\alpha \searrow 0$ then

$$m^\star = m + \frac{\alpha}{2\,f(m)} + o(\alpha)\,. \tag{4.47}$$

Similarly, when $t < m$ we get

$$m^\star = m - \frac{\alpha}{2\,f(m)} + o(\alpha)\,. \tag{4.48}$$

See Problem 18. So the influence function of $m(F)$ is

$$\begin{aligned}
\xi_F(t) &= \left[\frac{d}{d\alpha}\,m^\star\right]_{\alpha=0} \\
&= \left[\frac{d}{d\alpha}\left\{\,m + \mathrm{sgn}(t-m)\,\frac{\alpha}{2\,f(m)} + o(\alpha)\,\right\}\right]_{\alpha=0} \\
&= \frac{\mathrm{sgn}(t-m)}{2\,f(m)}\,.
\end{aligned} \tag{4.49}$$

In contrast to the influence function for the mean functional, this influence function is bounded and depends only on t through $\mathrm{sgn}(t-m)$. This property of the influence function of $m(F)$ is a reflection of the well-known insensitivity of the median to outlier contamination.

We can also use the von Mises calculus to calculate the asymptotic distribution of the sample median $m(\widehat{F}_n)$. This requires additional work to prove that the remainder term $\text{Rem}_F(\widehat{F}_n - F) = o_p(n^{-1/2})$. Suppose n is odd, so that we can write $n = 2k+1$. With probability one the order statistics $X_{(1)} < X_{(2)} < \cdots < X_{(2k+1)}$ will be distinct. For samples of odd size, we have $m(\widehat{F}_n) = X_{(k+1)}$. Then

$$\sqrt{n}\,\left[X_{(k+1)} - m\right] \stackrel{d}{\Longrightarrow} \mathcal{N}(0,\, v)$$

where

$$\begin{aligned} v &= \text{Var}_F\, \xi_F(X) \\ &= E_F\,[\,\xi_F(X)\,]^2 \\ &= \frac{1}{4\,[\,f(m)\,]^2}\,. \end{aligned}$$

4.14 Problems

1. Let $m \geq n \geq 1$, and suppose that X is a non-negative random variable.
   (a) Using Jensen's inequality, prove that $E(X^m)^n \geq E(X^n)^m$.
   (b) Using the inequality above, prove that
   $$E(X^m) \geq E(X^n)\, E(X^{m-n})\,.$$
   (c) Extend the previous result to the following. Suppose $X_1, \ldots, X_n$ are independent identically distribution random variables and that $1 \leq j_1 \leq \cdots \leq j_m \leq n$. Prove that
   $$|\, E(X_{j_1}\, X_{j_2}\, \cdots\, X_{j_m})\,| \leq E\,|X_1^m|\,.$$

2. Suppose $X \sim \mathcal{N}(\mu,\, 1)$, where $\mu > 0$, and we wish to use the delta method to approximate the mean and variance of $h(X) = \sqrt{X}$. (Here, we consider only a single random variable so that $n = 1$ in the formulas.) We assume that μ is reasonably large, so that $P(X > 0) \approx 1$. Without doing any detailed calculations, explain how well the delta method should work for different values of μ. Do we expect the approximation to be better for smaller μ or for larger μ? Justify your answer.

3. Let $X_1, \ldots, X_n$ be independent Poisson random variables with mean λ. Prove that
   $$E\,\sin \overline{X}_n = \sin\lambda - \frac{\lambda\,\sin\lambda}{2\,n} - \frac{\lambda\,\cos\lambda}{6\,n^2} + \frac{\lambda^2\,\sin\lambda}{8\,n^2} + O\left(\frac{1}{n^3}\right)\,. \quad (4.50)$$

4. Suppose X is $\mathcal{B}(n, \theta)$, where n is known.

   (a) Show that the maximum likelihood estimator for θ is $\widehat{\theta} = X/n$.

   (b) Show that $E(\widehat{\theta}) = \theta$ and that $\text{Var}(\widehat{\theta}) = \theta\,(1-\theta)/n$.

   (c) Since θ is unknown, we can estimate the value of $\text{Var}(\widehat{\theta})$ using the plug-in estimate $\widehat{\theta}\,(1-\widehat{\theta})/n$. Find the bias of this plug-in estimate by showing first that

$$E\left[\widehat{\theta}\,(1-\widehat{\theta})\right] = \frac{(n-1)\,\theta\,(1-\theta)}{n}\,.$$

   (d) Use the delta method for moments—particularly Proposition 6—to show that

$$\text{Var}\left[\widehat{\theta}\,(1-\widehat{\theta})\right] = \frac{(1-2\,\theta)^2\,\theta\,(1-\theta)}{n} + O(n^{-2})\,.$$

   (e) Finally, determine the exact value of the variance approximated in part (d) above. Use this to show that the order term $O(n^{-2})$ obtained in part (d) is correct.

5. Let $X_1, \ldots, X_n$ be independent exponential random variables with density

$$f(x) = \begin{cases} \lambda\, e^{-\lambda\, x} & \text{for } x \geq 0\,, \\ 0 & \text{for } x < 0\,. \end{cases}$$

   (a) Find the maximum likelihood estimator for $P_\lambda(X_1 \leq 2)$, and express this in the form $h(\overline{X}_n)$.

   (b) Prove that

$$E_\lambda\, h(\overline{X}_n) = P_\lambda(X_1 \leq 2) + \frac{2\, e^{-2\lambda}\, \lambda\,(1-\lambda)}{n} + O(n^{-2})\,,$$

   and

$$\text{Var}_\lambda\, h(\overline{X}_n) = \frac{4\,\lambda^2\, e^{-4\,\lambda}}{n} + O(n^{-2})\,.$$

6. Prove that the transformation $h(\overline{X}_n)$, with $h(x)$ as given in (4.16) stabilises the variance to order $O(n^{-1})$.

7. Prove Proposition 7. (Don't worry about the regularity assumptions of Propositions 4 and 6, because $h(x) = \sqrt{x}$ does not have bounded derivatives on the positive reals.)

8. Prove that the choice of h given in the differential equation (4.22), annihilates the coefficient on n^{-2} as required for (4.21). (This is a messy expansion. You may find Maple helpful.)

9. Verify formula (4.29) under appropriate assumptions on the moments of (X_j, Y_j) and on the derivatives of h.

10. Verify formula (4.30) under appropriate assumptions on the moments of (X_j, Y_j) and on the derivatives of h.

11. Suppose that

$$X_n = \mu + O_p(1/\sqrt{n}) \qquad \text{and} \qquad Y_n = \mu + O_p(1/\sqrt{n})$$

as $n \to \infty$, where $\mu \neq 0$.

(a) Prove that

$$\frac{Y_n}{X_n} = 1 + \frac{Y_n - X_n}{\mu} + O_p(n^{-1}).$$

(b) Suppose (X_n, Y_n) is a bivariate normal vector, with $E(X_n) = E(Y_n) = \mu$, $\text{Var}(X_n) = \text{Var}(Y_n) = n^{-1}\sigma^2$ and with covariance $\text{Cov}(X_n, Y_n) = n^{-1}\rho\sigma^2$. Show that to order $1/\sqrt{n}$ the ratio Y_n/X_n is normally distributed. Calculate the mean and variance of this asymptotic distribution.

(c) Generalise your results in part (a) above to the case where X_n and Y_n have different means μ_x and μ_y.

(d) Give an example to show that the conclusion of part (a) does not follow when $\mu = 0$.

12. Prove (4.31).

13. Prove (4.32).

14. Prove (4.33).

15. Suppose we are given a real-valued function $f(x)$ of a real variable which has no simple algebraic formula. Evaluation of $f(x)$ must be done numerically and involves a certain amount of error which is random. When f is approximated at a set of domain values $x_1, \ldots, x_k$ the result is a set of approximations

$$f(x_1) + \epsilon_1, \qquad f(x_2) + \epsilon_2, \qquad \ldots \qquad f(x_k) + \epsilon_k$$

where ϵ_j, $1 \leq j \leq k$ are small independent identically distributed errors with mean zero and variance σ^2.

In order to approximate the derivative of $f(x)$ which has no simple form, the numerical approximation is

$$f'(x) \approx \frac{Y_2 - Y_1}{h}$$

where $Y_1 = f(x) + \epsilon_1$ and $Y_2 = f(x+h) + \epsilon_2$.

(a) For any given value of h, calculate the expectation and variance of this numerical approximation to $f'(x)$ in terms of the moments of ϵ.

(b) Find the mean square error of this numerical approximation.

(c) Do a Taylor expansion of $f(x+h)$ about $f(x)$. Use this to find the value of h which minimises the mean square error of the numerical approximation. Which order of Taylor expansion is most insightful here?

(d) Suppose the domain variables also have independent errors. In this case, an evaluation of f at $x_1, \ldots, x_k$ produces

$$f(x_1 + \delta_1) + \epsilon_1, \qquad f(x_2 + \delta_2) + \epsilon_2, \qquad \ldots \qquad f(x_k + \delta_k) + \epsilon_k$$

where δ_j has mean zero and variance τ^2. Use the delta method to approximate the mean and variance of the numerical derivative to order $O(\tau^2)$ as $\tau \to 0$.

16. Prove that when $\theta = \beta$ in Theorem 1, the transformation $\nu = h(\beta)$ which satisfies (4.28) will satisfy (4.26).

17. Recall that the variance functional $\sigma^2(F)$ is defined by

$$\sigma^2(F) = \int_{-\infty}^{\infty} \int_{-\infty}^{\infty} \frac{(t_1 - t_2)^2}{2}\, dF(t_1)\, dF(t_2)\,.$$

Prove that the influence curve of the variance functional is

$$\xi_F(t) = [\,t - \mu(F)\,]^2 - \sigma^2(F)\,.$$

18. Prove formulas (4.47) and (4.48).

19. Let $X_1, \ldots, X_n$ be independent identically distributed random variables, and $Y_n = h_n(X_1, \ldots, X_n)$ a random variable which is a function of them. Suppose we wish to find the limiting distribution of Y_n

as $n \to \infty$. Define

$$\widetilde{Y}_n = \sum_{j=1}^{n} E(Y_n \,|X_j) - (n-1)\,E(Y_n)\,.$$

The random variable $\widetilde{Y}_n$ is called the *Hájek projection* of Y_n. Note that the Hájek projection is defined as a sum over independent random variables. So the Lindeberg-Feller central limit theorem can sometimes be applied to $\widetilde{Y}_n$ when it is not immediately applicable to Y_n.

(a) Prove that $E(\widetilde{Y}_n) = E(Y_n)$.

(b) Show that

$$\operatorname{Var}(Y_n) = \operatorname{Var}(\widetilde{Y}_n) + E(Y_n - \widetilde{Y}_n)^2\,.$$

(c) Suppose that

$$\operatorname{Var}(\widetilde{Y}_n) \sim \operatorname{Var}(Y_n)$$

as $n \to \infty$. Prove that

$$\frac{\widetilde{Y}_n - E(\widetilde{Y}_n)}{\sqrt{\operatorname{Var}(\widetilde{Y}_n)}} \stackrel{d}{\Longrightarrow} \mathcal{N}(0,\,1) \qquad \text{implies} \qquad \frac{Y_n - E(Y_n)}{\sqrt{\operatorname{Var}(Y_n)}} \stackrel{d}{\Longrightarrow} \mathcal{N}(0,\,1)\,.$$

Conclude that if the Lindeberg-Feller Central Limit Theorem applies to $\widetilde{Y}_n$, then Y_n is also asymptotically normal.

(d) The Hájek projection is widely used to prove the asymptotic normality of U-statistics. Let

$$U_n = \sum u_k(X_{j_1},\, X_{j_2},\, \ldots,\, X_{j_k})$$

where u_k is a symmetrical function of k variables and the sum is over all $1 \leq j_1 < j_2 < \ldots < j_k$. The random variable U_n is called a U-statistic of order k. Using the Hájek projection, state conditions under which U_n is asymptotically normal. Express its mean and variance in terms of the distribution of $u_k(X_1,\, \ldots,\, X_k)$, and its conditional mean given X_1.

CHAPTER 5

Optimality and likelihood asymptotics

5.1 Historical overview

5.1.1 *R. A. Fisher and maximum likelihood*

The method of maximum likelihood was formally introduced into statistics by Fisher (1922). Let $X_1, \cdots, X_n$ be a random sample from a distribution with density $f(x;\, \theta)$ governed by some parameter θ. A maximum likelihood estimator $\widehat{\theta}_n$ for θ is that value of θ which maximises the likelihood

$$L_n(\theta) = \prod_{j=1}^{n} f(X_j;\, \theta)\,,$$

or equivalently, maximises the log-likelihood

$$\ell_n(\theta) = \sum_{j=1}^{n} \ln\, f(X_j;\, \theta)\,.$$

We can write

$$\widehat{\theta}_n = \operatorname{argmax}_\theta L_n(\theta)\,,$$

provided $L_n(\theta)$ has a unique global maximum. More generally, the term "maximum likelihood estimator" is often used to describe an estimator which is a local maximum of the likelihood, but not necessarily the global maximum.

Fisher's 1922 paper also provided three criteria for good estimation, namely

- consistency,
- sufficiency,
- and efficiency.

A sample estimate is consistent in Fisher's original sense if it has the property that

> when applied to the whole population the derived statistic should be equal to the parameter.

Consistency, as Fisher originally defined it in 1922, is now called *Fisher consistency* to distinguish it from the modern definition. For example, any parameter defined by a functional $h(F)$ will have a sample analogue $h(\widehat{F}_n)$ that is Fisher consistent. Maximum likelihood estimators for independent identically distributed variates can be represented as M-functionals evaluated on the empirical distribution. So, it follows that maximum likelihood estimators are Fisher consistent.

Unlike the modern definition of consistency, Fisher consistency makes no statement about the behaviour of an estimator as the sample size goes to infinity. So, in many cases, there is no straightforward way to evaluate a sample estimate on the entire population to determine whether it is Fisher consistent. For example, it is not obvious how to check for the Fisher consistency of a kernel density estimator, because the kernel of such a density estimator has no population analogue. There are many estimators based upon random samples which are not Fisher consistent but are consistent in the modern sense that as $n \to \infty$, the estimator converges to the true value of the parameter. A few years later, Fisher (1925) redefined a consistent estimator as one which

> when calculated from an indefinitely large sample ... tends to be accurately equal to that of the parameter.

This definition is deceptively close to the previous one. However, its use of the terms "indefinitely large" and "tends to be" shows that it is much closer in meaning to the modern sense of consistency that an estimate is required to converge to the parameter for large samples.

In his 1922 paper, Fisher also defined the concepts of sufficiency and efficiency. About sufficiency, he wrote

> In mathematical language we may interpret this statement by saying that if θ be the parameter to be estimated, θ_1 a statistic which contains the whole of the information as to the value of θ, which the sample supplies, and θ_2 any other statistic, then the surface of distribution of pairs of values of θ_1 and θ_2, for a given value of θ, is such that for a given value of θ_1, the distribution of θ_2 does not involve θ.

The reader will recognise that this is the modern definition. Fisher also introduced the *factorisation criterion* for sufficiency. An estimator $T = t(X_1, \ldots, X_n)$ is sufficient if we can write the likelihood function in the form

$$L_n(\theta;\, X_1,\, \ldots,\, X_n) = A(\theta;\, T\,)\, B(X_1,\, \ldots,\, X_n)$$

for some choice of functions $A(\theta; T)$ and $B(X_1, \ldots, X_n)$, where $A(\theta; T)$ depends on $X_1, \ldots, X_n$ only through T, and $B(X_1, \ldots, X_n)$ does not depend on θ. Equivalently, T is sufficient if we can write the log-likelihood function in the form

$$\ell_n(\theta; X_1, \ldots, X_n) = a(\theta; T) + b(X_1, \ldots, X_n)$$

for some choice of functions $a(\theta; T)$ which depends on the data only through T, and $b(X_1, \ldots, X_n)$ which does not depend on θ.

Fisher (1922) stated that maximum likelihood estimators are always sufficient. However, this turned out to be overly optimistic, an error incurred by implicitly assuming that a sufficient statistic is one-dimensional. He correctly established that the maximum likelihood estimator can be expressed as a function of a sufficient statistic. Nevertheless, his conclusion that the maximum likelihood estimators are always sufficient is most reasonably interpreted as applying to the case where there is a sufficient one-dimensional statistic for all sample sizes, notably for the one-parameter exponential family. In fact, the maximum likelihood estimator is *not* sufficient in general. However, it has the property of *asymptotic sufficiency*, which is implied by the fact that the factorisation criterion is asymptotically valid in a small neighbourhood of θ as $n \to \infty$.

Among Fisher's three criteria, his statement and claims for the efficiency of maximum likelihood estimators have received the most attention. According to Fisher, the criterion of efficiency requires

> [t]hat in large samples, when the distributions of the statistics tend to normality, that statistic is to be chosen which has the least probable error.

Again, this is the modern definition, although we would probably prefer to call this *asymptotic efficiency* nowadays. Fisher (1922) incorrectly assumed that the sufficiency implied efficiency, and it is clear from his comments that he did not distinguish between asymptotic efficiency and asymptotic sufficiency. He argued that in contrast to the method of maximum likelihood, estimators produced by the method of moments are not asymptotically efficient in general. However, the task of demonstrating the asymptotic efficiency of maximum likelihood estimation was left for others.

For Fisher, an efficient estimator is an asymptotically normal estimator with asymptotically minimum "probable error." There is no reference to standard deviation or to variance in the definition. Although the expression "probable error" was ambiguous, Fisher's choice of this term is in the modern spirit, namely that we should be concerned with the dispersion of the asymptotic distribution of the estimator and not with the asymptotic dispersion of the distribution, when these two differ.*

Fisher was prescient in choosing to define efficiency only asymptotically and within the class of asymptotically normal estimators. Estimators which are normal and centred around the true value of the parameter may be compared in precision in the absolute sense, that the comparison may be made independently of the measure of dispersion. For example, this allows us to say that the sample median is less efficient in Fisher's sense than the sample mean when estimating the mean of the normal distribution. This fact is not dependent on any decision-theoretic criterion for error, because both the sample mean and the sample median are unbiased, and normal (asymptotically in the latter case).

Nevertheless, by 1925, Fisher recognised that a careful statement of efficiency could not avoid discussion of asymptotic bias and variance more explicitly. In Fisher (1925) he stated the following.

> In a large and important class of consistent statistics the random sampling distribution tends to normal ... and ... the variance ... falls off inversely to the size of the sample. in such cases, the characteristics of any particular statistic, for large samples, are completely specified by (i) its bias, and (ii) its variance. The question of bias is only of preliminary interest .... If we wish to make tests of significance for the deviation of ... [an estimator] from some hypothetical value, then ... [the bias] must fall off more rapidly than $n^{-\frac{1}{2}}$ ....

5.1.2 *H. Cramér and asymptotic efficiency*

Fisher's arguments led to considerable optimism that the method of maximum likelihood would be shown to be fully optimal. However, the reality was more complex. While it is customary today to speak of the asymptotic optimality of maximum likelihood, it should not be forgotten that this optimality is a consequence of some extensive regularity whose precise statement took decades to develop. The small sample optimality of the likelihood ratio for testing simple hypotheses was proved by Neyman and Pearson (1933). Wilks (1938) obtained the asymptotic properties of the likelihood ratio test statistic. Cramér (1946a) provided the standard regularity conditions for the asymptotic normality of the

* The probable error of $\widehat{\theta}_n$ is that value $\alpha > 0$ such that $P\left(|\,\widehat{\theta}_n - \theta\,| < \alpha\right) = 0.5$. Unlike the mean square error and the variance which are sensitive to events of small probability in the tail of the distribution, the probable error is determined by the centre of the distribution, where the central limit approximation for the estimator applies.

maximum likelihood estimator. Strictly speaking, these conditions do not directly verify that the global maximum of the likelihood is consistent and asymptotically normal. Instead, under Cramér's regularity, we may conclude that a point of stationarity of the likelihood function which is consistent will be asymptotically normal. Wald (1949) provided conditions under which the global maximum of the likelihood is consistent. More generally, there exists a local point of stationarity of the likelihood which is consistent and asymptotically normal. However, this point may be far from the global maximum of the likelihood when Wald's conditions fail. See Ferguson (1982) for a simple example. It is natural to assume that a consistent stationary point of the likelihood must be unique, and a local maximum. This is indeed the case, at least asymptotically, a fact that was proved by Huzurbazar (1948).

In the previous chapter, we discussed the asymptotic normality of the maximum likelihood estimator $\widehat{\theta}_n$ as an M-functional. Under Cramér's regularity assumptions for independent indentically distributed random variables $X_1, \ldots, X_n$, when θ_0 is the true value of the parameter, then the maximum likelihood estimator has asymptotic distribution

$$\sqrt{n}\,(\,\widehat{\theta}_n - \theta_0\,) \stackrel{d}{\Longrightarrow} \mathcal{N}(\,0,\, I(\theta_0)^{-1}\,) \tag{5.1}$$

where $I(\theta_0)$ is the expected (or Fisher) information at θ_0 defined by

$$I(\theta_0) = -E_{\theta_0}\, \frac{\partial^2}{\partial \theta^2}\, \ln\, f(X_j;\, \theta_0)\,. \tag{5.2}$$

The importance of this for the method of maximum likelihood is that $\widehat{\theta}_n$ is shown to be an asymptotic version of a uniformly minimum variance unbiased estimator for θ, a theory that is due to Aitken and Silverstone (1942), Fréchet (1943), Darmois (1945), Rao (1945), and Cramér (1946b).† In particular, under standard regularity, any statistic T_n, which is unbiased in the sense that

$$E_{\theta_0}(T_n) = \theta_0$$

for all θ_0 will also satisfy the inequality

$$\mathrm{Var}_{\theta_0}(T_n) \geq \frac{1}{n\, I(\theta_0)}\,.$$

So if $E_{\theta_0}(T_n) = \theta_0$ and $\mathrm{Var}_{\theta_0}(T_n) = 1/[n\, I(\theta_0)]$, then T_n will be a

† The quantity $[\,n\, I\,]^{-1}$ is called the Cramér-Rao lower bound or the Rao-Cramér lower bound. The bound is more appropriately known as the information bound or information inequality, out of respect for the work of other researchers, which was contemporary or marginally earlier than that of Rao and Cramér. However, there are many information inequalities. So we shall use the term Cramér-Rao inequality (or lower bound) to refer to that involving $I(\theta)$ and use the term information inequality to refer to the general class.

uniformly minimum variance unbiased estimator. The maximum likelihood estimator $\widehat{\theta}_n$ achieves this asymptotically in the sense that, under Cramér regularity, the result in (5.1) will hold. It is in this sense that we say that $\widehat{\theta}_n$ is "asymptotically efficient."

5.1.3 Le Cam asymptotics

Perhaps it was inevitable that a backlash would develop against Fisher's optimistic assessment of maximum likelihood. There were those who argued that maximum likelihood holds no special place among the class of good estimators, and that in certain cases maximum likelihood can be outperformed by other methods. Among those who took this viewpoint was Lucien Le Cam, who proposed a radically different type of asymptotic inference from that initiated by Fisher and developed by Cramér. Le Cam and others pointed out certain problems in the Cramér approach: for example, there exist estimators which are asymptotically normal, whose asymptotic variance is no larger than maximum likelihood and is strictly smaller at certain parameter values. These "superefficient" estimators highlight the inadequacies of the concept of asymptotic efficiency. One would expect that an asymptotically unbiased estimator should have asymptotic variance bounded by the Cramér-Rao inequality (as the maximum likelihood estimator does). However, this turns out not to be the case. Despite their apparent superefficiency, the estimators whose asymptotic variance is smaller than maximum likelihood are not particularly good estimators. The reason for this is that the points of superefficiency at which they perform so well are obtained by a reduction in performance within a neighbourhood of those points.

The theory of local asymptotics was developed in part to overcome the obstacles due to superefficiency. The word "local" has a technical meaning here that is distinct from its usual meaning in this book. Specifically,

> the word "local" is meant to indicate that one looks at parameter values ... so close to ... [the true parameter value] that it is not possible to separate .... [them] easily.‡

This idea is made precise with the concept of a *contiguity neighbourhood*, which is parametrised by a new contiguity parameter η for model. To understand the idea of a continuity neighbourhood, consider two sequences of probability distributions, say P_n and Q_n, $n \geq 1$. Let P_n and Q_n be respective joint probability distributions for $X_1, \ldots, X_n$. The sequence Q_n, $n \geq 1$ is said to be *contiguous* to P_n, $n \geq 1$ if every sequence

‡ Le Cam and Yang (2000, p. 117).

of random variables $V_n = v_n(X_1, \ldots, X_n)$ converging in probability to zero under P_n also converges in probability to zero under Q_n. If Q_n is contiguous to P_n and P_n is contiguous to Q_n, then we say that P_n and Q_n are contiguous alternatives.

Suppose P_n is a joint distribution for $X_1, \ldots, X_n$ governed by some fixed value of the parameter θ_0 and that Q_n is a joint distribution for $X_1, \ldots, X_n$ governed by a parameter sequence θ_n. For Q_n to be contiguous to P_n it is sufficient that when θ_0 is the true parameter value then

$$\ell_n(\theta_n) - \ell_n(\theta_0) \stackrel{d}{\Longrightarrow} W$$

where $E(e^W) = 1$. See Le Cam and Yang (2000, p. 36). Under Cramér's regularity assumptions, this will be true if

$$\theta_n = \theta_0 + \eta/\sqrt{n}$$

where η is any given real number. Here, the value η can be regarded as a parameter for a contiguity neighbourhood of the fixed parameter θ. For this "contiguity model" parametrised by η we may study the properties of superefficient estimators as determined by the various values of η. By shifting attention from fixed θ_0 to the contiguity parameter η, we can measure the extent to which some estimators borrow efficiency from their contiguity neighbourhoods in order to achieve superefficiency at certain points.

Another objection raised by Le Cam asymptotics to the Cramér regularity conditions is that in practice real data are often not obtained by random sampling—a sampling mechanism which is often called the "i.i.d. case." As Le Cam and Yang say[§]

> It must be pointed out that the asymptotics of the "standard i.i.d. case" are of little relevance to practical use of statistics, in spite of their widespread study and use. The reason for this is very simple: One hardly ever encounters fixed families ... with a number of observations that will tend to infinity.

In support of this observation, we should add that there are many models in which the relevant estimators are approximately normal and for which the standard "i.i.d. assumptions" do not hold. As we shall see, the key insight from Le Cam asymptotics is that models for which $\widehat{\theta}_n$ is asymptotically normal typically possess a property called *local asymptotic normality*. Loosely speaking, a model is locally asymptotically normal if the log-likelihood difference $\ell_n(\theta_n) - \ell_n(\theta_0)$ is asymptotically normal whenever θ_n is contiguous to θ_0.

[§] See Le Cam and Yang (2000, p. 175).

Under Cramér's assumptions, the log-likelihood in a contiguity neighbourhood of θ has the asymptotic form

$$\ell_n\left(\theta_0 + \eta/\sqrt{n}\right) - \ell_n(\theta_0) \;\sim\; \eta\, I(\theta_0)\left[\sqrt{n}\,(\widehat{\theta}_n - \theta_0)\right] - \frac{\eta^2\, I(\theta_0)}{2} \qquad (5.3)$$

when θ_0 is the true value of the parameter. As $\sqrt{n}\,(\widehat{\theta}_n - \theta_0)$ is asymptotically normal, it follows that the Cramér assumptions imply local asymptotic normality. However, as (5.3) shows, it is the local asymptotic normality of the model which "drives" the asymptotic normality of $\widehat{\theta}_n$. So the Cramér assumptions are not required.

By formulating model assumptions in terms of local asymptotic normality, we see that the class of asymptotically efficient estimators is quite large, containing much more than $\widehat{\theta}_n$. For example, there is nothing special about $\widehat{\theta}_n$ in (5.3). Any estimator T_n for which

$$T_n - \theta_0 \;\sim\; \widehat{\theta}_n - \theta_0 \qquad \text{as } n \to \infty$$

can be substituted for $\widehat{\theta}_n$ in the right-hand side. Such estimators T_n can be constructed for locally asymptotically normal models from *auxiliary estimators* which have the property that they lie in a contiguity neighbourhood of the true value θ_0 with high probability. If the log-likelihood is asymptotically quadratic in a contiguity neighbourhood, then one can

> fit a quadratic to the log likelihood around ... [the auxiliary estimator] and take as the new estimated value of θ the point that maximises the fitted quadratic.¶

The estimators obtained by this procedure are called centred variables. Centering an auxiliary estimator can be done with various types of quadratic fits to the log-likelihood function. Examples include the class of *efficient likelihood estimators* obtained by one-step Newton-Raphson iteration from an auxiliary estimator. Ironically, the process of justifying $\widehat{\theta}_n$ by asymptotic methods has succeeded but also downgraded its importance. It is undeniable that there are many estimators whose asymptotic properties are at least as good as maximum likelihood estimators when these properties are measured by decision-theoretic criteria. In addition to estimators derived by adjusting auxiliary estimators, the family of Bayes estimators also have equivalent asymptotic properties. Some estimators, such as Pitman estimators of location and scale parameters, may even outperform maximum likelihood for all sample sizes under some criterion such as quadratic loss, while being asymptotically equivalent.

Nevertheless, maximum likelihood remains *primus inter pares* within the

¶ Le Cam and Yang (2000, p. 124).

class of all such estimators, despite the criticisms of Le Cam asymptotics. The reason for this is easy to see. The method of maximum likelihood has proved to be highly reliable under regularity assumptions that can be clearly enunciated. On the other hand, it is not hard to construct efficient likelihood estimators which behave badly for standard models for a given sample size. Similarly, Bayes estimators can have poor properties for a given sample size if the prior distribution on the parameter space is misspecified. More than any other property it is the reliability of the method of maximum likelihood which has proved to be its greatest asset. We must recognise that asymptotic methodology has not completely justified Fisher's optimism for maximum likelihood. However, his optimism is amply supported.

5.2 The organisation of this chapter

In the sections that follow, the concepts and results described above shall be explained in greater detail. In Section 5.3 we shall look at properties of the likelihood, log-likelihood and related functions which arise from expansions around any given value of the parameter. The score function and the information function are introduced as coefficients in the expansion of the log-likelihood. The Bhattacharyya functions are introduced as coefficients in the expansion of the likelihood ratio. Some of the properties of these functions are developed under standard model regularity conditions. The family of information inequalities is introduced and the Cramér-Rao inequality is proved as a special case. Section 5.4 looks at the log-likelihood and uses Jensen's inequality to prove the consistency of maximum likelihood for finite paramater spaces. The problems of extending this result to infinite parameter spaces are addressed but not extensively explored. Section 5.5 proves the asymptotic normality of maximum likelihood estimators under the standard regularity assumptions first identified by Cramér.

In Section 5.6 we investigate the asymptotic efficiency of maximum likelihood estimators and other estimators as well. The classes of asymptotically unbiased and asymptotically normal estimators are introduced, motivated by the properties of maximum likelihood proved in the previous section. The asymptotic relative efficiency between two estimators is defined. It is argued that maximum likelihood is asymptotically efficient in the sense that its variance attains the Cramér-Rao lower bound as sample size goes to infinity. Nevertheless, it is also demonstrated that there exist superefficient estimators whose asymptotic efficiency is superior to maximum likelihood.

The remaining sections sketch some of the ideas of local asymptotics which address the problems of superefficiency. The local asymptotic risk function is introduced. It is shown by example that superefficient estimators which dominate maximum likelihood in terms of asymptotic variance do not dominate with respect to the local asymptotic risk. The statement that maximum likelihood is locally asymptotically minimax is briefly given in Section 5.7 and its full proof deferred until Section 5.9. The class of locally asymptotically normal models is defined in Section 5.8, and it is seen that the asymptotic normality of maximum likelihood estimators derives from this property.

Before we continue, a note on notation is in order. Throughout this chapter, θ shall denote any arbitrary value of the parameter selected from the parameter space Θ. The true value of the parameter for the purpose of calculating the distribution of a statistic shall be denoted by θ_0. The exception to this rule occurs in Definition 7, and associated material, where for the local asymptotic risk function at θ_0 and any given sample size n the true value of the parameter for the calculation of risk—expected loss—is $\theta_n = \theta_0 + \eta/\sqrt{n}$ and not θ_0 as is usually assumed. In this case, the value θ_n will be explicitly stated as the true value of the parameter. Occasionally, it will be necessary to compare different random variables under different distribution assumptions within the same expression. When this is the case, we shall write $(\,X\,)_H$ to denote that the random variable X is governed by the distribution assumption H.

5.3 The likelihood function and its properties

In this section, we shall examine some properties of maximum likelihood estimates that derive from the power series expansion of the likelihood function.

Let $X_1, X_2, \ldots, X_n$ be independent identically distributed random variables drawn from a distribution with density function $f(x;\,\theta)$, where θ is some unknown parameter. For the present discussion, we shall assume that θ takes values in some open subset Θ of the real line. We define the likelihood function $L_n(\theta)$ to be

$$L_n(\theta) = f(X_1;\,\theta)\,f(X_2;\,\theta)\,\cdots\,f(X_n;\theta)\,.$$

Here, and subsequently, we suppress the random variables $X_1, \ldots, X_n$ in the notation.

The log-likelihood function is defined as

$$\ell_n(\theta) = \ln\,L_n(\theta)$$

A maximum likelihood estimator for θ is defined to be any value

$$\widehat{\theta}_n = \widehat{\theta}_n(x_1, \ldots, x_n)$$

with the property that $L_n(\widehat{\theta}_n) \geq L_n(\theta)$ for all $\theta \in \Theta$. If $L_n(\theta)$ is differentiable with respect to θ, then

$$L'_n(\widehat{\theta}_n) = 0 \qquad \text{and} \qquad \ell'_n(\widehat{\theta}_n) = 0\,. \tag{5.4}$$

In particular, $\ell'_n(\theta) = u_n(\theta)$ is the score function. So equivalently,

$$u_n(\widehat{\theta}_n) = 0$$

is often solved as a way to find a maximum likelihood estimator. In general, however, this is only guaranteed to provide a stationary point of the likelihood.

The function

$$\begin{aligned} i_n(\theta) &= -\ell''_n(\theta) \\ &= -u'_n(\theta) \end{aligned}$$

is called the observed information function. For each $\theta \in \Theta$, the expectation of the observed information function is

$$E_\theta\, i_n(\theta) = n\, I(\theta)$$

where $I(\theta)$ is the expected information function. Evaluating the observed information function at the maximum likelihood estimate yields $i_n(\widehat{\theta}_n)$ which is called the *observed information* about θ.

A large amount of likelihood theory is dedicated to finding the approximate distribution and properties of $\widehat{\theta}_n$ as the sample size $n \to \infty$. In general, we cannot write down an explicit formula for $\widehat{\theta}_n$, and must rely on approximations based upon a power series expansion of $L_n(\theta)$ or $\ell_n(\theta)$. Suppose $\theta_0 \in \Theta$ is any given value of the parameter. Assuming that we can expand the likelihood about θ_0 in a Taylor series then

$$L_n(\theta) = L_n(\theta_0) + \sum_{j=1}^{\infty} L_n^{(j)}(\theta_0)\, \frac{(\theta - \theta_0)^j}{j!}\,.$$

Suppose $L_n(\theta_0) > 0$. Then we may subtract $L_n(\theta_0)$ from both sides and also divide by $L_n(\theta_0)$, to obtain

$$\psi(\theta_0, \theta) = \sum_{j=1}^{\infty} \psi_j(\theta_0)\, \frac{(\theta - \theta_0)^j}{j!}\,, \tag{5.5}$$

where the quantity

$$\psi(\theta_0, \theta) = \frac{L_n(\theta)}{L_n(\theta_0)} - 1 \tag{5.6}$$

that appears on the left-hand side of equation (5.5) is called a *centred likelihood ratio*, and the coefficients

$$\psi_j(\theta_0) = \frac{L_n^{(j)}(\theta_0)}{L_n(\theta_0)} \tag{5.7}$$

are called *Bhattacharyya functions.* The reader will note that the first Bhattacharyya function $\psi_1(\theta_0)$ is also the *score function* because

$$u_n(\theta) = \frac{\partial}{\partial\theta} \ln L_n(\theta) = \frac{L_n'(\theta)}{L_n(\theta)}.$$

At this stage, it will be convenient to use the multiple integral notation used in Definition 12, in Section 4.12.2 of the previous chapter. There are two important properties of centred likelihood ratios. The first property is that

$$\begin{aligned}
E_{\theta_0}\, \psi(\theta_0, \theta) &= \int^{[n]} \psi(\theta_0, \theta)\, L_n(\theta_0)\; d^n x \\
&= \int^{[n]} \left[\frac{L_n(\theta) - L_n(\theta_0)}{L_n(\theta_0)}\right] L_n(\theta_0)\, d^n x \\
&= \int L_n(\theta)\, d^n x - \int^{[n]} L_n(\theta_0)\; d^n x \\
&= 1 - 1 \\
&= 0\,.
\end{aligned}$$

The second property of $\psi(\theta_0, \theta)$ is that it is uncorrelated with any unbiased estimator of zero which has finite second moment when $\theta = \theta_0$.

Definition 1. *A statistic*

$$Z = z(X_1, \ldots, X_n)$$

is said to be an unbiased estimator of zero if

$$E_\theta\, Z = 0 \qquad \text{for all } \theta \in \Theta\,.$$

Let $\mathbf{Z}(\theta_0)$ *denote the class of all such unbiased estimators of zero* Z *for which* $E_{\theta_0}\, Z^2 < \infty$.

For example, the class $\mathbf{Z}(\theta_0)$ contains statistics such as $X_j - X_k$, provided the sample variables have finite second moment when $\theta = \theta_0$. More generally, contrast statistics of the form

$$Z = \sum_{j=1}^{n} a_j\, X_j\,, \qquad \text{where } \sum_{j=1}^{n} a_j = 0$$

will lie in $\mathbf{Z}(\theta_0)$ provided the second moment is finite.

Suppose $Z \in \mathbf{Z}(\theta_0)$ and that

$$E_{\theta_0}\left[\,\psi(\theta_0, \theta)\,\right]^2 < \infty\,.$$

Then

$$\begin{aligned} E_{\theta_0}[\,Z\,\psi(\theta_0,\,\theta)\,] &= \int^{[n]} z\,\psi(\theta_0, \theta)\,L_n(\theta_0)\,d^n x \\ &= \int^{[n]} z\,[\,L_n(\theta) - L_n(\theta_0)\,]\,d^n x \\ &= E_\theta\,Z - E_{\theta_0}\,Z \\ &= 0 - 0 \\ &= 0\,. \end{aligned}$$

In many models, the Bhattacharyya functions will also satisfy these two properties, namely

$$E_{\theta_0}\,\psi_j(\theta_0) = 0\,, \qquad \text{and} \qquad E_{\theta_0}\,[\,Z\,\psi_j(\theta_0)\,] = 0\,, \tag{5.8}$$

for all j, provided $Z \in \mathbf{Z}(\theta_0)$ and $E_{\theta_0}\,[\psi_j(\theta_0)]^2 < \infty$. However, there is some additional regularity required for this to be true. The two properties hold provided we can change the order of integration and differentiation in the sense that

$$\int^{[n]} \frac{\partial^j}{\partial \theta^j}\,[\;\sim\;]\,d^n x = \frac{\partial^j}{\partial \theta^j} \int^{[n]} [\;\sim\;]\,d^n x \tag{5.9}$$

where appropriate. For example,

$$\begin{aligned} E_{\theta_0}\,\psi_j(\theta_0) &= \int^{[n]} \left[\frac{L_n^{(j)}(\theta_0)}{L_n(\theta_0)}\right] L_n(\theta_0)\,d^n x \\ &= \int^{[n]} \left[\frac{\partial^j}{\partial \theta^j}\,L_n(\theta)\right]_{\theta=\theta_0} d^n x \\ &= \left[\frac{\partial^j}{\partial \theta^j} \int^{[n]} L_n(\theta)\,d^n x\right]_{\theta=\theta_0} \\ &= \frac{\partial^j}{\partial \theta_j}\,(1) \\ &= 0\,. \end{aligned}$$

The proof that $\psi_j(\theta_0)$ is uncorrelated with $Z \in \mathbf{Z}(\theta_0)$ is left to the reader as Problem 1.

Definition 2. *Let $\mathbf{S}(\theta_0)$ denote the set of all statistics*

$$T = t(X_1, \ldots, X_n)$$

such that $E_{\theta_0} T^2 < \infty$ *and such that*

$$E_{\theta_0} T = 0 \qquad \text{and} \qquad E_{\theta_0} (Z\,T) = 0 \tag{5.10}$$

for all $Z \in \mathbf{Z}(\theta_0)$.

The centred likelihood ratios lie in this set if they have finite second moment when $\theta = \theta_0$. Under the regularity assumptions used earlier, the Bhattacharyya functions also lie in this set. It is straightforward to see that $\mathbf{S}(\theta_0)$ is closed under linear combinations, and is a vectorspace. The importance of $\mathbf{S}(\theta_0)$ for the problem of estimation can be illustrated by the following proposition.

Definition 3. *Let* $\mathbf{T}(\theta_0)$ *denote the set of all statistics*

$$T = t(X_1, \ldots, X_n)$$

which are unbiased estimators for θ, *in the sense that*

$$E_\theta T = \theta \qquad \text{for all } \theta \in \Theta\,,$$

and such that $E_{\theta_0} T^2 < \infty$.

Proposition 1. *Let* θ_0 *be any given value of the parameter. Let*

$$T_1 = t_1(X_1, \ldots, X_n) \qquad \text{and} \qquad T_2 = t_2(X_1, \ldots, X_n)$$

be two statistics which both lie in $\mathbf{T}(\theta_0)$. *If* $T_1 - \theta_0 \in \mathbf{S}(\theta_0)$ *then*

$$\text{Var}_{\theta_0} T_1 \le \text{Var}_{\theta_0} T_2\,.$$

When this is true, T_1 *is said to be a* locally minimum variance unbiased estimator *(LMVUE) for* θ *at* θ_0.

Proof. Let $Z = T_2 - T_1$. Then for all $\theta \in \Theta$,

$$\begin{aligned} E_\theta Z &= E_\theta T_2 - E_\theta T_1 \\ &= 0\,. \end{aligned}$$

So Z is an unbiased estimator of zero. Therefore

$$\begin{aligned} \text{Var}_{\theta_0}(T_2) &= \text{Var}_{\theta_0}(T_1 + Z) \\ &= \text{Var}_{\theta_0} T_1 + \text{Var}_{\theta_0} Z \\ &\ge \text{Var}_{\theta_0} T_1 \end{aligned}$$

as required. ■

The statistic T_1 is a LMVUE at θ_0 in the sense that its variance is minimum at θ_0 among all unbiased estimators for θ. Now suppose $T_1 - \theta \in \mathbf{S}(\theta)$ for all $\theta \in \Theta$. Then for any unbiased estimator T_2 we will have

$$\mathrm{Var}_\theta\, T_1 \le \mathrm{Var}_\theta\, T_2 \qquad \text{for all } \theta \in \Theta\,.$$

When T_1 satisfies this property we say that T_1 is a uniformly minimum variance unbiased estimator (UMVUE). Unlike LMVUEs, the existence of UMVUEs cannot be guaranteed for general models.

Now consider any function $\psi(\theta_0) = \psi(\theta_0; X_1, \ldots, X_n)$ with the property that

$$\psi(\theta_0) \in \mathbf{S}(\theta_0)$$

Suppose T_1 and T_2 are both in $\mathbf{T}(\theta_0)$. Once again, $T_2 - T_1 \in \mathbf{Z}(\theta_0)$. Therefore,

$$\begin{aligned} \mathrm{Cov}_{\theta_0}[\,T_2,\,\psi(\theta_0)\,] &= \mathrm{Cov}_{\theta_0}[\,T_1,\,\psi(\theta_0)\,] + \mathrm{Cov}_{\theta_0}[\,T_2 - T_1,\,\psi(\theta_0)\,] \\ &= \mathrm{Cov}_{\theta_0}[\,T_1,\,\psi(\theta_0)\,] + 0 \\ &= \mathrm{Cov}_{\theta_0}[\,T_1,\,\psi(\theta_0)\,]\,. \end{aligned}$$

It follows from this that for any unbiased estimator T, we can write

$$\mathrm{Cov}_{\theta_0}[\,T,\,\psi(\theta_0)\,] = h_\psi(\theta_0)$$

where the function $h_\psi(\theta_0)$ does not depend upon the choice of $T \in \mathbf{T}(\theta_0)$. From this fact we can write the covariance inequality between T and $\psi(\theta_0)$ as

$$\begin{aligned} \mathrm{Var}_{\theta_0}\, T &\ge \frac{\mathrm{Cov}^2_{\theta_0}[\,T,\,\psi(\theta_0)\,]}{\mathrm{Var}_{\theta_0}\,\psi(\theta_0)} \\ &= \frac{[\,h_\psi(\theta_0)\,]^2}{E_{\theta_0}\,\psi^2(\theta_0)}\,. \end{aligned} \tag{5.11}$$

The right-hand side of this inequality does not depend on the choice of $T \in \mathbf{T}(\theta_0)$. Therefore, it is a universal lower bound on the variance of all unbiased estimators $T \in \mathbf{T}(\theta_0)$. Bounds of the variance of T of this type are known as *information inequalities*. Various special cases are of interest. The choice $\psi(\theta_0) = \psi(\theta_0, \theta)$ gives us

$$\begin{aligned} h_\psi(\theta_0) &= E_{\theta_0}[\,T\,\psi(\theta_0,\,\theta)\,] \\ &= \int^{[n]} t\left[\frac{L_n(\theta) - L_n(\theta_0)}{L_n(\theta_0)}\right] L_n(\theta_0)\, d^n x \\ &= \int^{[n]} t\,[\,L_n(\theta) - L_n(\theta_0)\,]\, d^n x \\ &= \theta - \theta_0\,. \end{aligned}$$

Therefore, in this case the information inequality is

$$\mathrm{Var}_{\theta_0}\, T \geq \frac{(\theta - \theta_0)^2}{E_{\theta_0}\, [\, \psi(\theta_0,\, \theta)\,]^2}\,.$$

We are free to choose the value of θ on the right-hand side, provided that $\theta \neq \theta_0$. Taking a supremum of the right-hand side yields the *Hammersley-Chapman-Robbins inequality*, which states that

$$\mathrm{Var}_{\theta_0}\, T \geq \sup_{\theta \neq \theta_0} \frac{(\theta - \theta_0)^2}{E_{\theta_0}\, [\, \psi(\theta_0,\, \theta)\,]^2}\,. \tag{5.12}$$

The score function $u_n(\theta_0)$ is also the first Bhattacharyya function, and therefore $u_n(\theta_0) \in \mathbf{S}(\theta_0)$ provided its second moment is finite. So it can also be used in (5.11) by setting $\psi(\theta_0) = u_n(\theta_0)$. With this choice, under the regularity allowing the interchance of derivatives and integrals,

$$\begin{aligned}
h_u(\theta_0) &= E_\theta\,[T\, u(\theta_0)] \\
&= \int^{[n]} t\, \frac{L_n'(\theta)}{L_n(\theta)}\, L_n(\theta)\, d^n x \\
&= \int^{[n]} t\, \frac{\partial}{\partial\theta}\, L_n(\theta)\, d^n x \\
&= \left[\frac{\partial}{\partial\theta} \int^{[n]} E_\theta\, T\right]_{\theta=\theta_0} \\
&= 1\,.
\end{aligned}$$

Therefore,

$$\mathrm{Var}_{\theta_0}\, T \geq \frac{1}{E_\theta\, [\, u_n(\theta_0)\,]^2}\,. \tag{5.13}$$

Again, under regularity permitting the interchange of derivatives and integrals,

$$\begin{aligned}
E_{\theta_0}\, [u_n(\theta_0)\,]^2 &= \int^{[n]} \left[\frac{L_n'(\theta_0)}{L_n(\theta_0)}\right]^2 L_n(\theta_0)\, d^n x \\
&= \int^{[n]} \frac{[\, L_n'\,]^2 - L_n\, L_n''}{[\, L_n\,]^2}\, L_n\, d^n x + \int^{[n]} L_n''\, d^n x \\
&= -\int^{[n]} [\ln L_n\,]''\, L_n\, d^n x + \int^{[n]} [\, L_n''\,]\, d^n x \\
&= -\int^{[n]} [\ln L_n\,]''\, L_n\, d^n x + \left[\int^{[n]} L_n\, d^n x\right]'' \\
&= -\int^{[n]} [\ln L_n\,]''\, L_n\, d^n x + [\, 1\,]''
\end{aligned}$$

$$= -\int^{[n]} [\ln L_n]'' \, L_n \, d^n x \, .$$

This last integral is $-E_{\theta_0} \ell_n''(\theta_0)$. The expected information function is

$$I(\theta) = -E_\theta \frac{\partial^2}{\partial \theta^2} \ln f(X;\theta)$$

and therefore

$$\begin{aligned} E_{\theta_0} [u_n(\theta_0)]^2 &= -E_{\theta_0} \ell_n''(\theta_0) \\ &= n\, I(\theta_0) \, . \end{aligned}$$

We may conclude the following.

Proposition 2. *Let $I(\theta_0)$ be the expected information at θ_0. We assume sufficient regularity to interchange derivatives and integrals as above. Then*

$$\mathrm{Var}_{\theta_0} T \geq \frac{1}{n\, I(\theta_0)} \, . \tag{5.14}$$

for all $T \in \mathbf{T}(\theta_0)$.

This result is usually called the *Cramér-Rao inequality.*

5.4 Consistency of maximum likelihood

At this stage, let us formalise some assumptions.

Assumption 1. *We adopt the following assumptions.*

1. *Let $X_1, X_2, \ldots, X_n$ be independent and identically distributed, each with density $f(x;\, \theta)$ depending on some real parameter θ.*
2. *We assume that the set*

$$C = \{ x \; : \; f(x;\, \theta) > 0 \}$$

   does not depend on θ. (The sample will lie with probability one in this set where the likelihood will be strictly positive and the log-likelihood finite.)
3. *We shall suppose that θ is identifiable in the sense that when $\theta_1 \neq \theta_2$ then the distribution of X_j under θ_1 is distinct from the distribution of X_j under θ_2.*

We can write the likelihood of $X_1, \ldots, X_n$ as

$$L_n(\theta) = f(X_1;\, \theta)\, f(X_2;\, \theta) \cdots f(X_n;\, \theta)\,,$$

and the log-likelihood as

$$\ell_n(\theta) = \sum_{j=1}^{n} \ln f(X_j;\, \theta)\,.$$

So by the strong law of large numbers,

$$P_{\theta_0}\left[\frac{1}{n}\,\ell_n(\theta) \to E_{\theta_0} \ln f(X;\, \theta)\right] = 1\,. \tag{5.15}$$

When $\theta \neq \theta_0$,

$$\begin{aligned} E_{\theta_0} \ln f(X;\, \theta) - E_{\theta_0} \ln f(X; \theta_0) &= E_{\theta_0} \ln \frac{f(X;\, \theta)}{f(X;\, \theta_0)} \\ &= \int \ln \frac{f(x;\, \theta)}{f(x;\, \theta_0)}\, f(x;\, \theta_0)\, dx \\ &< \ln \int \frac{f(x;\, \theta)}{f(x;\, \theta_0)}\, f(x;\, \theta_0)\, dx \\ &= \ln(1) = 0\,. \end{aligned}$$

This is equivalent to

$$E_{\theta_0} \ln f(X;\, \theta) < E_{\theta_0} \ln f(X; \theta_0) \qquad \text{for } \theta \neq \theta_0\,. \tag{5.16}$$

We may use the strict version of Jensen's inequality above because the logarithm function is strictly convex, and the ratio $f(x;\, \theta)/f(x; \theta_0)$ is a nonconstant random variable. Combining (5.15) and (5.16), we conclude that for any $\theta \neq \theta_0$,

$$P_{\theta_0}\left[\ell_n(\theta_0) > \ell_n(\theta) \text{ for all sufficiently large } n\right] = 1\,. \tag{5.17}$$

However, the finite intersection of events of probability one is also of probability one. Therefore, if $\theta_0, \theta_1, \ldots, \theta_k$ are distinct parameter values, and k is fixed we obtain

$$P_{\theta_0}\left[\ell_n(\theta_0) > \max_{j \geq 1} \ell_n(\theta_j) \text{ for all sufficiently large } n\right] = 1\,. \tag{5.18}$$

This result can also be expressed as follows.

Proposition 3. *Let $\Theta = \{\theta_0, \theta_1, \ldots, \theta_k\}$ be a finite parameter space satisfying the assumptions above. Then*

$$P_{\theta_0}\left[\widehat{\theta}_n = \theta_0 \text{ for all sufficiently large } n\right] = 1\,,$$

as $n \to \infty$.

However, the infinite intersection of events of probability one is not necessarily of probability one. So the conclusion of Proposition 3 fails if the finite parameter space Θ is replaced by one with infinitely many points. To prove consistency of $\widehat{\theta}_n$ for an infinite parameter space it is necessary to ensure that the strong law of large numbers holds uniformly over $\theta \in \Theta$. In an infinite parameter space, there may be a sequence of parameter values θ_j which converge to any θ_0 as $j \to \infty$. If this is the case, then it is usually too much to ask for (5.18) to hold. Instead, we might ask for a uniform version of the strong law of large numbers when θ is bounded away from θ_0. Provided that for all $\epsilon > 0$ we can show that

$$P_{\theta_0}\left[\ell_n(\theta_0) > \max_{|\theta-\theta_0|>\epsilon} \ell_n(\theta) \text{ for all sufficiently large } n\right] = 1\,, \tag{5.19}$$

then we can establish that

$$P_{\theta_0}\left[\,|\,\widehat{\theta}_n - \theta_0\,| \le \epsilon \text{ for all large } n\right] = 1\,.$$

When this result holds for all $\epsilon > 0$, then we say that $\widehat{\theta}_n$ is *(strongly) consistent.* This is equivalent to

$$P_{\theta_0}\left[\widehat{\theta}_n \to \theta_0 \text{ as } n \to \infty\right] = 1\,.$$

In some models, the strong consistency may fail because of the failure of (5.19) above. For such cases, it may still be possible to show that a local maximum of the likelihood is consistent even when $\widehat{\theta}_n$—the global maximum—is not. For the consistency of a local maximum of a single real parameter it is sufficient that (5.19) be replaced by

$$P_{\theta_0}\left[\ell_n(\theta_0) > \max\{\ell_n(\theta_0-\epsilon),\, \ell_n(\theta_0+\epsilon)\} \text{ for large } n\right] = 1\,, \tag{5.20}$$

for all $\epsilon > 0$. For fixed $\epsilon > 0$, this last condition follows under the same general conditions as (5.18), because it is based upon a finite submodel of three parameter points, namely $\theta_0 - \epsilon$, θ_0, $\theta_0 + \epsilon$.

5.5 Asymptotic normality of maximum likelihood

In the previous chapter, we used the von Mises calculus to show that many statistical functionals are asymptotically normal. Maximum likelihood estimators, as M-functionals, are therefore asymptotically normal when the regularity conditions of the von Mises calculus are satisfied. However, it is useful to revisit the problem by using more direct methods to prove this asymptotic normality.

Let $u_n(\theta) = \ell'(\theta)$ and $i_n(\theta) = -\ell''(\theta)$ be the score function and observed information function for θ, respectively. Suppose that the following conditions are satisfied.

Assumption 2. *Suppose the following properties hold.*

1. *Let $X_1, X_2, \ldots, X_n$ be independent and identically distributed, each with density $f(x;\,\theta)$ depending on some real parameter $\theta \in \Theta$, where Θ is a (nonempty) open interval of real numbers.*
2. *We assume that the set*
$$C = \{\, x \,:\, f(x;\,\theta) > 0 \,\}$$
does not depend on θ. (Note that $X_j \in C$ with probability one.)
3. *For all $x \in C$, the third derivative*
$$\frac{\partial^3}{\partial\,\theta^3}\, f(x;\,\theta)$$
exists and is a continuous function of θ, for all $\theta \in \Theta$.
4. *Furthermore, it is possible to interchange derivatives and integrals so that*
$$\frac{\partial^k}{\partial\,\theta^k} \int_C f(x;\,\theta)\,dx = \int_C \frac{\partial^k}{\partial\,\theta^k}\, f(x;\,\theta)\,dx$$
for all $1 \le k \le 3$.
5. *The expected information*
$$I(\theta) = -E_\theta\, \frac{\partial^2}{\partial \theta^2}\, \ln\, f(X_j;\,\theta)$$
is strictly positive and finite for all $\theta \in \Theta$.
6. *There exists a real-valued function $b(x)$ which does not depend on θ, such that*
$$\left| \frac{\partial^3}{\partial\,\theta^3}\, \ln\, f(x;\,\theta) \right| \le b(x)$$
for all $x \in C$, and such that $E_\theta\, b(X) < \infty$, for all $\theta \in \Theta$.

Throughout the argument which follows let us suppose that the properties listed in Assumption 2 hold. Now suppose that θ_0 is the true value of the parameter θ. Suppose in addition that the likelihood is maximised at some value $\widehat{\theta}_n \in \Theta$. Then with probability one, *i.e.*, for $X_1, \ldots, X_n \in C$,

$$L_n(\widehat{\theta}_n) > 0\,, \qquad L_n'(\widehat{\theta}_n) = 0\,, \qquad L_n''(\widehat{\theta}_n) \le 0\,.$$

Typically, $L_n''(\widehat{\theta}_n) < 0$. We can evaluate the score function and the observed information function at $\widehat{\theta}_n$ to get

$$\begin{aligned} u_n(\widehat{\theta}_n) &= L_n'(\widehat{\theta}_n) \Big/ L_n(\widehat{\theta}_n) \\ &= 0\,, \end{aligned}$$

and

$$\begin{aligned} i_n(\widehat{\theta}_n) &= -u_n'(\widehat{\theta}_n) \\ &= -\frac{L''(\widehat{\theta}_n)}{L(\widehat{\theta}_n)} + \left[\frac{L'(\widehat{\theta}_n)}{L(\widehat{\theta}_n)}\right]^2 \\ &\geq -\frac{L''(\widehat{\theta}_n)}{L(\widehat{\theta}_n)} \\ &\geq 0\,. \end{aligned}$$

Next, let us suppose that $\widehat{\theta}_n$ is consistent. That is,

$$\widehat{\theta}_n - \theta_0 = o_p(1) \tag{5.21}$$

for large n. Also, from Assumption 2.3, the function $\ell_n(\theta)$ is three times differentiable with respect to θ for $x \in C$. We can expand $u_n(\widehat{\theta}_n)$ about θ_0 using Taylor's expansion with Lagrange form of the remainder to get

$$\begin{aligned} 0 &= u_n(\widehat{\theta}_n) \\ &= u_n(\theta_0) - i_n(\theta_0)\,(\widehat{\theta}_n - \theta_0) - \frac{i_n'(\theta_1)}{2}\,(\widehat{\theta}_n - \theta_0)^2 \end{aligned}$$

for some value θ_1 which lies between θ_0 and $\widehat{\theta}_n$. Therefore,

$$(\widehat{\theta}_n - \theta_0)\left[i_n(\theta_0) + i_n'(\theta_1)\,(\widehat{\theta}_n - \theta_0)/2\right] = u_n(\theta_0)\,.$$

Thus we can write

$$\sqrt{n}\,(\widehat{\theta}_n - \theta_0) = \frac{u_n(\theta_0)/\sqrt{n}}{i_n(\theta_0)/n + i_n'(\theta_1)\,(\widehat{\theta}_n - \theta_0)/(2\,n)} \tag{5.22}$$

We can investigate the terms in the denominator. First,

$$\begin{aligned} \frac{i_n(\theta_0)}{n} &= -\frac{1}{n}\sum_{j=1}^{n} \frac{\partial^2}{\partial\theta^2} \ln f(X_j;\,\theta_0) \\ &\stackrel{a.s.}{\to} -E_{\theta_0}\frac{\partial^2}{\partial\theta^2} \ln f(X;\,\theta_0) \\ &= I(\theta_0)\,. \end{aligned}$$

The almost sure convergence above follows from the strong law of large numbers. We shall only use $i_n(\theta_0)/n \stackrel{P}{\to} I(\theta_0)$ below, a fact that is implied by almost sure convergence. As we assumed earlier,

$$\widehat{\theta}_n - \theta_0 \stackrel{P}{\to} 0\,.$$

Also

$$\begin{aligned} |\, i'_n(\theta_1)/(2\,n)\,| &= \left| \frac{1}{2\,n} \sum_{j=1}^{n} \frac{\partial^3}{\partial\theta^3} \ln f(X_j;\,\theta_1) \right| \\ &\le \frac{1}{2n} \sum_{j=1}^{n} b(X_j) \\ &\xrightarrow{P} \frac{1}{2} E_{\theta_0}\, b(X) < \infty\,. \end{aligned}$$

So the denominator of (5.22) has limit

$$i_n(\theta_0)/n + i'_n(\theta_1)\,(\widehat{\theta}_n - \theta_0)/(2\,n) \xrightarrow{P} I(\theta_0)\,, \tag{5.23}$$

where $0 < I(\theta_0) < \infty$. In the numerator of (5.22), we may write

$$u_n(\theta_0) = \sum_{j=1}^{n} \frac{\partial}{\partial\theta} \ln f(X_j;\,\theta_0)\,.$$

Under the assumption permitting the interchange of derivatives and integrals,

$$E_{\theta_0} \left[\frac{\partial}{\partial\theta} \ln f(X;\,\theta_0) \right] = 0 \qquad \text{and} \qquad \mathrm{Var}_{\theta_0} \left[\frac{\partial}{\partial\theta} \ln f(X;\,\theta_0) \right] = I(\theta_0)\,. \tag{5.24}$$

See Problem 2. So we can apply the central limit theorem to get

$$\begin{aligned} \frac{u_n(\theta_0)}{\sqrt{n}} &= \frac{1}{\sqrt{n}} \sum_{j=1}^{n} \frac{\partial}{\partial\theta} \ln f(X_j;\,\theta_0) \\ &\stackrel{d}{\Longrightarrow} \mathcal{N}(0,\, I(\theta_0))\,. \end{aligned} \tag{5.25}$$

Putting (5.25) and (5.23) together gives us

$$\sqrt{n}\,(\widehat{\theta}_n - \theta_0) \stackrel{d}{\Longrightarrow} \mathcal{N}\left(0,\, \frac{1}{I(\theta_0)}\right)\,, \tag{5.26}$$

as $n \to \infty$.

5.6 Asymptotic comparison of estimators

While this argument seems to fulfill Fisher's expectations for the method of maximum likelihood, there are some nagging issues that must not be overlooked. First, uniformly minimum variance unbiased estimators are only optimal—in the sense of minimising variance—in the class of estimators whose bias is exactly zero. Typically, maximum likelihood estimators do not lie inside this class. However, maximum likelihood estimators are asymptotically unbiased in the sense of the following definition.

Definition 4. *Let $T_n = t_n(X_1, \ldots, X_n)$ be a sequence of estimators for a parameter θ. Suppose there exists a* nondegenerate *probability distribution*|| *H and a sequence of constants c_n such that*

$$c_n\,(T_n - \theta_0) \stackrel{d}{\Longrightarrow} H$$

as $n \to \infty$, whenever θ_0 is the true value of the parameter.

- *If the distribution H has expectation zero, then we say that T_n is* asymptotically unbiased. *If H has nonzero expectation, then we say that T_n is* asymptotically biased.
- *If $c_n = \sqrt{n}$, and H is a normal distribution, then we say that T_n is* asymptotically normal.

In this definition, we understand H to be a nondegenerate probability distribution in the sense that it does not assign probability one to a constant (its mean). Asymptotic unbiasedness and asymptotic normality are of course asymptotic versions of unbiasedness and normality, respectively. Unlike its finite sample counterpart, asymptotic unbiasedness is parametrisation equivariant, in the sense that if T_n, $n \geq 1$ is asymptotically unbiased estimator for θ, and $h(x)$ is a smooth real-valued function with non-vanishing derivative, then $h(T_n)$, $n \geq 1$ is also an asymptotically unbiased estimator for $h(\theta)$. Similarly, if T_n is asymptotically normal, then so is $h(T_n)$. Both of these results are a consequence of the delta method for distributions. This is essentially the same as part 1 of Proposition 11 in Chapter 4.

The concept of asymptotic unbiasedness should not be confused with the concept of *unbiasedness in the limit.* The latter states that $\lim E_\theta\, T_n = \theta$ as $n \to \infty$, or equivalently that the bias of T_n goes to zero.

In regular models, maximum likelihood estimators are both asymptotically unbiased and asymptotically normal, as (5.26) shows. However, when this regularity fails, the estimator can be both asymptotically biased and nonnormal. For example, let $X_1, \ldots, X_n$ be a random sample from $\mathcal{U}(0, \theta)$, where $\theta > 0$. In this case, the maximum likelihood estimator is the largest order statistic, namely $\widehat{\theta}_n = X_{(n)}$, which converges faster than $O(n^{-\frac{1}{2}})$ to θ. The correct scaling sequence is $c_n = n$, and it can be shown that if θ_0 is the true value of θ, then

$$n\,(\theta_0 - X_{(n)}) \stackrel{d}{\Longrightarrow} \mathcal{E}(\theta_0^{-1})\,.$$

|| The probability distribution H can depend upon the value of θ_0, although this dependence is suppressed in the notation.

This follows from the fact that for all $t > 0$,

$$\begin{aligned} P_{\theta_0}\left\{ n\left[\theta_0 - X_{(n)}\right] > t \right\} &= P_{\theta_0}\left\{ X_{(n)} < \theta_0 - \frac{t}{n} \right\} \\ &= P_{\theta_0}\left\{ X_{(n)} < \theta_0 \left[1 - \frac{t/\theta_0}{n}\right] \right\} \\ &= \left[1 - \frac{t/\theta_0}{n}\right]^n \to e^{-t/\theta_0} . \end{aligned}$$

As the limiting distribution $\mathcal{E}(\theta_0^{-1})$ has a nonzero mean, it follows that $\widehat{\theta}_n$ is asymptotically biased. This example differs from those usually encountered in statistics because the support of the distribution is a function of the parameter, a condition explicitly ruled out by Cramér's regularity. Note that although $X_{(n)}$ is asymptotically unbiased in this example, it is unbiased in the limit.

In general, c_n and H are uniquely determined by the sequence T_n, $n \geq 1$ in the following limiting sense. Suppose

$$c_n\,(T_n - \theta_0) \stackrel{d}{\Longrightarrow} H_1 \qquad \text{and} \qquad d_n\,(T_n - \theta_0) \stackrel{d}{\Longrightarrow} H_2$$

as $n \to \infty$, where H_1 and H_2 are nondegenerate as in Definition 4. Then there exists some constant c such that $H_2(t) = H_1(t/c)$ for all t and such that $d_n/c_n \to c$ as $n \to \infty$. The proof of this result is left to the reader as Problem 3.

Within the class of estimators with some common scaling sequence c_n, we may be able to compare asymptotic efficiencies. Let T_n and T_n^* be two sequences of estimators based on some common random sample such that

$$c_n\,(T_n - \theta_0) \stackrel{d}{\Longrightarrow} H \qquad \text{and} \qquad c_n\,(T_n^* - \theta_0) \stackrel{d}{\Longrightarrow} H^*$$

To compare T_n with T_n^* we can compare H with H^*. If H is more closely concentrated around zero than H^*, say, then we can conclude that T is more efficient than T^*. In general, this is a difficult comparison, because distributions are not naturally totally ordered in terms of their concentration about zero.

In some cases, H and H^* differ only in a scale factor. For example, this is true when H and H^* are both normal distributions centred at zero. Under these conditions, T will be asymptotically superior to T^* if the dispersion of H is smaller than the dispersion of H^*, a fact that does not depend on the choice of dispersion parameter. Nevertheless, a quantitative measurement of the relative efficiency of T to T^* will still depend on the choice of dispersion parameter. So, it is helpful to reformulate the comparison in the follow way. Suppose $k(1) \leq k(2) \leq$

$k(3) \le \cdots$ is an increasing sequence of positive integers such that $k(n)/n$ has a limit as $n \to \infty$, and such that

$$c_n\,(T_n - \theta_0) \stackrel{d}{\Longrightarrow} H \qquad \text{and} \qquad c_n\,(T^*_{k(n)} - \theta_0) \stackrel{d}{\Longrightarrow} H\,, \tag{5.27}$$

where H is a nondegenerate distribution. We define the asymptotic relative efficiency of T to T^* to be this limit, namely

$$e(T,\,T^*) = \lim_{n\to\infty} \frac{k(n)}{n}\,.$$

The idea behind this definition is that when the asymptotic relative efficiency of T_n to T^*_n is $e(T,\,T^*) = 3$, say, then using T with a sample of size n to estimate θ is asymptotically equivalent to using T^* with a sample three times as large. Clearly, in such a case T is to be favoured over T^*.

Now suppose T_n and T^*_n are asymptotically normal random variables with

$$\sqrt{n}\,(T_n - \theta_0) \stackrel{d}{\Longrightarrow} \mathcal{N}(0,\,v) \qquad \text{and} \qquad \sqrt{n}\,(T^*_n - \theta_0) \stackrel{d}{\Longrightarrow} \mathcal{N}(0,\,v^*)\,.$$

In this case we can define

$$k(n) = \left\lfloor \frac{n\,v^*}{v} \right\rfloor$$

where $\lfloor x \rfloor$ is the floor of x, or greatest integer less than or equal to x. Then the two limits can be written as

$$\sqrt{n}\,(T_n - \theta_0) \stackrel{d}{\Longrightarrow} \mathcal{N}(0,\,v) \qquad \text{and} \qquad \sqrt{n}\,(T^*_{k(n)} - \theta_0) \stackrel{d}{\Longrightarrow} \mathcal{N}(0,\,v)\,.$$

Therefore

$$\begin{aligned} e(T,\,T^*) &= \lim_{n\to\infty} \frac{\lfloor n\,v^*/v \rfloor}{n} \\ &= \frac{v^*}{v}\,. \end{aligned}$$

So, for asymptotically normal random variables with scaling $c_n = \sqrt{n}$, the asymptotic relative efficiency can be expressed as the ratio of the variances. Therefore, within the class of asymptotically unbiased asymptotically normal estimators, a sequence T_n, $n \ge 1$ will be asymptotically efficient relative to all others if its asymptotic variance is minimised.

The Cramér-Rao inequality of Proposition 2 provides a lower bound for the variance of an unbiased estimator. A sequence of unbiased estimators T_n, $n \ge 1$ which attains this lower bound for all n may be said to be efficient (in a finite sample sense), and will have a variance which is proportional to n^{-1}. Provided T_n is also normal or approximately normal, the variance provides a natural stochastic ordering of the dispersion of T_n about θ. However, maximum likelihood estimators are generally not unbiased and typically do not attain the lower bound except in an asymptotic sense. The following definition is an asymptotic version of the concept of finite sample efficiency.

Definition 5. *Within the class of asymptotically unbiased, asymptotically normal estimators, an estimator T_n, $n \geq 1$ satisfying*

$$\sqrt{n}\,(T_n - \theta_0) \stackrel{d}{\Longrightarrow} \mathcal{N}(\,0,\, 1/I(\theta_0)\,)$$

is said to be asymptotically efficient.

In particular, under standard regularity, maximum likelihood estimators are asymptotically efficient.

This result would seem to be the fulfillment of Fisher's original claim that maximum likelihood estimators are (asymptotically) efficient. However, Definition 5 raises some concerns that must be addressed. Definition 5 says nothing about estimators which are asymptotically nonnormal or asymptotically biased. Do maximum likelihood estimators have optimality properties in a larger class of estimators?

Another problem appears upon inspection. The Cramér-Rao lower bound of Proposition 2 only refers to estimators which are exactly unbiased. We have no immediate reason to believe that an estimator which is asymptotically unbiased must satisfy an asymptotic version of the same inequality. Is it possible for an estimator T to be asymptotically efficient according to Definition 5, and yet for another estimator T^* to exist such that

- $e(T, T^*) \leq 1$ for all θ, and
- $e(T, T^*) < 1$ for one or more values of θ?

This is critical for making sense out of Definition 5. Therefore, it was a surprise when the existence of "superefficient" estimators was discovered.

Definition 6. *An asymptotically unbiased, asymptotically normal estimator T_n, $n \geq 1$ is said to be* superefficient *at a given parameter value θ_1 if for all $\theta_0 \in \Theta$, $\sqrt{n}\,(T_n - \theta_0) \stackrel{d}{\Longrightarrow} \mathcal{N}(\,0,\, v(\theta_0)\,)$, where*

- *$v(\theta_0) \leq 1/I(\theta_0)$ for all θ_0, and in particular,*
- *$v(\theta_1) < 1/I(\theta_1)$.*

The following example of a superefficient estimator is due to J. L. Hodges, Jr. as reported in Le Cam (1953). Surprisingly, superefficient estimators exist for the normal location model $\mathcal{N}(\theta,\, 1)$, where the maximum likelihood estimator attains the lower bound of the Cramér-Rao inequality. Suppose $X_1,\, \ldots,\, X_n$ are independent identically distributed $\mathcal{N}(\theta,\, 1)$ random variables. The expected information function is $I(\theta) = 1$ for all $\theta \in \Theta$. The maximum likelihood estimator for θ is the sample average $\overline{X}_n$ which is unbiased and attains the Cramér-Rao lower bound for all sample sizes n. As $n \to \infty$,

$$\sqrt{n}\,(\overline{X}_n - \theta_0) \overset{d}{\Longrightarrow} \mathcal{N}(0,\, 1)\,, \tag{5.28}$$

so that $\overline{X}_n$ is asymptotically efficient. Next, let $0 < a < 1$ be any given real number. Define

$$T_n = \begin{cases} \overline{X}_n & \text{when } \left|\overline{X}_n\right| \geq n^{-1/4} \\ a\,\overline{X}_n & \text{when } \left|\overline{X}_n\right| < n^{-1/4}\,. \end{cases}$$

Then it can be shown that

$$\sqrt{n}\,(T_n - \theta_0) \overset{d}{\Longrightarrow} \begin{cases} \mathcal{N}(0,\, 1) & \text{when } \theta_0 \neq 0\,, \\ \mathcal{N}(0,\, a^2) & \text{when } \theta_0 = 0\,. \end{cases} \tag{5.29}$$

The limit in (5.29) can be compared with the limit in (5.28). We see that both estimators are asymptotically unbiased and asymptotically normal. Curiously, the variance of the asymptotic distribution of T_n is better than $\overline{X}_n$: the variance of the asymptotic distribution of T_n equals that of $\overline{X}_n$ when $\theta_0 \neq 0$ and is smaller when $\theta_0 = 0$.

To prove the limit in (5.29), note that when $\theta_0 \neq 0$, then $\overline{X}_n \to \theta_0 \neq 0$. So

$$P_{\theta_0}\left(\left|\overline{X}_n\right| < n^{-1/4}\right) \to 0$$

This implies that $P_{\theta_0}(T_n = \overline{X}_n) \to 1$, as $n \to \infty$. So when $\theta_0 \neq 0$, it follows that $\sqrt{n}\,(T_n - \theta_0)$ has the same asymptotic distribution as $\sqrt{n}\,(\overline{X}_n - \theta_0)$, which is asymptotically $\mathcal{N}(0,\, 1)$.

On the other hand, when $\theta_0 = 0$, $\overline{X}_n$ will converge to zero in probability. Furthermore

$$\begin{aligned} P_{\theta_0=0}\left(\left|\overline{X}_n\right| < n^{-\,1/4}\right) &= P_{\theta_0=0}\left(\left|\sqrt{n}\,\overline{X}_n\right| < n^{+\,1/4}\right) \\ &= P_{\theta_0=0}\left(\left|\mathcal{N}(0,\, 1)\right| < n^{+\,1/4}\right) \\ &\to 1\,. \end{aligned}$$

Therefore $P_{\theta_0=0}(T_n = a\,\overline{X}_n) \to 1$ as $n \to \infty$. So when $\theta_0 = 0$, it follows that $\sqrt{n}\,(T_n - \theta_0)$ has the same asymptotic distribution as $\sqrt{n}(a\,\overline{X}_n - \theta_0)$, which is asymptotically $\mathcal{N}(0,\, a^2)$.

The superefficient estimator T_n is not a serious contender as a good estimator of θ. Although T_n formally outperforms $\overline{X}_n$, it is the efficiency criterion of Definition 5 which comes under suspicion, and not the sample mean $\overline{X}_n$. Estimators such as T_n acquire a point of superefficiency by "borrowing" efficiency from neighbouring parameter values. As a result, superefficiency at a point such as $\theta = 0$ above leads to poor behaviour in a neighbourhood of that point.

How can we fix this problem? Several solutions have been proposed. Le Cam (1953) and Bahadur (1964) noted that the set of points of superefficiency must be of Lebesgue measure zero for regular models. Thus, in a certain sense, superefficiency only occurs on a negligible set of parameter values. Can we ignore sets of measure zero in the parameter space? However negligible the points of superefficiency are measure-theoretically, there is no reason to suppose that these points are *statistically* negligible. Even if superefficiency occurs only at a single point within a continuum of possibilities, there may be good reasons to believe that this point is a credible value for the parameter. For example, null hypotheses are typically low dimensional sets with parameter spaces, and usually have Lebesgue measure zero. Despite this fact, a null hypothesis is rarely statistically negligible.

Another way of solving the problem is to find some criterion that can be imposed which eliminates superefficient estimators. If all superefficient estimators are irregular in some sense, then we might restrict to the class of regular estimators. Suppose that we are given two functions, say $g(\theta)$ and $h(\theta)$, which are both continuous and defined on some open set Θ. Let N be a subset of Θ of Lebesgue measure zero. If $g(\theta) \geq h(\theta)$ for all $\theta \in \Theta - N$, then the continuity of g and h can be used to show that $g(\theta) \geq h(\theta)$ on the set N as well. Note that in (5.29), the variance of the limiting normal distribution is not continuous at $\theta = 0$. The argument using function g and h shows that this discontinuity is a consequence of superefficiency at $\theta = 0$. Therefore, superefficient estimators can be abolished if we restrict consideration to estimators T_n for which

$$\sqrt{n}\,(T_n - \theta_0) \stackrel{d}{\Longrightarrow} \mathcal{N}(\,0,\, v(\theta_0)\,)$$

where v is continuous on Θ. The restriction to estimators that are regular in this sense provides a mathematical solution to the problem of superefficient estimators. Is it statistically sensible? It is rather obvious that estimators are often optimal in a sufficiently small class. Indeed, every estimator is optimal within the class consisting only of itself! To be a statistical solution to the problem of superefficiency, the restriction to regular estimators must be justified by statistical and not mathematical considerations. This is difficult, because there is no obvious statistical advantage in requiring that $v(\theta)$ be continuous.

Another solution due to Rao (1962) is to redefine the efficiency of an asymptotically unbiased estimator T_n as the limiting correlation between T_n and the score function $u_n(\theta)$, or more generally

$$\liminf_{n\to\infty} \text{Corr}_{\theta_0}[\,T_n,\, u_n(\theta_0)\,]\,.$$

Under this criterion, the maximum likelihood estimator is efficient for regular models because we can write

$$\widehat{\theta}_n - \theta_0 \quad \sim \quad \frac{u_n(\theta_0)}{n\,I(\theta_0)}$$

as in Section 5.5.** Under this criterion, both $\overline{X}_n$ and Hodges' superefficient estimator are efficient.

Nice as this idea is, some problems remain. First, the redefined criterion for efficiency gives us no reason to reject Hodges' estimator, as it is fully efficient under the new criterion. What is still needed is a theory which shows why such superefficient estimators are defective. Secondly, we must ask whether we have ended up with the cart in front of the horse, so to speak. Rao's criterion essentially says that an estimator is efficient if it is "score-like." As the score function and $\widehat{\theta}_n$ are asymptotically related, this is much the same thing as saying that an estimator is efficient to the degree that it resembles $\widehat{\theta}_n$. The problem with Rao's criterion is that it is only indirectly related to the performance of the estimator T_n. However efficiency is defined, it should say something directly about the accuracy of T_n as an estimator for θ.

5.7 Local asymptotics

5.7.1 Local asymptotic risk

As we noted in the previous section, estimators can achieve superefficiency at a point in the parameter space by borrowing efficiency from neighbouring parameter values. This "borrowing" is sufficiently mild that it does not affect the asymptotic efficiency of neighbouring parameter values. However, superefficiency at one point reduces efficiency in a

** The idea behind this clever way of redefining efficiency was also instrumental in the development of the theory of optimal estimating equations, which places the criterion of efficiency on the estimating equation rather than the estimator obtained by solving the equation.

neighbourhood of the point. The idea of local asymptotics, due to Hájek and Le Cam, based upon earlier suggestions of C. Stein, H. Rubin and H. Chernoff, is to measure the efficiency of an estimator at a point through its local behaviour around that point. In the discussion which follows we must leave out many of the technical details necessary for the full power of the theory. Our primary concern here will be to provide some idea of its basic definitions and concepts. Proofs and technical details will be considered later.

For the moment, let us consider the family of estimators such that

$$\sqrt{n}\,(T_n - \theta_0) \stackrel{d}{\Longrightarrow} H$$

for some nondegerate distribution H which can depend on θ_0. To assess the behaviour of T_n close to θ_0, we could allow θ to vary in a contiguity neighbourhood of θ_0. Define

$$\theta_n = \theta_0 + \frac{\eta}{\sqrt{n}} \tag{5.30}$$

for some constant value η. We shift from θ_0 as the true value of the parameter to θ_n. Here θ_0 is fixed, and the model is parametrised by the new "parameter" η. Now suppose that $X_1, X_2, \ldots, X_n$ are a random sample from a distribution with density $f(x;\,\theta_n)$, and that $T_n = t_n(X_1, \ldots, X_n)$. Suppose there is some nondegenerate distribution H_η such that

$$\sqrt{n}\,(T_n - \theta_n) \stackrel{d}{\Longrightarrow} H_\eta \tag{5.31}$$

where H_η may depend upon η.

Definition 7. *Suppose that $X_1, \ldots, X_n$ have a distribution governed by the parameter value θ_n, where $\theta_n = \theta_0 + \eta/\sqrt{n}$ is a sequence contiguous to θ_0. Let $T_n = t_n(X_1, \ldots, X_n)$. We define $R(\eta, T)$, called the* local asymptotic risk *of T_n, to be the second moment of the distribution H_η. That is*

$$R(\eta, T) = \int x^2\, dH_\eta(x)\,, \qquad \text{where } \sqrt{n}\,(T_n - \theta_n) \stackrel{d}{\Longrightarrow} H_\eta\,, \tag{5.32}$$

provided this limit exists for some nondegenerate H_η.

The reader should note carefully that the limit distribution of $\sqrt{n}\,(T_n - \theta_n)$ is *not* calculated under the assumption that θ_0 is the true value of the parameter, as has previously been the case. In this respect, the calculation of the local asymptotic risk function differs from previous calculations.

The mean square of the limit distribution H_η can often be calculated as

$$R(\eta, T) = \lim_{n\to\infty} E_{\theta_n}\left[n\,(T_n - \theta_n)^2\right]. \tag{5.33}$$

However, this will only be true under an assumption of uniform integrability which ensures that the second moment of the limit distribution is the same as the limit of the sequence of second moments. Generally, when these differ, the local asymptotic risk is defined as the mean square of the limit distribution H_η, as defined by (5.32).

5.7.2 Example: Hodges' superefficient estimator

We may compare the asymptotic performance of two estimators T_n, $n \geq 1$ and T_n^*, $n \geq 1$ in a neighbourhood of θ, by comparing their respective local asymptotic risk functions. Let us consider Hodges' superefficient estimator of Section 5.6 above. Once again, the random variables $X_1, \ldots, X_n$ are independent $\mathcal{N}(\theta, 1)$. We compare Hodges' estimator

$$T_n = \begin{cases} \overline{X}_n & \text{when } \left|\overline{X}_n\right| \geq n^{-1/4} \\ a\,\overline{X}_n & \text{when } \left|\overline{X}_n\right| < n^{-1/4}. \end{cases}$$

where $0 < a < 1$, with the sample mean $T^* = \overline{X}_n$. When $\theta_0 \neq 0$, the sequence $\theta_n = \theta_0 + \eta/\sqrt{n}$ will be nonzero for all sufficiently large n. Also $P_{\theta_n}(|\overline{X}_n| \geq n^{-1/4}) \to 1$ as $n \to \infty$. So $\sqrt{n}\,(T_n - \theta_n) \sim \sqrt{n}\,(\overline{X}_n - \theta_n)$, which is $\mathcal{N}(0, 1)$. Therefore

$$R(\eta, T) = R(\eta, T^*) = 1$$

for all η, when $\theta_0 \neq 0$. The case where $\theta_0 = 0$ is a more interesting comparison. Once again $R(\eta, T^*) = 1$ for all η. It is left as an exercise to the reader to show that

$$R(\eta, T) = a^2 + \eta^2\,(a - 1)^2.$$

From these formulas, we see that $R(\eta, T) < R(\eta, T^*)$ if and only if

$$|\eta| < \sqrt{\frac{1+a}{1-a}}. \tag{5.34}$$

See Problem 4. Although T_n is superefficient, its local asymptotic risk does not uniformly dominate the sample mean $\overline{X}_n$. By calculating the local asymptotic risk, we can quantify the extent that a superefficient estimator "borrows efficiency" from neighbouring parameter values at its points of superefficiency. In the example above, this borrowing is reflected in the fact that $R(\eta, T) > R(\eta, T^*)$ for large values of $|\eta|$.

5.7.3 *Example: exponential with lag parameter*

The next example we shall consider is one in which the standard regularity assumptions fail. Suppose $X_1, X_2, \ldots . X_n$ is a random sample from an exponential distribution with lag θ and mean $1+\theta$. That is, the random variables have density function

$$f(x;\, \theta) = \begin{cases} e^{\theta - x} & \text{for } x \geq \theta \\ 0 & \text{otherwise}. \end{cases}$$

It can be checked that the maximum likelihood estimator for θ is the smallest order statistic

$$T_n = \min(X_1,\, X_2,\, \ldots,\, X_n)\,.$$

The reader may check that T_n has an exponential distribution with lag θ and mean $\theta + n^{-1}$. An obvious alternative to this estimator is the bias corrected estimator $T_n^* = T_n - n^{-1}$. How do the local asymptotic risks of these two estimators compare?

In this example we must use a scaling sequence of the form $c_n = n$ rather than the usual $\sqrt{n}$ scaling. The concept of a contiguity neighbourhood must be correspondingly modified. Define

$$\theta_n = \theta_0 + \frac{\eta}{n}\,.$$

With this choice, we find that

$$n\,(T_n - \theta_n) \overset{d}{\Longrightarrow} \mathcal{E}(1)\,,$$

and therefore that

$$n\,(T_n - n^{-1} - \theta_n) \overset{d}{\Longrightarrow} \mathcal{E}(1) - 1\,.$$

We can compute the mean square of these two limiting distributions, to get

$$R(\eta,\, T) = 2 \qquad \text{and} \qquad R(\eta, T^*) = 1\,, \tag{5.35}$$

for all η.

5.7.4 *Example: normal with constrained mean*

In our third example, we revisit the $\mathcal{N}(\theta,\, 1)$ model. Consider a random sample $X_1,\, \ldots,\, X_n$ of independent identically distributed random variables from $\mathcal{N}(\theta,\, 1)$ with the constraint on the parameter that $\theta \geq 0$. How can we estimate θ if we know that it is nonnegative? Two reasonable estimators for θ are

$$T_n = \overline{X}_n \qquad \text{and} \qquad T_n^* = \max\left(\overline{X}_n,\, 0\right)\,.$$

The estimator T_n is unbiased. However, a side effect of its unbiasedness is that it takes on values which lie outside the parameter space. The estimator T_n^* is the maximum likelihood estimator for θ and always lies in Θ. However, it is biased, particularly when θ is close to zero. Which estimator has better local asymptotic risk? As before, we have $R(\eta, T) = 1$ for all η. To compute $R(\eta, T^*)$, we break the argument into two cases.

Let $\theta_0 > 0$. For a contiguous sequence $\theta_n = \theta_0 + \eta/\sqrt{n}$,

$$\begin{aligned} P_{\theta_n}\left(T_n^* = \overline{X}_n\right) &= P_{\theta_n}\left(\overline{X}_n \geq 0\right) \\ &= P_{\theta_n}\left[\sqrt{n}\,(\overline{X}_n - \theta_n) \geq -\sqrt{n}\,\theta_0 - \eta\right] \\ &= 1 - \Phi\left(-\sqrt{n}\,\theta_0 - \eta\right) \\ &= \Phi\left(\sqrt{n}\,\theta_0 + \eta\right)\,, \end{aligned}$$

which goes to one as $n \to \infty$.[††] Therefore, $T_n^* - \theta_n \sim \overline{X}_n - \theta_n$, so that

$$\sqrt{n}\,(T_n^* - \theta_n) \stackrel{d}{\Longrightarrow} \mathcal{N}(0, 1)$$

as $n \to \infty$. Therefore $R(\eta, T) = R(\eta, T^*) = 1$ for all η in this case.

Let $\theta_0 = 0$. Then

$$\begin{aligned} \sqrt{n}\,(T_n^* - \theta_n) &= \begin{cases} \sqrt{n}\,\left(\overline{X}_n - \theta_n\right) & \text{for } \sqrt{n}\,(\overline{X}_n - \theta_n) \geq -\eta \\ -\eta & \text{for } \sqrt{n}\,(\overline{X}_n - \theta_n) < -\eta\,. \end{cases} \\ &\stackrel{d}{=} \max(W, -\eta) \end{aligned}$$

where W has a standard normal distribution. So for $\theta_0 = 0$,

$$R(\eta, T^*) = \int_{-\eta}^{\infty} w^2\,\phi(w)\,dw \;+\; \eta^2\,[1 - \Phi(\eta)]\,.$$

This function satisfies $R(0, T^*) = \frac{1}{2}$ and $R(\eta, T^*) < 1$ for all $\eta > 0$.

5.7.5 Local asymptotic admissibility

In the exponential lag example above, the local asymptotic risk of the bias corrected maximum likelihood estimator uniformly dominated the local asymptotic risk of the maximum likelihood estimate. It turns out that this situation is atypical. It arises from the fact that the maximum likelihood estimator is not asymptotically unbiased for this model. An estimator T_n is said to be *locally asymptotically inadmissible* if there

[††] We remind the reader that Φ is the distribution function of the standard normal, and ϕ its density function.

exists an estimator T_n^* such that $R(\eta, T^*) \leq R(\eta, T)$ for all η and such that there exists some η_1 for which $R(\eta_1, T^*) < R(\eta_1, T)$. An estimator which is not locally asymptotically inadmissible is said to be locally asymptotically admissible. In our exponential lag example, the maximum likelihood estimator is locally asymptotically inadmissible.

Local asymptotic admissibility can be used to narrow the class of estimators. However, a stronger criterion is needed to narrow the class further. Minimaxity is one possibility. For any estimator T_n, $n \geq 1$ with local asymptotic risk $R(\eta, T)$, define

$$R(T) = \sup_{\eta} R(\eta, T)\,. \tag{5.36}$$

An estimator T_n is said to be *locally asymptotically minimax* if $R(T)$ is minimum, in the sense that $R(T) \leq R(T^*)$ for any other estimator T_n^*. Proving local asymptotic minimaxity can be technically challenging, but is assisted by the following proposition.

Proposition 4. *Suppose T_n, $n \geq 1$ is a locally asymptotically admissible estimator such that $R(\eta, T)$ is a constant function of η. Then T_n is locally asymptotically minimax.*

Proof. Let T_n^* be any other estimator. By assumption T_n is locally asymptotically admissible. Therefore there exists an η_0 such that $R(\eta_0, T) \leq R(\eta_0, T^*)$. However, $R(\eta, T)$ is a constant function of η. Therefore

$$\begin{aligned} R(T) &= R(\eta_0, T) \\ &\leq R(\eta_0, T^*) \\ &\leq \sup_{\eta} R(\eta, T^*) = R(T^*)\,. \end{aligned}$$

Therefore T_n is locally asymptotically minimax. ∎

As we saw earlier, Hodges' superefficient estimator for the $\mathcal{N}(\theta, 1)$ model does not have constant risk, whereas the maximum likelihood estimator $\overline{X}_n$ does. In general, in models satisfying the Cramér assumptions, the maximum likelihood estimator has constant local asymptotic risk because the limiting distribution of $\sqrt{n}\,(\widehat{\theta}_n - \theta_n)$ does not depend on η. The same regularity conditions also imply that $\widehat{\theta}_n$ is locally asymptotically admissible. Combining these properties and applying Proposition 4, we conclude that $\widehat{\theta}_n$ is locally asymptotically minimax, a property not shared with superefficient estimators in regular models.

5.8 Local asymptotic normality

The theory of local asymptotic normality was introduced to avoid the restrictive assumptions of Cramér's asymptotic theory. As we have seen, Cramér's assumptions which require random variables to be independent and identically distributed, a condition that is often not met in many models for which the maximum likelihood estimate is known to be asymptotically normal.

Let θ_0 be the true value of the parameter, and assume for the moment that the regularity conditions of Assumption 2 in Section 5.5 hold. Then as $n \to \infty$, in a contiguity neighbourbood of θ_0, a Taylor expansion gives

$$\begin{aligned} \ell_n\left(\theta_0 + \frac{\eta}{\sqrt{n}}\right) - \ell_n(\theta_0) &\sim \ell'_n(\theta_0)\,\frac{\eta}{\sqrt{n}} + \frac{1}{2}\,\ell''_n(\theta_0)\,\frac{\eta^2}{n} \\ &\sim \eta\,\frac{u_n(\theta_0)}{\sqrt{n}} - \frac{\eta^2}{2}\,I(\theta_0)\,. \end{aligned} \tag{5.37}$$

In addition, using formula (5.22), we obtain

$$\begin{aligned} \sqrt{n}\,(\widehat{\theta}_n - \theta_0) &\sim \frac{u_n(\theta_0)/\sqrt{n}}{i_n(\theta_0/n} \\ &\sim \frac{u_n(\theta_0)/\sqrt{n}}{I(\theta_0)}\,. \end{aligned}$$

This can be rewritten as

$$\frac{u_n(\theta_0)}{\sqrt{n}} \sim I(\theta_0)\left[\sqrt{n}\,(\widehat{\theta}_n - \theta_0)\right]. \tag{5.38}$$

Plugging (5.38) into (5.37), we get

$$\ell_n\left(\theta_0 + \frac{\eta}{\sqrt{n}}\right) - \ell_n(\theta_0) \sim \eta\,I(\theta_0)\left[\sqrt{n}\,(\widehat{\theta}_n - \theta_0)\right] - \frac{\eta^2\,I(\theta_0)}{2} \tag{5.39}$$

as $n \to \infty$. The use of $\widehat{\theta}_n$ in (5.39) is not essential. Many other efficient estimators will do. For example, if $T_n - \theta_0 \sim \widehat{\theta}_n - \theta_0$, then T_n is asymptotically equivalent to $\widehat{\theta}_n$. In this case, (5.39) will remain true in probability if $\widehat{\theta}_n$ is replaced by T_n. Similar remarks also hold for the quadratic term and its coefficient involving $I(\theta_0)$. This leads us to the following definition.

Definition 8. *A model is said to be* locally asymptotically quadratic *at θ_0 if the following two conditions hold when $n \to \infty$ and θ_0 is the true value of the parameter.*

1. *There exist two random sequences $S_n(\theta_0)$ and $I_n(\theta_0)$ which do not depend on η such that*

$$\ell_n\left(\theta_0 + \frac{\eta}{\sqrt{n}}\right) - \ell_n(\theta_0) = \eta\, S_n(\theta_0) - \frac{\eta^2\, I_n(\theta_0)}{2} + \rho_n(\eta,\, \theta_0)\,,$$

where $\rho_n(\eta, \theta_0) \xrightarrow{P} 0$.

2. *The sequence $I_n(\theta_0)$ is positive and bounded in probability away from zero.*‡‡

The statement that $\rho_n(\eta, \theta_0) \xrightarrow{P} 0$ can be interpreted in various ways. Three possible interpretations are

- For every real value η, the sequence $\rho_n(\eta, \theta_0)$ converges in probability to zero.
- For every real value η, and for every sequence η_n converging to η, the sequence $\rho_n(\eta_n, \theta_0)$ converges in probability to zero.
- The sequence $\rho_n(\eta, \theta_0)$ converges in probability to zero uniformly for η in every bounded set. That is,

$$\sup_{|\eta| \le m} |\, \rho_n(\eta, \theta_0)\, | \xrightarrow{P} 0$$

  for every $m > 0$.

Of these three interpretations, the second is the standard definition. However, the choice of an appropriate condition will depend upon the particular application.

Since $\theta_0 + \eta/\sqrt{n}$ is contiguous to θ_0, it follows that

$$P_{\theta_0}\left[\rho_n(\eta,\, \theta_0) \xrightarrow{P} 0\right] \qquad \text{implies} \qquad P_{\theta_0 + n/\sqrt{n}}\left[\rho_n(\eta,\, \theta_0) \xrightarrow{P} 0\right].$$

So, if the log-likelihood is locally asymptotically quadratic under θ_0 it is locally asymptotically quadratic under contiguous alternatives. However, the asymptotic distributions of $S_n(\theta_0)$ and $I_n(\theta_0)$ under θ_0 will differ from their respective asymptotic distributions under the contiguous sequence $\theta_0 + n/\sqrt{n}$.

More informally, the model is locally asymptotically quadratic at θ_0 if the log-likelihood is approximately quadratic throughout the contiguity neighbourhood, so that

$$\ell_n\left(\theta_0 + \frac{\eta}{\sqrt{n}}\right) - \ell_n(\theta_0) \;\sim\; \eta\, S_n(\theta_0) - \frac{\eta^2\, I_n(\theta_0)}{2}\,, \tag{5.40}$$

‡‡ This is the same as saying that $[I_n(\theta_0)]^{-1}$ is $O_p(1)$.

In such models, the log-likelihood may be approximated by a quadratic function whose coefficients depend on S_n and I_n. In models where these quantities do not depend heavily on θ_0 it may be possible to use (S_n, I_n) as an approximately sufficient pair for estimation.

Returning to (5.39), we note that $\sqrt{n}\,(\widehat{\theta}_n - \theta_0)$ is often asymptotically normal, and that the coefficient on the quadratic term is not random. This leads to the following definition.

Definition 9. *A model is said to be* locally asymptotically normal *at θ_0 if it is locally asymptotically quadratic as in Definition 8 above, and if additionally the following conditions hold when $n \to \infty$ and θ_0 is the true value of the parameter.*

1. *The random variables $I_n(\theta_0)$ converge in probability to some nonrandom $I(\theta_0) > 0$ which does not depend on η.*
2. *The random variables $S_n(\theta_0)$ converge in distribution to $\mathcal{N}(\,0,\, I(\theta_0)\,)$.*

Clearly from the definition, a model which is locally asymptotically normal is locally asymptotically quadratic. Furthermore, the regularity conditions of Assumption 2 in Section 5.5 are strong enough to ensure local asymptotic normality, although the latter condition is more general than the former. Local asymptotic normality turns out to be the right tool for proving that a class of asymptotically efficient estimators are asymptotically normal. Every such model admits a quadratic approximation to the log-likelihood within a contiguity neighbourhood of the true value of the parameter. Therefore, such models admit the construction of a variety of estimators obtained by maximising a quadratic approximation to the likelihood. For example, suppose θ_n^* is an auxiliary estimator for θ which lies with high probability within a contiguity neighbourhood of the true value of the parameter as $n \to \infty$. Thus

$$\theta_n^* = \theta_0 + O_p\left(1/\sqrt{n}\right) .$$

Next, we compute the log-likelihood at three parameter points within this contiguity neighbourhood. For example, we might compute

$$\ell_n(\theta_n^*), \qquad \ell_n\left(\theta_n^* + c/\sqrt{n}\right), \qquad \ell_n\left(\theta_n^* - c/\sqrt{n}\right),$$

where $c \neq 0$. Provided the log-likelihood is sufficiently smooth, it can be uniformly approximated by the quadratic function within a neighbourhood which includes these three values. It can be checked that the quadratic in η approximating the log-likelihood is

$$\ell_n\left(\theta_n^* + \eta/\sqrt{n}\right) \;\sim\; A_n + \eta\, B_n + \eta^2\, C_n$$

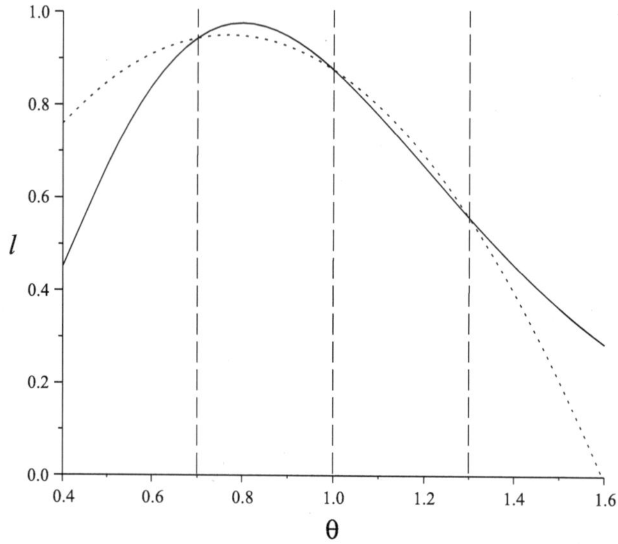

Figure 5.1 *A quadratic fit to the likelihood using three points, each marked with a vertical line. In the toy example above, an auxiliary estimator* $\theta_n^* = 1.0$ *is updated by fitting a quadratic (the dotted curve) to the log-likelihood (the unbroken curve) using* $c/\sqrt{n} = 0.3$. *The updated estimator* θ_n^{**} *is the maximum of this quadratic fit.*

where

$$A_n = \ell_n(\theta_n^*)$$

$$B_n = \frac{1}{2\,c}\left[\ell_n(\theta_n^* + c/\sqrt{n}) - \ell_n(\theta_n^* - c/\sqrt{n})\right]$$

$$C_n = \frac{1}{2\,c^2}\left[\ell_n(\theta_n^* + c/\sqrt{n}) + \ell_n(\theta_n^* - c/\sqrt{n}) - 2\,\ell_n(\theta_n^*)\right].$$

If the log-likelihood is strictly concave within the contiguity neighbourhood, then $C_n < 0$ and the quadratic will be maximised at

$$\theta_n^{**} = \theta_n^* - \frac{B_n}{2\,C_n}. \tag{5.41}$$

See Figure 5.1. As $c \to 0$, this becomes

$$\theta_n^{**} = \theta_n^* - \frac{\ell_n'(\theta_n^*)}{\ell_n''(\theta_n^*)}$$

which can be recognised as the value obtained by applying one step of

Newton-Raphson iteration to solve the equation $\ell(\widehat{\theta}_n) = 0$ starting from θ_n^*.

We can eliminate the linear term in the quadratic approximation by completing the square about θ_n^{**}. So, within the contiguity neighbourhood, the quadratic approximation to the log-likelihood can be written as

$$\ell_n(\theta_n^{**} + \eta/\sqrt{n}) \sim \ell_n(\theta_n^{**}) - \frac{\eta^2 I_n}{2}$$

where $I_n = -2\, C_n$. Replacing η in this approximation by $\sqrt{n}\,(\theta_0 - \theta_n^{**})$ and by $\eta + \sqrt{n}\,(\theta_0 - \theta_n^{**})$ and taking the difference of these two expressions, we get

$$\ell_n\left(\theta_0 + \frac{\eta}{\sqrt{n}}\right) - \ell_n(\theta_0) \sim \eta\,\left[\sqrt{n}\, I_n(\theta_0)\,(\theta_n^{**} - \theta_0)\right] - \frac{\eta^2\, I_n(\theta_0)}{2}\,. \tag{5.42}$$

The coefficients of formula (5.42) can be compared with the coefficients of formula (5.40). Assuming that the model is locally asymptotically normal, the coefficients of the quadratic in (5.42) have limits

$$I_n(\theta_0) \xrightarrow{P} I(\theta_0)$$

and

$$\sqrt{n}\, I_n\,(\theta_n^{**} - \theta_0) \stackrel{d}{\Longrightarrow} \mathcal{N}(\,0,\, I(\theta_0)\,)$$

as $n \to \infty$.

5.9 Local asymptotic minimaxity

In this section, we shall sketch a proof of the local asymptotic minimaxity of the maximum likelihood estimator $\widehat{\theta}_n$ under the assumption of local asymptotic normality.

Assumption 3. *Suppose that in a contiguity neighbourhood of θ_0 we can write*

$$\ell_n\left(\theta_0 + \frac{\eta}{\sqrt{n}}\right) - \ell_n(\theta_0) \sim \eta\, I(\theta_0)\,\sqrt{n}\,(\widehat{\theta}_n - \theta_0) - \eta^2\,\frac{I(\theta_0)}{2}\,, \tag{5.43}$$

where

$$\sqrt{n}\,(\widehat{\theta}_n - \theta_0) \stackrel{d}{\Longrightarrow} \mathcal{N}[\,0,\, 1/I(\theta_0)\,] \tag{5.44}$$

under θ_0, that is, when θ_0 is the true value of the parameter.

In the argument which follows, $\widehat{\theta}_n$ can be replaced by any other estimator θ_n^{**} provided that the corresponding assumptions to (5.43) and

(5.44) are satisfied with $\widehat{\theta}_n$ replaced by θ_n^{**}. To prove that $\widehat{\theta}_n$ is locally asymptotically minimax, we shall prove that $\widehat{\theta}_n$ is locally asymptotically admissible with constant local asymptotic risk.

Suppose θ_n^* is any estimator of θ. Define

$$T_n = \sqrt{n}\,(\widehat{\theta}_n - \theta_0) \qquad \text{and} \qquad T_n^* = \sqrt{n}\,(\theta_n^* - \theta_0)\,.$$

Then the joint distribution of $(T_n,\, T_n^*)$ has a joint limiting distribution whose behaviour is described by the following proposition.

Proposition 5. *Assume (5.43) and (5.44). Suppose that under θ_0 the pair $(T_n,\, T_n^*)$ converges to some distribution with joint distribution function $H_0(x, y)$. Then, under the sequence of contiguous alternatives $\theta_n = \theta_0 + \eta/\sqrt{n}$ the pair $(T_n,\, T_n^*)$ converges in distribution to some limit with joint distribution function $H_\eta(x,\, y)$, satisfying*

$$dH_\eta(x,\, y) = \exp\left(\eta\, I\, x - \eta^2\, \frac{I}{2}\right)\, dH_0(x,\, y)\,, \tag{5.45}$$

where $I = I(\theta_0)$.

Proof. Define $g(x, \eta) = \exp(\eta\, I\, x - \eta^2\, I/2)$. This function is continuous in the variable x, and therefore as $n \to \infty$, the joint distribution of $[\,T_n, T_n^*, g(T_n, \eta)\,]$ under θ_0 converges to the distribution of $[\,X, Y, g(X, \eta)\,]$ under H_0. So by (5.43),

$$\left(T_n,\, T_n^*,\, \exp\left[\ell_n\left(\theta_0 + \frac{\eta}{\sqrt{n}}\right) - \ell_n(\theta_0)\right]\right)_{\theta_0} \stackrel{d}{\Longrightarrow} [\,X,\, Y,\, g(X,\, \eta)\,]_{H_0}\,. \tag{5.46}$$

Now suppose $h(x,\, y)$ is a bounded continuous function of x and y. Using the likelihood tilting trick that we have used before, we can write

$$\begin{aligned} E_{\theta_n} h(T_n,\, T_n^*) &= E_{\theta_0}\left[h(T_n,\, T_n^*)\, \frac{L_n(\theta_n)}{L_n(\theta_0)}\right] \\ &= E_{\theta_0}\left\{h(T_n,\, T_n^*)\, e^{\ell_n(\theta_0 + \eta/\sqrt{n}) - \ell_n(\theta_0)}\right\}\,. \end{aligned} \tag{5.47}$$

The next step is to use (5.46) in (5.47) to obtain

$$E_{\theta_0}\left\{h(T_n,\, T_n^*)\, e^{\ell_n(\theta_0+\eta/\sqrt{n}) - \ell_n(\theta_0)}\right\} \to E_{H_0}[h(X, Y)\, g(X, \eta)]\,. \tag{5.48}$$

In order to prove convergence in expectation as in (5.48) from convergence of the joint distribution in (5.44) it is necessary to verify uniform integrability of

$$h(T_n,\, T_n^*)\, \exp[\ell_n(\theta_0 + \eta/\sqrt{n}) - \ell_n(\theta_0)]$$

under θ_0. First note that the sequence

$$V_n = \exp[\ell_n(\theta_0 + \eta/\sqrt{n}) - \ell_n(\theta_0)]$$

is uniformly integrable. To verify this, we check that the sequence of random variables V_n satisfies the following conditions.

- $(V_n)_{\theta_0} \overset{d}{\Longrightarrow} (g(X, \eta))_{H_0}$.
- $E_{\theta_0} V_n = E_{H_0} g(X, \eta)$. (See Problem 6.)
- $V_n, g(X, \eta) \geq 0$.

It is left to the reader to check that these conditions imply that the random variables V_n are uniformly integrable. From this fact, and the boundedness of $h(x, y)$ the reader can also show that $h(T_n, T_n^*) V_n$ is uniformly integrable. Therefore (5.48) follows.

Putting (5.47) and (5.48) together gives us

$$E_{\theta_n} h(T_n, T_n^*) \to E_{H_0}[h(X, Y) g(X, \eta)], \tag{5.49}$$

Statement (5.49) is true for every bounded continuous function $h(x, y)$. Since the limit of the left-hand side exists for all such functions, it follows that (T_n, T_n^*) has a limiting distribution H_η. So, we can write

$$\begin{aligned} E_{\theta_n} h(T_n, T_n^*) &\to \int\int h(x, y)\, dH_\eta(x, y) \\ &= \int\int h(x, y)\, g(x, \eta)\, dH_0(x, y) \end{aligned}$$

which implies the statement of the proposition. ∎

Proposition 5 implies that the maximum likelihood estimator—or any other estimator satisfying (5.43) and (5.44)—is asymptotically sufficient.

Proposition 6. *Under the assumptions given in (5.43) and (5.44), the estimator* $\widehat{\theta}_n$ *has constant local asymptotic risk given by* $R(\eta, \widehat{\theta}) = 1/I(\theta_0)$ *for all* η.

Proof. In Proposition 5, we set $T_n^* = T_n$. Then

$$(T_n, T_n)_{\theta_0} \overset{d}{\Longrightarrow} (X, X)_{H_0} \qquad \text{and} \qquad (T_n, T_n)_{\theta_\eta} \overset{d}{\Longrightarrow} (X, X)_{H_\eta}.$$

From (5.44), we see that X is $\mathcal{N}(0, 1/I(\theta_0))$ under H_0. Therefore, $(T_n)_{\theta_\eta}$ converges to a distribution with density

$$\exp\left(\eta\, I\, x - \eta^2 \frac{I}{2}\right) \times \sqrt{\frac{I}{2\pi}} \exp\left(-\frac{1}{2} I\, x^2\right)$$

This reduces to

$$\sqrt{\frac{I}{2\,\pi}}\,\exp\left[-\frac{I\,(x-\eta)^2}{2}\right],$$

namely, the density of $\mathcal{N}(\eta,\,1/I)$. It is left to the reader to prove from this that the local asymptotic risk of $\widehat{\theta}_n$ is $R(\eta,\,\widehat{\theta}) = 1/I(\theta_0)$ for all η. ∎

The next proposition requires some results from the theory of weak convergence. We refer the reader to Billingsley (1999) for many of the basic definitions and results in this area.

Proposition 7. *Assume that $\widehat{\theta}_n$ satisfies (5.43) and (5.44). Then $\widehat{\theta}_n$ is locally asymptotically admissible.*

Proof. From Proposition 6, we see that $R(\eta,\widehat{\theta}) = 1/I(\theta_0)$. Suppose that θ_n^* is any other estimator such that $R(\eta,\theta^*) \le 1/I(\theta_0)$ for all η. It is sufficient to prove that $R(\eta,\theta^*) = 1/I(\theta_0)$ for all η.

Let $(T_n,\,T_n^*)$ be defined as in Proposition 5. First, note that

$$R(0,\widehat{\theta}),\; R(0,\theta^*) \le 1/I\,.$$

This implies that T_n and T_n^* are both bounded in mean square in the limit. This in turn implies that $(T_n,\,T_n^*)_{\theta_0}$ has a subsequence $(T_{n_j},\,T_{n_j}^*)_{\theta_0}$ which converges to some limiting distribution H_0. Proposition 5 then implies that $(T_{n_j},\,T_{n_j}^*)_{\theta_n}$ has limiting distribution H_η as defined by (5.45). We are now in a position to use the form of the limiting family given in (5.45). The following results follow from (5.45).

1. The limiting distribution of $(\,T_n\,)_{\theta_0}$ is $\mathcal{N}(0,\,1/I)$. Formula (5.45) implies that in the limit $(\,T\,)_{H_\eta}$ is $\mathcal{N}(\eta,\,1/I)$ as noted earlier. This is a normal location model.
2. Under H_η, the statistic T_n is a sufficient statistic for the parameter η.

So, we can write

$$(\,T_n,\,T_n^*\,)_{H_\eta} \stackrel{d}{=} (\,T_n\,)_{\mathcal{N}(\eta,1/I)} \times (T_n^*\,|\,T_n\,)$$

where the conditional component does not involve the parameter η. This is a normal location model with an additional random component attached which does not depend on the parameter η. In the normal location model (plus random noise represented by the second component in the factorisation) the usual unbiased estimator of the parameter—here represented by T—is admissible. Here, we have a limiting admissible estimator whose risk is dominated by another estimator T^*. This can only occur if $P_{H_\eta}(T = T^*) = 1$ so that $E_{H_\eta}(T^*)^2 = E_{H_\eta} T^2$, for all η.

We conclude from this argument that $R(\eta, \theta^*) = 1/I$ as required. ∎

Theorem 2. *Under assumptions (5.43) and (5.44), the estimator $\widehat{\theta}_n$ is locally asymptotically minimax.*

Proof. By Proposition 7 it is admissible, and by Proposition 6 it has constant local asymptotic risk. Together these imply the result. ∎

Our proof of Theorem 2 requires the admissibility of the sample mean for the normal location model. It is well known that in higher dimensions, the sample mean of the data is inadmissible for the normal location parameter. For this reason, our proof does not extend to maximum likelihood estimation in general dimensions. Nevertheless, the sample mean is minimax in all dimensions for the normal model. The generalisation of Theorem 2 remains true in higher dimensions as well.

5.10 Various extensions

Many of the results in this chapter extend rather routinely into higher dimensions by replacing each object with its higher dimensional counterpart. Random variables extend to random vectors, with joint density functions used to define likelihood. Single parameter models extend to k parameter models, with the likelihood function $L_n(\theta)$ defined for the p-dimensional vector θ. Although the likelihood and log-likelihood remain real-valued functions of the vector θ, the score function $u_n(\theta)$ becomes the vector-valued function

$$u_n(\theta) = \left(\frac{\partial}{\partial \theta_j} \ell_n(\theta) \right)$$

whose components are the partial derivatives of the log-likelihood. The observed information function is a $p \times p$ dimensional matrix

$$i_n(\theta) = \left(- \frac{\partial^2}{\partial \theta_j \, \partial \theta_k} \ell_n(\theta) \right)$$

with expected information matrix $I(\theta_0) = E_\theta \, i_1(\theta)$. Positivity of $I(\theta)$ in the single parameter case is replaced by positive definiteness in the multiparameter case. The regularity conditions extend so that

$$\sqrt{n}\,(\widehat{\theta}_n - \theta_0) \stackrel{d}{\Longrightarrow} \mathcal{N}_p(\,0,\, I^{-1}(\theta_0)\,)$$

where $\mathcal{N}_p(\mu, \Sigma)$ is the multivariate normal in dimension p with vector mean μ and covariance matrix Σ, and $I^{-1}(\theta_0)$ is the matrix inverse of $I(\theta_0)$. A full derivation of these multivariate extensions can be found in many standard texts on mathematical statistics including Lehmann and Casella (1998).

Local asymptotic normality and local asymptotic minimaxity extend also, but the proof given in this chapter does not extend. For a more general development of the theory, the reader is referred to Le Cam and Yang (2000). Locally asymptotically quadratic models and locally asymptotically normal models are defined through the quadratic approximation

$$\ell_n\left(\theta_0 + \frac{\eta}{\sqrt{n}}\right) - \ell_n(\theta_0) \sim \eta^t\, S_n(\theta_0) - \frac{1}{2}\,\eta^t\, I(\theta_0)\,\eta$$

where η is a p-dimensional column vector, and the column vector $S_n(\theta_0)$ converges in distribution to $\mathcal{N}_p(0,\, I(\theta_0))$. In turn, the condition of local asymptotic normality can be extended still further to *local asymptotic mixed normality (LAMN)*. Such models are locally asymptotically quadratic as above. However, in this case $I(\theta_0)$ is a random variable and $S(\theta_0)$ is $\mathcal{N}_p(0,\, I(\theta_0))$ conditionally on $I(\theta_0)$.

Another important result in local asymptotics is the Hájek convolution theorem—known also as the Hájek-Inagaki convolution theorem—for regular estimators of θ. Roughly speaking, an estimator T_n is regular if the distribution H_η of Definition 7 does not depend upon η.[§§] The convolution theorem states that for locally asymptotically normal models, the asymptotic distribution of regular T_n can be written as a convolution, namely

$$\sqrt{n}\,(T_n - \theta_0) \stackrel{d}{\Longrightarrow} W + Y$$

where W and Y are independent and $W \stackrel{d}{=} \mathcal{N}(0,\, I^{-1}(\theta_0))$. For regular $\widehat{\theta}_n$ the random variable Y is degenerate, being the constant zero. This convolution theorem is another way of expressing the idea that $\widehat{\theta}_n$ and equivalent estimators asymptotically reach the Cramér-Rao lower bound, because the sum $W + Y$ of two independent random variables can be thought of as more widely dispersed than either. So, all regular estimators for θ are at least as widely dispersed as $\widehat{\theta}_n$. The reader is

[§§] Regularity, in the sense defined here, is simply *local asymptotic location equivariance* of the distribution of T_n. More precisely, the distribution of $\sqrt{n}\,(T_n - \theta_n)$ is asymptotically invariant with respect to the parameter η when θ_n is in a contiguity neighbourhood of θ_0. Asymptotic invariants of this kind are also known as asymptotic pivotals, and have played an important role in the theory of hypothesis testing and interval estimation. Regular estimators have constant local asymptotic risk functions. Superefficient estimators are not regular, a result that follows from the fact that their local asymptotic risk functions are not constant.

referred to Beran (1999) for a summary of the definitions and results and for discussion of different types of regularity imposed on estimators. For a discussion of its relationship to local asymptotic minimaxity, see Le Cam and Yang (2000). The original work is by Hájek (1970) and independently using a more restrictive condition by Inagaki (1970). See also Inagaki (1973).

The theory of local asymptotics has also been applied to study the properties of shrinkage estimators. As James and Stein (1961) discovered, the sample mean for the estimation of μ is inadmissible in dimension $p \geq 3$. In particular, if $X_1, \ldots, X_n$ are independent identically distributed $\mathcal{N}_p(\mu, 1_{p\times p})$, where $1_{p\times p}$ is the identity matrix, then the sample mean $\overline{X}_n$ has larger mean square error for every μ than the shrinkage estimator of the form

$$T_n = \left(1 - \frac{p-2}{n\,||\,\overline{X}_n\,||^2}\right)\overline{X}_n\,.$$

This paradoxical result runs counter to our intuition that $\overline{X}_n$ is the best estimator for μ under all reasonable criteria. The key insight provided by the theory of local asymptotics is that James-Stein shrinkage estimators are similar in their asymptotic behaviour to Hodges' superefficient estimators. Just as the Hodges' superefficient estimator shrinks $\overline{X}_n$ towards zero in dimension one, so James-Stein estimators shrink $\overline{X}_n$ towards zero in higher dimensions. It turns out that neither superefficient estimators nor James-Stein estimators are regular in the sense defined above. In both cases, regularity fails at zero. For James-Stein estimators, the limiting distribution has the form

$$\sqrt{n}\left(T_n - \theta_0 - \frac{\eta}{\sqrt{n}}\right) \stackrel{d}{\Longrightarrow} H_\eta\,.$$

When $\theta_0 \neq 0$, the distribution H_η does not depend on η. However, when $\theta_0 = 0$, the limiting distribution is

$$\sqrt{n}\left(T_n - \frac{\eta}{\sqrt{n}}\right) \stackrel{d}{\Longrightarrow} \left(1 - \frac{p-2}{||\,W\,||^2}\right)W - \eta\,,$$

where $W \stackrel{d}{=} \mathcal{N}_p(0, 1_{p\times p})$.

5.11 Problems

1. Let ψ_j be the jth Bhattacharyya function, and let Z be an element of $\mathbf{Z}(\theta_0)$. Prove that

$$E_{\theta_0}\,[\,Z\,\psi(\theta_0)\,] = 0$$

under appropriate regularity allowing the interchange of derivatives and integrals.

2. Prove (5.24).

3. Suppose

$$c_n(T_n - \theta_0) \stackrel{d}{\Longrightarrow} H_1 \qquad \text{and} \qquad d_n(T_n - \theta_0) \stackrel{d}{\Longrightarrow} H_2$$

as $n \to \infty$, where H_1 and H_2 are nondegenerate. Prove that there exists some constant c such that $H_2(t) = H_1(t/c)$ for all t and such that $d_n/c_n \to c$ as $n \to \infty$.

4. Derive formula (5.34) to complete the calculation of the local asymptotic risk for Hodges' superefficient estimator.

5. In the example given in Section 5.7.3, it was stated that the maximum likelihood estimator for the lag parameter θ of an exponential distribution based on a random sample $X_1, \ldots, X_n$ is $T_n = \min(X_1, \ldots, X_n)$.
   (a) Prove this result.
   (b) Prove that T_n has an exponential distribution with lag θ and mean $\theta + n^{-1}$.

6. In the proof of Proposition 5, the fact that $E_{\theta_0}(V_n) = E_{H_0}\, g(X, \eta)$ was used. Prove this, and complete the other omitted steps in the proof of the proposition.

7. Complete the proof of Proposition 6, and show that $\widehat{\theta}_n$ has constant local asymptotic risk under conditions (5.43) and (5.44).

8. A box of 100 good widgets has been accidentally contaminated by a certain number of defective widgets. Let θ be the number of defective widgets in the box, where θ is an unknown nonnegative integer. A sample of n widgets is chosen without replacement from the box. When each widget in the sample is tested, it is discovered that x of the widgets are good and y are defective, where $x + y = n$. Let $L_n(\theta)$ be the likelihood function for the unknown parameter θ based upon this sample of size n.
   (a) Show that $L_n(\theta)$ satisfies the equation

$$L_n(\theta + 1) = \frac{(\theta + 1)(101 + \theta - n)}{(\theta + 1 - y)(101 + \theta)} L_n(\theta)$$

(b) Using this equation, find a formula for the maximum likelihood estimator $\widehat{\theta}_n$.

9. Data which follow a count distribution often have the feature that the zero counts are not recorded. For example, this will occur for data where an incident (that is, an event leading to a positive count) is required in order for the data to be collected. Such data are called *zero-truncated.* In this problem, we shall consider a zero-truncated Poisson distribution. A random variable X is a zero-truncated Poisson with parameter θ provided

$$P(X = x) = \frac{\theta^x e^{-\theta}}{(1 - e^{-\theta})\, x!}$$

for $x = 1, 2, 3, \ldots$.

(a) Verify that this is a probability mass function on the strictly positive integers, and determine the mean of the distribution in terms of θ.

(b) Suppose $X_1, X_2, \ldots, X_n$ are independent and identically distributed random variables having this zero-truncated count distribution. Prove that the maximum likelihood estimator for θ satisfies the equation

$$\widehat{\theta}_n = \overline{X}_n \left(1 - e^{-\widehat{\theta}_n}\right) .$$

10. In this problem, we shall consider a model where the likelihood function is not differentiable. Suppose $X_1, X_2, \ldots, X_n$ are independent and identically distributed random variables having a double exponential distribution. The density function for this distribution has the form

$$f(x : \theta) = \frac{1}{2} \exp\left(-|x - \theta|\right) \qquad \text{for } -\infty < x < \infty ,$$

where $-\infty < \theta < \infty$.

(a) Find expressions for $\ell_n(\theta)$, $u_n(\theta)$ and $i_n(\theta)$. State clearly in the case of $u_n(\theta)$ and $i_n(\theta)$ where these functions are defined.

(b) Suppose that n is an odd integer, so that $n = 2\,m - 1$, for integer m. Prove that $\widehat{\theta}_n = X_{(m)}$, the middle order statistic of the sample.

(c) Suppose that n is an even integer, so that $n = 2\,m$. Prove that $\widehat{\theta}_n$ is not uniquely defined in this case. Rather the likelihood $L_n(\theta)$ is maximised at any value of θ which lies between the two middle order statistics $X_{(m)}$ and $X_{(m+1)}$.

(d) Does the identity $E_{\theta_0}\, u_n(\theta_0) = 0$ hold?

(e) Does the identity $\text{Var}_{\theta_0} u_n = E_{\theta_0} i_n(\theta_0)$ hold? If not, how should the expected information function be defined?

(f) Show that for odd sample sizes, $\widehat{\theta}_{2m-1}$ is unbiased and has finite variance.

(g) Show that the variance of $\widehat{\theta}_1$ is strictly greater than the Cramér-Rao lower bound, appropriately defined.

11. The *log-relative likelihood function* is defined as $r_n(\theta) = \ell_n(\theta) - \ell_n(\widehat{\theta}_n)$. When θ is real-valued, a quadratic approximation for $r_n(\theta)$ in a contiguity neighbourhood of $\widehat{\theta}_n$ typically has the form

$$r_n\left(\widehat{\theta}_n + \frac{c}{\sqrt{n}}\right) = -\frac{1}{2}\frac{\widehat{i}_n}{n}c^2 + O_p\left(\frac{c^3}{n\sqrt{n}}\right),$$

where $\widehat{i}_n = i_n(\widehat{\theta}_n)$.

(a) Provide regularity assumptions suitable for the derivation of this quadratic approximation.

(b) Consider two independent experiments A and B which produce relative likelihoods $r_n^A(\theta)$ and $r_n^B(\theta)$ and maximum likelihood estimates $\widehat{\theta}_n^A$ and $\widehat{\theta}_n^B$, respectively, for some common parameter θ. Similarly, let

$$\widehat{i}_n^A = i_n^A(\widehat{\theta}_n^A) \qquad \text{and} \qquad \widehat{i}_n^B = i_n^B(\widehat{\theta}_n^B)$$

be the respective observed informations for experiments A and B. Let $r_n(\theta)$, $\widehat{\theta}_n$ and $\widehat{i}_n$ be the corresponding quantities for the combined experiment (A, B).

Suppose that $\widehat{\theta}_n^A$ and $\widehat{\theta}_n^B$ both lie in a contiguity neighbourhood of $\widehat{\theta}_n$. Prove that

$$\widehat{\theta}_n \approx \frac{\widehat{i}_n^A}{\widehat{i}_n^A + \widehat{i}_n^B}\,\widehat{\theta}_n^A + \frac{\widehat{i}_n^B}{\widehat{i}_n^A + \widehat{i}_n^B}\,\widehat{\theta}_n^B,$$

and provide an expression for the order of the error.

12. Suppose $X_1, X_2, \ldots, X_n$ are independent identically distributed continuous random variables from a distribution whose density $f(x)$ vanishes off the interval $0 \le x \le 1$, and is bounded so that $0 \le f(x) \le 2$. Suppose also that $f(x)$ is continuous. We can think of the density function as an unknown nonparametric infinite dimensional parameter. Let $\widehat{f}_n$ be a maximum likelihood estimator for f.

(a) Prove that $\widehat{f}$ is a maximum likelihood estimator for f if and only if

- $\int_0^1 \widehat{f}(x)\,dx = 1$,
- $\widehat{f}(x)$ is a continuous function of x, and
- $\widehat{f}(X_j) = 2$ for all $j = 1, 2, \ldots, n$.

(b) Use the result above to show that there exist infinitely many maximum likelihood estimators for the density f in this model.

(c) For nonparametric models, it is often unreasonable to expect that $\widehat{f}_n \to f$ as $n \to \infty$. However, in many nonparametric models it is nevertheless true that $\widehat{F}_n(t) \to F(t)$ for all t. For any given choice of maximum likelihood estimator $\widehat{f}_n$, define

$$F(t) = \int_0^t f(x)\,dx \qquad \text{and} \qquad \widehat{F}_n(t) = \int_0^t \widehat{f}_n(x)\,dx\,,$$

for all $0 \le t \le 1$. Show that for the model developed above, the function $\widehat{F}_n$ does not necessarily converge to F.

(d) Show also that consistent choices of $\widehat{F}_n$ exist.

13. For locally asymptotically normal models, what is the limiting distribution of

$$\frac{\ell_n\left(\theta_0 + \dfrac{1}{\sqrt{n}}\right) - \ell_n(\theta_0) + \dfrac{I(\theta_0)}{2}}{\sqrt{I(\theta_0)}}$$

as $n \to \infty$?

14. Let $X_1, X_2, \ldots, X_n$ be independent identically distributed random variables from $\mathcal{N}(0, \sigma^2)$. For each value of the constant a, determine the local asymptotic risk of

$$T_n = \frac{1}{n+a} \sum_{j=1}^n X_j^2$$

as an estimator for σ^2.

15. Consider a random sample $X_1, \ldots, X_n$ of independent identically distributed $\mathcal{N}(\theta, 1)$ random variables, where θ is constrained to be integer-valued. For any given θ_0, characterise the collection of all sequences θ_n which are contiguous to θ_0 (and also constrained to be integer-valued). Justify your answer.

CHAPTER 6

The Laplace approximation and series

6.1 A simple example

The evaluation of integrals is a core analytical problem in many branches of statistics. When we evaluate a marginal distribution of a statistic, a moment, or a posterior density, we are evaluating integrals either implicitly or explicitly. In some cases, the difficulties involved in integration can turn an otherwise promising solution to a problem into a computational quagmire.

To manage difficult integrals we resort to a variety of methods. In some cases, by appropriately restricting the class of models under consideration, the required integrals can be solved using standard textbook techniques. For example, standard texts often list the moment generating functions for the binomial, the Poisson, the normal, and other families of distributions of exponential type. These distributions have generating functions whose integral representations can be computed in closed form. Some integrals, particularly multiple integrals in high dimensions, have been attacked by Monte Carlo methods. In the previous chapter, we found that series expansions can provide sensible approximations to the values of some integrals.

The *Laplace approximation*, which is the main topic of this chapter, has some affinity to the series expansions that we considered previously. The basic idea is very simple. In certain circumstances it is reasonable to approximate the logarithm of an integrand by a quadratic function so that the integrand itself is approximated by a "bell-shaped" curve, *i.e.*, a function which can be written as a constant multiple of a normal (Gaussian) probability density function.

To understand the circumstances in which such an approximation is possible, let us consider the functions plotted in Figure 6.1. On the left-hand side are plotted the functions

$$g_n(x) = (1 - x^2)^n\,, \quad n = 1, 2, 4, 8\,.$$

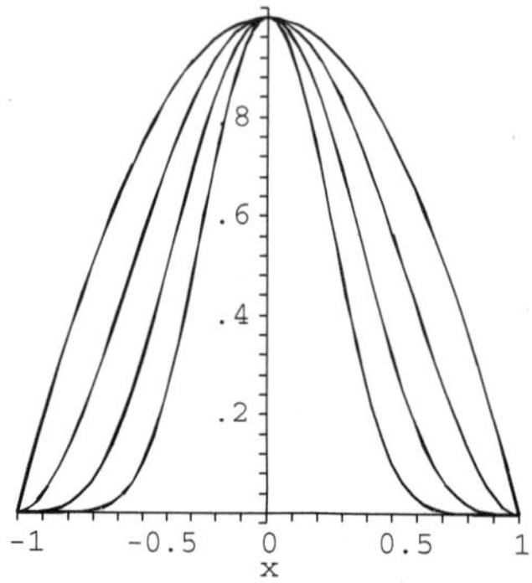

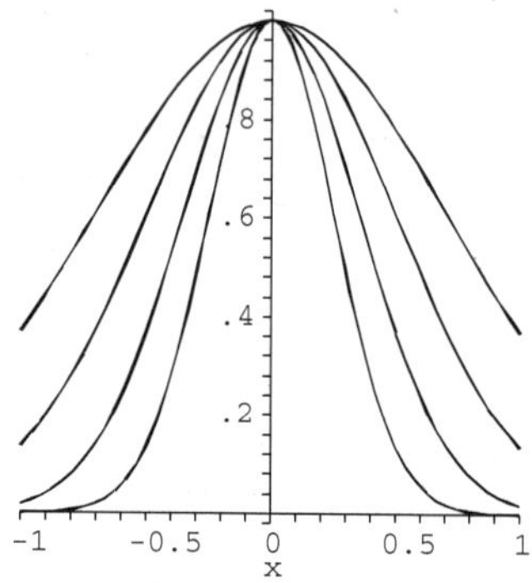

Figure 6.1 *The effect of power transformations on the shape of a function*

On the right-hand graph are plotted the bell-curve approximations to these functions, which are

$$f_n(x) = \exp(-n\,x^2)$$

found by matching the functions and the first two derivatives at $x = 0$. As can be seen from the graph, $g_n(x)$ is much closer in shape to the bell curve $h_n(x)$ when n is large and x is close to zero. This is easy to understand because, letting $x = y/\sqrt{n}$, and keeping y constant, we obtain

$$\left(1 - \frac{y^2}{n}\right)^n = \exp(-\,y^2)\,\left[1 + O\left(n^{-1}\right)\right], \text{ as } n \to \infty\,.$$

This is a special case of Problem 16 at the end of Chapter 1.

In view of these approximations, we might seek to approximate an integral whose integrand is raised to a high power by the integral over an appropriately fitted bell curve. Then we can write

$$\begin{aligned}
\int_{-1}^{1} (1 - x^2)^n\,dx &= \frac{1}{\sqrt{n}} \int_{-\sqrt{n}}^{\sqrt{n}} \left(1 - \frac{y^2}{n}\right)^n dy \\
&= \frac{1}{\sqrt{n}} \int_{-\sqrt{n}}^{\sqrt{n}} \exp(-y^2)\,\left[1 + O\left(n^{-1}\right)\right]\,dy \\
&\sim \frac{1}{\sqrt{n}} \int_{-\infty}^{\infty} \exp(-y^2)\,\left[1 + O\left(n^{-1}\right)\right]\,dy \\
&= \sqrt{\frac{\pi}{n}}\,\left[1 + O\left(n^{-1}\right)\right].
\end{aligned}$$

The resulting approximation to the integral is the Laplace approximation, which is the subject of this chapter.

6.2 The basic approximation

One of our first tasks in this chapter is to determine the precise conditions under which the Laplace approximation holds. First of all, let us assume that $f(x) > 0$ over the region of integration. When this is the case, we can write $f(x) = \exp[h(x)]$, and apply a quadratic approximation to $h(x)$ about its maximum. As an initial exploration of these ideas, let us consider the approximation of an integral of the form

$$\int_{-\infty}^{\infty} e^{n\,h(x)}\,dx\,. \tag{6.1}$$

We would like to calculate this integral approximately, as $n \to \infty$. Suppose $h(x)$ attains its maximum at some point $x = x_0$. Under a suitable change of coordinates, we can make $x_0 = 0$, which we now assume. In addition, we shall assume that $h(0) = 0$. This restriction is not an important one from the point of view of our investigation. If it were not the case, we could remove a factor $e^{n\,h(0)}$ from the integral. Suppose $h(x)$ can be expressed locally about $x = 0$ by its Taylor expansion. Expanding out the function $h(x)$, we see that

$$\exp[n\,h(x)] = \exp\left[n\,h(0) + x\,n\,h'(0) + \frac{x^2\,n\,h''(0)}{2} + o(x^2)\right], \tag{6.2}$$

where the order term is understood to be as $x \to 0$, and not in n. By assumption, we have $h(0) = 0$. In addition, we also have $h'(0) = 0$ because $h(x)$ is maximised at $x = 0$. Additionally, since $x = 0$ is a maximum, we must have $h''(0) \le 0$. Typically, this second derivative will be strictly negative. If this is the case, we see that

$$\int_{-\infty}^{\infty} e^{n\,h(x)}\,dx = \int_{-\infty}^{\infty} \exp\left[\frac{x^2\,n\,h''(0)}{2} + o(x^2)\right]\,dx\,, \tag{6.3}$$

to lowest non-zero order about $x = 0$. Note that if we ignore the error term $o(x^2)$ in the expansion, this integral can be evaluated analytically. We can recognise the integrand as a density function for a normal random variable without the constant–a so-called "bell curve" mentioned in the previous section. The mean of this normal distribution is zero, and the variance is $-[n\,h''(0)]^{-1}$. This leads us to the following approximation, which is not yet fully justified, namely

$$\int_{-\infty}^{\infty} \exp[n\,h(x)]\,dx \quad \sim \quad \int_{-\infty}^{\infty} \exp\left[\frac{x^2\,n\,h''(0)}{2}\right]\,dx$$

$$= \sqrt{\frac{2\,\pi}{-n\,h''(0)}} \tag{6.4}$$

as $n \to \infty$.

A serious objection to our argument so far is that the quadratic approximation provided by Taylor's theorem is only valid locally about $x = 0$. On the other hand, an integral approximation requires a global approximation to the integrand where the error is uniformly small over the range of integration. However, as we shall prove more carefully later, asymptotically as $n \to \infty$, the dominant contribution to the value of the integral is provided by values of the integrand locally about its global maximum. More generally, we can prove the following result.

Proposition 1. *Let $-\infty \le a < b \le \infty$. Let $h(x)$ be defined on the open interval (a, b). Suppose*

1. *the function $h(x)$ is differentiable throughout (a, b), is uniquely maximised at some point $x_0 \in (a, b)$, and that $h''(x_0)$ exists and is strictly negative;*
2. *there exist constants $\eta > 0$ and $\delta > 0$ such that $h(x) < h(x_0) - \eta$ for all $x \in (a, b)$ such that $|x - x_0| \ge \delta$; and that*
3. *the integral below exists for $n = 1$.*

Then

$$\int_a^b e^{n\,h(x)}\,dx \sim e^{n\,h(x_0)}\sqrt{\frac{2\,\pi}{-n\,h''(x_0)}} \tag{6.5}$$

as $n \to \infty$.

We shall defer the proof of this result until a later stage. For the moment, we shall be content to state the relevant results and observe them in action. Proposition 1 also has an extension (which is not quite a generalisation) to the next result.

Proposition 2. *Suppose that the conditions of Proposition 1 hold. (The integrability condition of Proposition 1 requires appropriate modification to the absolute integrability of the integral below.) Suppose additionally that $g(x)$ is a continuous function defined on the open interval (a, b) such that $g(x_0) \neq 0$. Then*

$$\int_a^b e^{n\,h(x)}\,g(x)\,dx \sim e^{n\,h(x_0)}\,g(x_0)\sqrt{\frac{2\,\pi}{-n\,h''(x_0)}} \tag{6.6}$$

Pierre-Simon Laplace (1749–1827)

Astronomer and mathematician, Laplace made important contributions to probability theory as well as the approximations of integrals.

> "The theory of probabilities is at bottom nothing but common sense reduced to calculus; it enables us to appreciate with exactness that which accurate minds feel with a sort of instinct for which ofttimes they are unable to account."
>
> Laplace in his Introduction to *Théorie Analytique des Probabilités* (1812).

as $n \to \infty$.

Once again, we shall defer the proof. The right-hand sides of formulas (6.5) and (6.6) are both known as the Laplace approximation. The Laplace approximation is often known as the saddle-point approximation by physicists. However, statisticians have reserved the latter term for a related approximation which is useful for approximating the distribution of sums of independent random variables. This can sometimes lead to some confusion in terminology. However, the two approximations are related as we shall see later.

One dissatisfying aspect of the approximation in (6.6) is that no account is taken of the higher derivatives of $g(x)$ about $x = x_0$. To improve the approximation, we can expand the functions $g(x)$ and $h(x)$ in power series about the point $x = x_0$. Suppose we write

$$\begin{aligned} g(x) &= \alpha_0 + \alpha_1 (x - x_0) + \alpha_2 (x - x_0)^2 + \cdots \\ h(x) &= \beta_0 + \beta_2 (x - x_0)^2 + \beta_3 (x - x_0)^3 + \cdots \end{aligned}$$

where $\beta_1 = 0$ because $h(x)$ is maximised at x_0. We can write the required integral in the form

$$\int_a^b e^{n\,h(x)}\, g(x)\, dx = e^{n\,\beta_0} \int_a^b e^{n\,\beta_2\,(x-x_0)^2} \left\{ g(x)\, e^{n\,\beta_3\,(x-x_0)^3+\cdots} \right\} dx.$$

The next step is to expand the expression $\{\text{——}\}$ to obtain

$$g(x)\, e^{n\,\beta_3\,(x-x_0)^3+\cdots} = \left[\alpha_0 + \alpha_1(x - x_0) + \alpha_2(x - x_0)^2 + \cdots\right] \times$$

$$\left\{ 1 + \left[n\,\beta_3\,(x - x_0)^3 + \cdots\right] + \frac{\left[n\,\beta_3\,(x - x_0)^3 + \cdots\right]^2}{2} + \cdots \right\}.$$

While it is a rather tedious task, this expression can be expanded out. Powers of $x - x_0$ and powers of n appear in the expansion, but in association with each other, so that each term has the form $n^k\,(x - x_0)^{m+3\,k}$ times some coefficient which does not depend upon n and x. Indeed, each term in the expansion can be represented in this form for a unique choice of k and m. Thus we can write

$$g(x)\, e^{n\,\beta_3\,(x-x_0)^3+\cdots} = \sum_{k=0}^{\infty} \sum_{m=0}^{\infty} \gamma_{k,\,m} \left[n\,(x - x_0)^3\right]^k (x - x_0)^m,$$

and this expansion can be taken as defining a doubly indexed array of coefficients $\gamma_{k,\,m}$ which do not depend upon n and x. Later in this chapter, we shall derive the following.

Proposition 3. *Let* $-\infty \le a < b \le \infty$, *and let* $x_0 \in (a,\, b)$.

1. *Suppose there exists $\delta > 0$, where $a < x_0 - \delta < x_0 + \delta < b$ such that $g(x)$ and $h(x)$ are analytic functions on the interval*

$$(x_0 - \delta,\ x_0 + \delta)$$

whose power series about x_0 converge absolutely. By choosing δ sufficiently small, we can ensure that both power series converge on the closed interval $[x_0 - \delta,\ x_0 + \delta]$ as well.

2. *Suppose that $g(x_0) \neq 0$, $h'(x_0) = 0$ and $h''(x_0) < 0$. Let x_0 be the unique global maximum of the function $h(x)$.*

3. *Suppose that for every $\delta > 0$ and every integer $m > 0$,*

$$\int_a^{x_0 - \delta} e^{n\,h(x)}\, g(x)\, dx = e^{n\,h(x_0)}\, O(n^{-m}) \tag{6.7}$$

and

$$\int_{x_0 + \delta}^b e^{n\,h(x)}\, g(x)\, dx = e^{n\,h(x_0)}\, O(n^{-m}) \tag{6.8}$$

as $n \to \infty$.

Then the integral of Proposition 2 above can be expanded as an asymptotic series so that

$$\int_a^b e^{n\,h(x)}\, g(x)\, dx \quad \sim \quad \frac{\exp[n\,h(x_0)]}{\sqrt{n}} \left(a_0 + \frac{a_1}{n} + \frac{a_2}{n^2} + \frac{a_3}{n^3} + \cdots \right) \tag{6.9}$$

or equivalently

$$\int_a^b e^{n\,h(x)}\, g(x)\, dx = \frac{\exp[n\,h(x_0)]}{\sqrt{n}} \left[a_0 + \frac{a_1}{n} + \cdots + \frac{a_k}{n^k} + O\left(\frac{1}{n^{k+1}}\right) \right]$$

$$\text{for all } k \geq 0 \tag{6.10}$$

where

$$a_j = \sum_{k=0}^{2j} \gamma_{k,\,2j-k} \frac{\Gamma\left(j + k + \frac{1}{2}\right)}{(-\beta_2)^{j+k+1/2}}. \tag{6.11}$$

In particular, we have

$$\gamma_{0,\,0} = \alpha_0 = g(x_0)\,. \tag{6.12}$$

Therefore

$$a_0 = g(x_0) \sqrt{\frac{2\,\pi}{-h''(x_0)}} \tag{6.13}$$

as we found earlier. So formula (6.6) is obtained from the special case of

(6.9) where only the first term of the series is included. The next term in the asymptotic series evaluates to

$$a_1 = \sqrt{\frac{2\pi}{-h''}} \left[-\frac{g''}{2\,h''} + \frac{h'''\,g'}{2\,(h'')^2} + \frac{h^{(iv)}\,g}{8\,(h'')^2} - \frac{5\,(h''')^2\,g}{24\,(h'')^3} \right] \tag{6.14}$$

where the functions g, h and their derivatives are to be evaluated at x_0. The next coefficient is

$$a_2 = \sqrt{\frac{2\,\pi}{-h''}} \left[\frac{1}{8}\frac{g^{(iv)}}{(h'')^2} - \frac{1}{48}\frac{g\,h^{(vi)}}{(h'')^3} - \frac{5}{12}\frac{g'''\,h'''}{(h'')^3} - \frac{5}{16}\frac{g''\,h^{(iv)}}{(h'')^3} \right.$$
$$- \frac{1}{8}\frac{g'\,h^{(v)}}{(h'')^3} + \frac{35}{384}\frac{g\,(h^{(iv)})^2}{(h'')^4} + \frac{35}{48}\frac{g''\,(h''')^2}{(h'')^4} + \frac{7}{48}\frac{g\,h'''\,h^{(v)}}{(h'')^4}$$
$$\left. + \frac{35}{48}\frac{g'\,h'''\,h^{(iv)}}{(h'')^4} - \frac{35}{48}\frac{(h''')^3\,g'}{(h'')^5} - \frac{35}{64}\frac{(h''')^2\,g\,h^{(iv)}}{(h'')^5} + \frac{385}{1152}\frac{g\,(h''')^4}{(h'')^6} \right].$$

The verification of formulas (6.13) and (6.14) is left to the reader in Problems 5 and 6.

Beyond this, the formulas for the coefficients escalate in complexity.

6.3 The Stirling series for factorials

We can write the factorial function in integral form as

$$n! = \int_0^\infty x^n\,e^{-x}\,dx.$$

We might initially consider setting $h(x) = \ln(x)$ and $g(x) = e^{-x}$. However, this will not work because the function $h(x)$ is not bounded above. To apply the Laplace approximation, we need to make a substitution. Let $y = n^{-1}\,x - 1$ or equivalently, let $x = n\,(1+y)$. Then

$$\begin{aligned} n! &= \int_{-1}^\infty [n\,(1+y)]^n\,e^{-n\,(1+y)}\,n\,dy \\ &= n^{n+1}\,e^{-n} \int_{-1}^\infty \left[(1+y)\,e^{-y}\right]^n\,dy. \end{aligned} \tag{6.15}$$

Now $(1+y)\,e^{-y}$ has its maximum where

$$\frac{d}{dy}\left[(1+y)\,e^{-y}\right] = 0\,.$$

This is solved by $y = 0$. Thus it seems reasonable to define $h(y)$ by setting $e^{h(y)} = (1+y)\,e^{-y}$. Then $h(0) = 0$, $h'(0) = 0$, and $h''(0) = -1$.

Therefore

$$\int_{-1}^{\infty} \left[(1+y)\,e^{-y}\right]^n\,dy \quad \sim \quad \sqrt{\frac{2\,\pi}{-n\,h''(0)}}$$
$$= \quad \sqrt{\frac{2\,\pi}{n}},$$

from Proposition 1. Inserting this asymptotic formula into equation (6.15), we obtain

$$n! \;\sim\; \sqrt{2\,\pi\,n}\,n^n\,e^{-n} \tag{6.16}$$

which is a version of Stirling's approximation. Higher order corrections to this formula can be obtained by using Proposition 3. In particular, we can take more terms of the asymptotic series to get

$$n! \;\sim\; \sqrt{2\pi n}\,n^n\,e^{-n}\left[1+\frac{1}{12\,n}+\frac{1}{288\,n^2}-\frac{139}{51840\,n^3}-\frac{571}{2488320\,n^4}+\cdots\right] \tag{6.17}$$

which is known as the *Stirling series.* Problem 7 asks the reader to check that this expansion is correct up to the term of order n^{-2}, using the formulas for a_0, a_1, and a_2. Further expansion of this asymptotic series yields the series discussed in Problem 20 of Chapter 2. The Stirling series provides a highly accurate approximation to $n! = \Gamma(n+1)$ without the need for anything more powerful than a hand calculator.*

6.4 Laplace expansions in Maple

The calculations leading to Proposition 3 are straightforward to code in Maple. However, when the integral can be evaluated analytically, much of the intermediate work can be skipped by appealing to the command *asympt*. For example, the command

$> \; expand(simplify(asympt(\Gamma(n+1),\, n,\, 3)))$

* Legendre used this series to calculate the common logarithms of $\Gamma(x)$, $1 < x < 2$, to twelve decimal places. See Problem 20 at the end of Chapter 2. Although the Stirling series for $\Gamma(x)$ is not particularly accurate for small values of x, this problem can be resolved using a trick. For example, we can write

$$\Gamma(x+n+1) = (x+n)\,(x+n-1)\,\cdots\,x\,\Gamma(x).$$

For sufficiently large n, $\Gamma(x+n+1)$ can be approximated reasonably accurately using Stirling's series. From this, an approximation for the right-hand side and $\Gamma(x)$ can be obtained.

leads to the output

$$\frac{\sqrt{2}\,\sqrt{\pi}}{e^n\sqrt{\frac{1}{n}}\left(\frac{1}{n}\right)^n}+\frac{1}{12}\,\frac{\sqrt{2}\,\sqrt{\pi}}{n\,e^n\sqrt{\frac{1}{n}}\left(\frac{1}{n}\right)^n}+\frac{O\left(\frac{\sqrt{\frac{1}{n}}}{n}\right)}{e^n\left(\frac{1}{n}\right)^n}$$

which can certainly be recognised as the asymptotic expansion for $n! = \Gamma(n+1)$, albeit in less than tidy form. What looks like obtuseness on the part of the Maple routine is simply a conservative approach to simplification. Without any guarantee that n is positive, the Maple routine chooses to be very careful with principal branches. If the investigator can anticipate the format of the answer in simplest form, then additional tidying up is possible. For example, the command

$$> expand\left(simplify\left(asympt\left(\frac{\Gamma(n+1)}{\sqrt{2\cdot\pi\cdot n}\cdot n^n\cdot exp(-n)},\, n\right)\right)\right)$$

can be used, leading to the output

$$1+\frac{1}{12\,n}+\frac{1}{288\,n^2}-\frac{139}{51840\,n^3}-\frac{571}{2488320\,n^4}+O\left(\frac{1}{n^5}\right).$$

The symbolic manipulations invoked by the command *asympt* do not parallel the manipulations that led to Proposition 3. So, we should not expect *asympt* to provide us with general Laplace expansions of integrals which cannot be evaluated analytically. Attempts to do so will yield the error message

```
Error, (in asympt) unable to compute series
```

if Maple cannot evaluate the integral.

Formal calculations can be performed in Maple more directly using the *powseries* package.

6.5 Asymptotic bias of the median

Consider a random sample $X_1, X_2, \ldots, X_n$ drawn from a continuous distribution that has median θ, density function $f(x)$ and distribution function $F(x)$. For simplicity, we shall assume that the sample size is odd, so that we may write $n = 2\,m+1$ for some integer m. Let

$$X_{(1)} < X_{(2)} < \cdots < X_{(2\,m+1)}$$

denote the order statistics of the sample. The middle order statistic $M = X_{(m+1)}$ is the median of the sample, which is often used as an estimator for the median θ when this is unknown. Let us consider the problem of approximating the moments of the sample median. Let M have probability density function $f_M(x)$. The r-th moment of the median has integral representation

$$\begin{aligned} E(M^r) &= \int_{-\infty}^{\infty} x^r\, f_M(x)\, dx \\ &= \int_{-\infty}^{\infty} x^r\, \frac{(2m+1)!}{(m!)^2} f(x)\, \{F(x)\,[1-F(x)]\}^m\, dx \\ &= \frac{(2m+1)!}{(m!)^2} \int_{-\infty}^{\infty} \{x^r f(x)\}\, \{F(x)\,[1-F(x)]\}^m\, dx. \end{aligned} \tag{6.18}$$

The factorials involving m in expression (6.18) can be approximated by Stirling's approximation that we derived in the previous example. The other factor involving m in (6.18) is the integral. In this case, m appears as an exponent inside the integral. This suggests that we can use a Laplace approximation as $m \to \infty$. Using Stirling's approximation we obtain

$$\begin{aligned} \frac{(2\,m+1)!}{(m!)^2} &= (2\,m+1) \times \frac{(2\,m)!}{(m!)^2} \\ &= 2\,m \cdot \left[1 + \frac{1}{2m}\right] \times \frac{1}{\sqrt{\pi m}} \cdot 2^{2m} \cdot \left[1 - \frac{1}{8m} + O(m^{-2})\right] \\ &= 2^{2\,m+1} \cdot \sqrt{\frac{m}{\pi}} \cdot \left[1 + \frac{3}{8\,m} + O(m^{-2})\right]. \end{aligned} \tag{6.19}$$

Problem 8 asks the reader to verify these steps. A lower order approximation that we shall also use later is simply

$$\frac{(2\,m+1)!}{(m!)^2} = 2^{2\,m+1} \cdot \sqrt{\frac{m}{\pi}} \cdot \left[1 + O(m^{-1})\right]. \tag{6.20}$$

Formulas (6.19) and (6.20) provide the necessary asymptotic approximations to the combinatorial coefficient in (6.18).

Turning next to the integral in (6.18), let us set $g(x) = x^r f(x)$ and $h(x) = \ln F(x) + \ln[1 - F(x)]$. The function $h(x)$ is maximised at the point x_0 where $h'(x_0) = 0$. This is equivalent to the equation

$$\frac{f(x_0)}{F(x_0)} - \frac{f(x_0)}{1 - F(x_0)} = 0.$$

Assuming that $f(x_0) \neq 0$, the only solution to this equation is $F(x_0) = \frac{1}{2}$, which means that the maximum is obtained at the median $x_0 = \theta$.

Evaluating about the median, we find that $h(\theta) = -2 \cdot \ln 2$, $h'(\theta) = 0$ and

$$\begin{aligned} h''(\theta) &= -8\,[f(\theta)]^2\,, && (6.21)\\ g(\theta) &= \theta^r\, f(\theta)\,. && (6.22)\end{aligned}$$

Plugging the values for $h(\theta)$, $h''(\theta)$ and $g(\theta)$ into formula (6.6) for $r = 1$ gives

$$\int_{-\infty}^{\infty} \{x\, f(x)\}\, \{F(x)\,[1 - F(x)]\}^m\, dx = 2^{-(2\,m+1)} \cdot \theta \cdot \sqrt{\frac{\pi}{m}} \cdot [1 + O(m^{-1})]\,. \qquad (6.23)$$

We can combine formula (6.23) with (6.20) to obtain

$$\begin{aligned} E(M) &= \frac{(2m+1)!}{(m!)^2} \cdot \int_{-\infty}^{\infty} \{x\, f(x)\}\, \{F(x)\,[1 - F(x)]\}^m\, dx \\ &= \left\{2^{2m+1} \sqrt{\frac{m}{\pi}} \left[1 + O\left(\frac{1}{m}\right)\right]\right\} \cdot \left\{2^{-(2m+1)}\, \theta \sqrt{\frac{\pi}{m}} \left[1 + O\left(\frac{1}{m}\right)\right]\right\} \\ &= \theta\, \left[1 + O\left(m^{-1}\right)\right]. \end{aligned}$$

We can conclude from this that the bias of M is $O(m^{-1})$ as $m \to \infty$. On the face of it, this is a rather modest accomplishment for so much work. However, we can do better by calculating the next term in the Laplace expansion. After some algebra, we find that

$$\begin{aligned} h'''(\theta) &= -24\, f'(\theta)\, f(\theta) && (6.24)\\ h^{(iv)}(\theta) &= -32\, f''(\theta)\, f(\theta) - 192\,[f(\theta)]^4 - 24\,[f'(\theta)]^2 && (6.25)\\ g'(\theta) &= r\,\theta^{r-1}\, f(\theta) + \theta^r\, f'(\theta) && (6.26)\\ g''(\theta) &= r\,(r-1)\,\theta^{r-2}\, f(\theta) + 2\,r\,\theta^{r-1}\, f'(\theta) + \theta^r f''(\theta). && (6.27)\end{aligned}$$

Evaluating the Laplace expansions for the integral to the next order, and using (6.19), we can obtain

$$\begin{aligned} E(M) &= \theta - \frac{f'(\theta)}{16\,m\,[f(\theta)]^3} + O\left(\frac{1}{m^2}\right) \\ &= \theta - \frac{f'(\theta)}{8\,n\,[f(\theta)]^3} + O\left(\frac{1}{n^2}\right). && (6.28)\end{aligned}$$

Problem 9 asks the reader to verify this.

It is also interesting to compute the asymptotic variance of M. This can be indirectly obtained using $r = 2$ in formula (6.18) to find $E(M^2)$. As the asymptotic variance of M is of order n^{-1} we must use Proposition 3, and the expansion in (6.9), including both a_0 and a_1. Calculating the required terms of the expansions leads to the well known result that

$$\begin{aligned}\mathrm{Var}(M) &= \frac{1}{8\,m\,[f(\theta)]^2} + O\left(\frac{1}{m^2}\right)\\ &= \frac{1}{4\,n\,[f(\theta)]^2} + O\left(\frac{1}{n^2}\right).\end{aligned}\tag{6.29}$$

The details are left to the reader.

6.6 Recurrence properties of random walks

Suppose X_1, X_2, X_3, ... is a sequence of infinitely many independent identically distributed random variables with common characteristic function $\chi(t)$, $-\infty < t < \infty$. Let us assume that each X_j is a continuous random variable that is symmetric in distribution about zero. That is, X_j and $-X_j$ have the same distribution. Define $Y_0 = 0$, and

$$Y_n = X_1 + \cdots + X_n\,, \quad n = 1, 2, 3, \ldots. \tag{6.30}$$

Then the sequence Y_n, $n \geq 0$ is said to be a *symmetric random walk* starting at zero.

Now let $\epsilon > 0$ be given. We will say that the random walk Y_n, $n \geq 0$ has an *ϵ-return* to the origin at time n if $|Y_n| < \epsilon$. In this section, we shall consider the expected number of ϵ-returns to the origin, and whether this is finite or infinite. As is well known, there are only two possibilities: either the expected number of returns is finite for all $\epsilon > 0$, in which case the random walk is said to be *transient*, or the expected number of returns is infinite for all $\epsilon > 0$, in which case the random walk is said to be *recurrent*. Since the expected number of returns is given by

$$\sum_{n=1}^{\infty} P(-\epsilon < Y_n < \epsilon) \tag{6.31}$$

It can be shown that when a random walk is recurrent as defined here, then there will be infinitely many ϵ-returns to the origin with probability one. It is straightforward to show that when a random walk is transient the number of ϵ-returns to the origin will be finite with probability one.

The characteristic function of Y_n is $[\chi(t)]^n$. So, by Fourier inversion of the characteristic function of Y_n, the density function can be determined to be

$$f(y) = \frac{1}{2\,\pi} \int_{-\infty}^{\infty} e^{-i\,t\,y}\,[\chi(t)]^n\,dt\,. \tag{6.32}$$

Therefore

$$P(-\epsilon < Y_n < \epsilon) \;=\; \frac{1}{2\,\pi} \int_{-\epsilon}^{\epsilon} \int_{-\infty}^{\infty} e^{-i\,t\,y}\,[\chi(t)]^n\,dt\,dy$$

$$= \frac{1}{2\pi} \int_{-\infty}^{\infty} \int_{-\epsilon}^{\epsilon} e^{-ity} [\chi(t)]^n \, dy \, dt$$

$$= \frac{1}{2\pi} \int_{-\infty}^{\infty} \frac{e^{it\epsilon} - e^{-it\epsilon}}{it} [\chi(t)]^n \, dt \,. \qquad (6.33)$$

In this expression, the integrand is understood to be equal to 2ϵ when $t = 0$.

Since the distribution of each Y_j is symmetric about zero it follows that $\chi(t)$ is real-valued, and that the inversion integral reduces to

$$P(-\epsilon < Y_n < \epsilon) = \frac{1}{\pi} \int_{-\infty}^{\infty} \frac{\sin(t\,\epsilon)}{t} [\chi(t)]^n \, dt \,. \qquad (6.34)$$

The integral in (6.34) now has a real integrand and is in a general form to which the Laplace approximation is applicable. We set

$$g(t) = \frac{\sin(t\,\epsilon)}{t}, \quad g(0) = \epsilon \,,$$

and

$$h(t) = \ln \chi(t) \,.$$

Although this is undefined where $\chi(t)$ is negative, it is defined within a neighbourhood of $t = 0$. Moreover, since $\chi(t)$ is the characteristic function of a continuous distribution, it can be shown that

$$|\chi(t)| < \chi(0) = 1 \quad \text{for all } t \neq 0 \,. \qquad (6.35)$$

Therefore, $h(t)$ is maximised at $t = 0$ where $h(0) = 0$.

Provided $\sigma^2 = \mathrm{Var}(X_j) < \infty$, we will also have a second derivative for $h(t)$ at $t = 0$ with $h''(0) < -\sigma^2 < 0$. Applying the basic Laplace approximation to (6.34) we obtain

$$P(-\epsilon < Y_n < \epsilon) \sim \frac{\epsilon}{\sigma} \sqrt{\frac{2}{\pi\, n}} \,. \qquad (6.36)$$

This is precisely what the central limit theorem says it should be. However, this is a local central limit rather than the usual normal limit for the standardised distribution function.

So, the expected number of ϵ-returns to the origin for the random walk Y_n, $n \geq 0$ can be approximated by summing this expression over n. The quality of this approximation is unclear. However, the sum of the right-hand side over the values of $n \geq 1$ is divergent. So, the sum of the left-hand side will be divergent as well.†

The case where $\sigma^2 = \infty$ still remains to be considered. We can further

† This follows easily from the Limit Comparison Test for absolute convergence.

subdivide this into two cases, namely where X_j has finite expectation and where X_j has infinite expectation. When X_j has finite expectation, so that $E(X_j) = 0$, the random walk $Y_n, n \geq 0$ is recurrent with infinitely many ϵ-returns to the origin for all $\epsilon > 0$. See Feller (1971, pp. 200–204). When the expectation is infinite, however, the recurrence properties of the random walk are less straightforward.

6.7 Proofs of the main propositions

The proofs of Propositions 1, 2 and 3 have been separated out from the earlier discussion in order to focus on the results and their applications. Readers who wish to skip the derivations can omit this section without difficulty. Nevertheless, the regularity assumptions of the main propositions cannot be taken for granted in applications.

Proof of Proposition 1. The function $h(x)$ is assumed to be twice differentiable, and to attain its maximum in the interval (a, b) uniquely at the point x_0, where $h''(x_0) < 0$. It is sufficient to prove the proposition for the special case $x_0 = 0$, which we now assume. If this is not the case, a suitable change of coordinates for the variable of integration can be used.

It is also sufficient to prove the proposition for the case $h(0) = 0$. If this is not the case, then removing a factor $e^{n\,h(0)}$ from the integral reduces the integral to the required condition. Henceforth, we also assume that $h(0) = 0$, for simplicity.

We shall also restrict to the case where $a = -\infty$ and $b = \infty$, although this restriction is more consequential than the other ones. As we shall see, the Laplace method works by showing that the contribution of the integrand to the value of the integral is asymptotically negligible outside a neighbourhood of $x_0 = 0$. The proof below can be modified for the case where either a or b is finite without much difficulty.

Since $h(0) = 0$ is the unique maximum of $h(x)$, we must have $h(x) < 0$ for all $x \neq 0$. It follows from this that

$$e^{h(x)} \geq e^{2\,h(x)} \geq e^{3\,h(x)} \geq \cdots 0\,,$$

with equality when $x = 0$ and strict inequality when $x \neq 0$. By assumption, we also have

$$\int_{-\infty}^{\infty} e^{h(x)}\,dx < \infty\,. \tag{6.37}$$

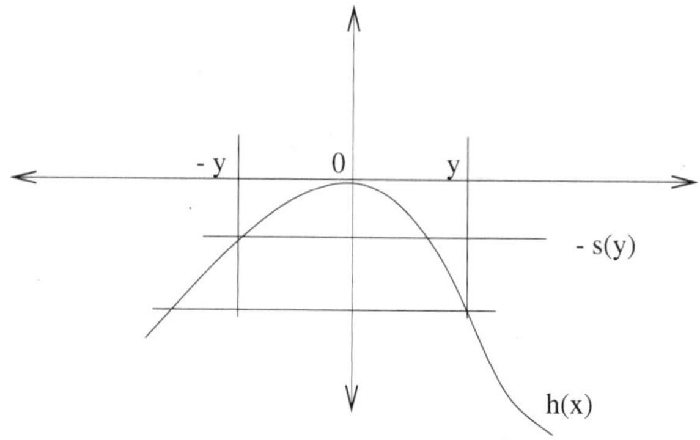

Figure 6.2 *The assumptions of Proposition 1 force the function $h(x)$ to be bounded away from 0 by a distance $s(y)$, for all values of x such that $|x| \geq y$.*

It follows from (6.37) that

$$\int_{-\infty}^{\infty} e^{n\,h(x)}\,dx < \infty \quad \text{for all } n \tag{6.38}$$

because the integrands in (6.38) are nonnegative and dominated by the integrand of (6.37).

Next, by assumption, there exist $\delta,\ \eta > 0$ such that $h(x) < -\eta$ whenever $|x| \geq \delta$. It follows from this that for all $y > 0$, there exists an $s(y) > 0$ such that

$$h(x) \leq -s(y) \quad \text{for all } x \text{ such that } |x| \geq y\,. \tag{6.39}$$

See Figure 6.2. Now, for any such $y > 0$, we can split our integral into three pieces, namely

$$\int_{-\infty}^{\infty} e^{n\,h(x)}\,dx \;=\; \underbrace{\int_{-\infty}^{-y} e^{n\,h(x)}\,dx}_{A} + \underbrace{\int_{-y}^{y} e^{n\,h(x)}\,dx}_{B} + \underbrace{\int_{y}^{\infty} e^{n\,h(x)}\,dx}_{C}\,. \tag{6.40}$$

Out task will be to show that integrals A and C are asymptotically negligible, with the integral B providing the required asymptotic formula. We have

$$\begin{aligned} A + C &= \int_{-\infty}^{-y} e^{n\,h(x)}\,dx + \int_{y}^{\infty} e^{n\,h(x)}\,dx \\ &= \int_{-\infty}^{-y} e^{(n-1)\,h(x)}\,e^{h(x)}\,dx + \int_{y}^{\infty} e^{(n-1)\,h(x)}\,e^{h(x)}\,dx \end{aligned}$$

$$
\begin{aligned}
&\le \int_{-\infty}^{-y} e^{-(n-1)\,s(y)}\, e^{h(x)}\, dx + \int_{y}^{\infty} e^{-(n-1)\,s(y)}\, e^{h(x)}\, dx \\
&= e^{-(n-1)\,s(y)} \left[\int_{-\infty}^{-y} e^{h(x)}\, dx + \int_{y}^{\infty} e^{h(x)}\, dx \right] \\
&< e^{-(n-1)\,s(y)} \int_{-\infty}^{\infty} e^{h(x)}\, dx = O\left(e^{-n\,\alpha}\right) \qquad (6.41)
\end{aligned}
$$

for some $\alpha > 0$.

To approximate integral B, we approximate $h(x)$ by $h''(0)/2$ in the interval $(-y,\, y)$. By Taylor's theorem

$$
\begin{aligned}
h(x) &= h(0) + x\,h'(0) + \frac{x^2\,h''(0)}{2} + o(x^2) \\
&= \frac{x^2\,h''(0)}{2} + o(x^2) \quad \text{as } x \to 0. \qquad (6.42)
\end{aligned}
$$

So for all $\epsilon > 0$, there is a $y > 0$ such that

$$
\left| h(x) - \frac{h''(0)\,x^2}{2} \right| \le \epsilon\,x^2 \quad \text{for all } -y \le x \le y\,. \qquad (6.43)
$$

Equivalently,

$$
\frac{h''(0)\,x^2}{2} - \epsilon\,x^2 \le h(x) \le \frac{h''(0)\,x^2}{2} + \epsilon\,x^2 \quad \text{for } -y \le x \le y\,.
$$

Therefore

$$
\int_{-y}^{y} e^{\frac{1}{2}\,n\,x^2\,[h''(0)-2\,\epsilon]}\, dx \le B \le \int_{-y}^{y} e^{\frac{1}{2}\,n\,x^2\,[h''(0)+2\,\epsilon]}\, dx\,. \qquad (6.44)
$$

Now in a way similar to (6.41),

$$
\int_{-y}^{y} e^{\frac{1}{2}\,n\,x^2\,[h''(0)-2\,\epsilon]}\, dx = \int_{-\infty}^{\infty} e^{\frac{1}{2}\,n\,x^2\,[h''(0)-2\,\epsilon]}\, dx + O(e^{-n\,\beta}) \qquad (6.45)
$$

for some $\beta > 0$, and

$$
\int_{-y}^{y} e^{\frac{1}{2}\,n\,x^2\,[h''(0)+2\,\epsilon]}\, dx = \int_{-\infty}^{\infty} e^{\frac{1}{2}\,n\,x^2\,[h''(0)+2\,\epsilon]}\, dx + O(e^{-n\,\gamma}) \qquad (6.46)
$$

for some $\gamma > 0$. Plugging (6.45), and (6.46) into (6.44), we obtain

$$
\begin{aligned}
&\int_{-\infty}^{\infty} e^{\frac{1}{2}\,n\,x^2\,[h''(0)-2\,\epsilon]}\, dx + O(e^{-n\,\beta}) \le B \\
&\qquad \le \int_{-\infty}^{\infty} e^{\frac{1}{2}\,n\,x^2\,[h''(0)+2\,\epsilon]}\, dx + O(e^{-n\,\gamma}). \qquad (6.47)
\end{aligned}
$$

The inequalities of (6.47) can then be combined with (6.41). Inserting both into (6.40) yields the bounds

$$\int_{-\infty}^{\infty} e^{\frac{1}{2}\,n\,x^2\,[h''(0)-2\,\epsilon]}\,dx + O(e^{-n\,\beta}) \leq \int_{-\infty}^{\infty} e^{n\,h(x)}\,dx + O(e^{-n\,\alpha})$$
$$\leq \int_{-\infty}^{\infty} e^{\frac{1}{2}\,n\,x^2\,[h''(0)+2\,\epsilon]}\,dx + O(e^{-n\,\gamma}). \tag{6.48}$$

The outermost integrals of (6.48) can now be integrated in closed form to give

$$\sqrt{\frac{2\,\pi}{n\,(-h''(0)+2\,\epsilon)}} + O(e^{-n\,\beta}) \leq \int_{-\infty}^{\infty} e^{n\,h(x)}\,dx + O(e^{-n\,\alpha})$$
$$\leq \sqrt{\frac{2\,\pi}{n\,(-h''(0)-2\,\epsilon)}} + O(e^{-n\,\gamma}). \tag{6.49}$$

Since ϵ is arbitrary, we can let it go to zero. Therefore, we obtain the desired asymptotic result

$$\int_{-\infty}^{\infty} e^{n\,h(x)}\,dx \sim \sqrt{\frac{2\,\pi}{-n\,h''(0)}} \tag{6.50}$$

as required. ■

Proof of Proposition 2. To prove the second proposition, we can follow the basic method used above, with the exception that where dx appears within the integral, we now have $g(x)\,dx$. As before, the integral is broken into three pieces, which can be labelled A, B, and C analogously to the integrals of the previous proof. The bounds on integrals A and C are much as before. To bound the integral B, it is necessary to simultaneously bound both

$$\left|h(x) - \frac{h''(0)\,x^2}{2}\right| \text{ and } |g(x)-g(0)|\,.$$

For all $\epsilon > 0$, we can choose $y > 0$ such that for $-y \leq x \leq y$,

$$\left|h(x) - \frac{h''(0)\,x^2}{2}\right| \leq \epsilon\,x^2 \text{ and } |g(x)-g(0)| \leq \epsilon\,. \tag{6.51}$$

Then the integral

$$B = \int_{-y}^{y} g(x)\,e^{n\,h(x)}\,dx$$

can be bounded on each side by integrals of the form

$$\int_{-y}^{y} e^{\frac{1}{2}\,n\,x^2\,[h''(0)\pm 2\,\epsilon]}\,[g(0)\pm\epsilon]\,dx.$$

Problem 10 asks the reader to fill in the details. ■

Proof of Proposition 3. The proof of Proposition 3 is far more technical than the previous two. The reader who is interested in studying the details will find a terse but rigorous proof in de Bruijn (1981).

6.8 Integrals with the maximum on the boundary

An important case that is not covered by the theory so far occurs when the maximum of $h(x)$ is on the boundary of the interval of integration. The following result can be proved.

Proposition 4. *Suppose that the following conditions hold:*

1. *We have $h(x) < h(a)$ for all $a < x < b$, and for all $\delta > 0$*

$$\inf\{h(a)-h(x)\ :\ x\in[a+\delta,\,b)\} > 0\,.$$

2. *The functions $h'(x)$ and $g(x)$ are continuous in a neighbourhood of $x = a$. Furthermore, the functions $h(x)$ and $g(x)$ have power series expansions about $x = a$ of the form*

$$\begin{aligned} g(x) &= \alpha_0+\alpha_1\,(x-a)+\alpha_2\,(x-a)^2+\cdots, \qquad \alpha_0\neq 0,\\ h(x) &= \beta_0+\beta_1\,(x-a)+\beta_2\,(x-a)^2+\cdots, \qquad \beta_1\neq 0. \end{aligned}$$

3. *The following integral converges absolutely for sufficiently large n.*

Then there exist coefficients a_n, $n\geq 1$ such that

$$\int_a^b g(x)\,e^{n\,h(x)}\,dx \;\sim\; e^{n\,h(a)}\left\{\frac{a_1}{n}+\frac{a_2}{n^2}+\frac{a_3}{n^3}+\cdots\right\}. \tag{6.52}$$

Furthermore, the formulas for the first two coefficients are

$$a_1 = -\frac{\alpha_0}{\beta_1},\qquad a_2=\frac{1}{\beta_1^2}\left[\alpha_1-\frac{2\,\beta_2\,\alpha_0}{\beta_1}\right].$$

Proof. We shall only sketch an outline of the proof here, and shall leave the details to other sources. In particular, we shall skip over the role of the regularity conditions in Proposition 4, as well as the proof that the series is asymptotic. The reader who wishes to see the details of this should consult Wong (2001). Here we shall concentrate on the formal derivation of the expansion for the particular case where the integral has the form

$$\int_0^\infty g(x)\, e^{n\, h(x)}\, dx\,.$$

so that $a = 0$ and $b = \infty$. The terms of the expansion can be obtained from the method of integration by parts which was introduced in Chapter 2. We write

$$g_1(x) = g(x)\, e^{n\beta_2 x^2 + n\beta_3 x^3 + \cdots}\,.$$

Then

$$\begin{aligned}
\int_0^\infty g(x) e^{nh(x)} dx &= e^{n\beta_0} \int_0^\infty g_1(x) e^{n\beta_1 x} dx \\
&= e^{n\beta_0} \left\{ \left[\frac{g_1(x) e^{n\beta_1 x}}{n\beta_1} \right]_0^\infty - \frac{1}{n\beta_1} \int_0^\infty g_1'(x) e^{n\beta_1 x} dx \right\} \\
&= e^{n\beta_0} \left\{ -\frac{\alpha_0}{n\beta_1} - \frac{1}{n\beta_1} \int_0^\infty g_1'(x) e^{n\beta_1 x} dx \right\}
\end{aligned}$$

Next we let $g_2(x) = g_1'(x)$, and proceed as before, evaluating the integral involving $g_2(x)$ using integration by parts. Evaluation of this second term to emerge gives us

$$\int_0^\infty g_2(x)\, e^{n\beta_1 x}\, dx = -\frac{\alpha_1}{n\,\beta_1} - \frac{1}{n\,\beta_1} \int_0^\infty g_2'(x)\, e^{n\beta_1 x}\, dx.$$

We set $g_3(x) = g_2'(x)$, and proceed to get

$$\int_0^\infty g_3(x)\, e^{n\beta_1 x}\, dx = -\frac{2\,\alpha_2 + 2\,n\,\alpha_0\,\beta_2}{n\,\beta_1} - \frac{1}{n\,\beta_1} \int_0^\infty g_3'(x)\, e^{n\beta_1 x}\, dx$$

and so on. Successive uses of integration by parts yield the higher order terms. Note that evaluating the series to order $O(n^{-m})$ will generally require more than m successive uses of integration by parts. ■

It is left to the reader to find a formula for a_3, the next coefficient in the series.

6.9 Integrals of higher dimension

The basic method behind the Laplace approximation extends to multiple integrals in higher dimensions. In such cases, we seek an asymptotic expression or series for a k-fold integral of the form

$$\int \cdots \int_{\mathcal{D}} g(x_1, \ldots, x_k)\, e^{n h(x_1, \ldots, x_k)}\, dx_1 \cdots dx_k\,, \tag{6.53}$$

where $\mathcal{D}$ is some domain in $\mathbb{R}^k$, and the function $h(x_1, \ldots, x_k)$ is maximised in the interior of $\mathcal{D}$.

We shall suppose that h has a unique maximum at

$$x_0 = (x_{01},\, x_{02},\, \ldots,\, x_{0k})\,.$$

and that all second derivatives of h exist and are continuous in a neighbourhood of the origin. For convenience, we shall write $x = (x_1, \ldots, x_k)$, and $dx = dx_1 \cdots dx_k$, and so on.

Then the extension of the Laplace approximation formula to the integral in (6.53) is

$$\int_{\mathcal{D}}^{[k]} g(x)\, e^{n h(x)}\, dx \sim \left(\frac{2\,\pi}{n}\right)^{k/2} g(x_0)\, (-\det H)^{-1/2}\, e^{n h(x_0)} \tag{6.54}$$

where H is the $k \times k$ Hessian matrix defined as

$$H = \left(\frac{\partial^2 h}{\partial x_j\, \partial x_m}\right)_{x=x_0}\,. \tag{6.55}$$

If $g(x)$ and $h(x)$ have power series expansions about x_0, this formula can be extended to an asymptotic series expansion of the form

$$\int_{\mathcal{D}}^{[k]} g(x)\, e^{n h(x)}\, dx \sim \frac{e^{n h(x_0)}}{n^{k/2}} \left(a_0 + \frac{a_1}{n} + \frac{a_2}{n^2} + \cdots\right). \tag{6.56}$$

The coefficients in this expansion increase in complexity rapidly in both n and k. So, unless the integrand has some symmetry, the direct evaluation of the coefficients in terms of the power series expansions of g and h will be quite laborious. The value of a_0 is determined by formula (6.54) above. We shall postpone the calculation of a_1 until the next section where we shall consider a generalisation of the basic result to integrals of products of functions.

A useful alternative to this approach is to convert the k-fold integral into a one-dimensional integral. To do this, we write

$$F(y) = \int_{\mathcal{D} \cap \mathcal{E}(y)}^{[k]} g(x)\, dx$$

where

$$\mathcal{E}(y) = \{x \;:\; h(x) - h(x_0) \geq -y^2/2\}\,.$$

Let $f(y) = F'(y)$. We can write‡

$$\begin{aligned} \int_{\mathcal{D}}^{[k]} g(x)\, e^{nh(x)}\, dx &= e^{n\,h(x_0)} \int_0^\infty e^{-ny^2/2}\, dF(y) \\ &= e^{n\,h(x_0)} \int_0^\infty f(x)\, e^{-n\,y^2/2}\, dy\,. \end{aligned} \tag{6.57}$$

The resulting integral is similar to the family of integrals studied in Propositions 2–4. The main difference is that

$$F(y) \sim y^k\, (-\det H)^{-1/2}\, g(x_0)\, V_k$$

as $y \searrow 0$, where V_k is the volume of the unit ball in k dimensions. Also

$$f(y) \sim k\, y^{k-1}\, (-\det H)^{-1/2}\, g(x_0)\, V_k\,.$$

The leading term of the asymptotic expansion of (6.53) becomes

$$e^{n\,h(x_0)}\, k\, (-\det H)^{-1/2}\, g(x_0)\, V_k \int_0^\infty y^{k-1}\, e^{-n\,y^2/2}\, dy\,.$$

This reduces to

$$e^{n\,h(x_0)}\, k\, (-\det H)^{-1/2}\, g(x_0)\, V_k \cdot 2^{(k-2)/2}\, \Gamma\left(\frac{k}{2}\right) n^{-k/2}\,.$$

Since

$$V_k = \frac{\pi^{k/2}}{\Gamma\left(\frac{k}{2}+1\right)}$$

this reduces to the expression that we obtained earlier in (6.54).

One final warning is in order. Asymptotic expansions in which $k \to \infty$ as well as $n \to \infty$ will have a different character altogether. Generally speaking, asymptotic expansions of high-dimensional integrals will require more terms to achieve the same accuracy in approximation as dimension one. For example, if the k-fold integral were to factorise into the product of k one-dimensional identical integrals, then the relative error in the first term would be of order

$$\begin{aligned} [1 + O(n^{-1})]^k - 1 &= [1 + k\, O(n^{-1})] - 1 \\ &= O\left(\frac{k}{n}\right). \end{aligned}$$

‡ Readers will note that this identity is essentially the same idea as the theorem in mathematical statistics that is commonly known as the "law of the unconscious statistician."

6.10 Integrals with product integrands

The integral of a function which is raised to some high power can be considered as a special case of the integral of a product of functions. In this section, we shall consider a generalisation of the Laplace method to integrals of products. In probability theory, such integrals are commonplace in the calculation of marginal distributions of random variables which are functions of data involving independent variates such as arise from random sampling. The integrals are of particular interest in Bayesian statistics, where they arise when calculating posterior expectations or other Bayes estimates.

Let us consider an integral of the form

$$\begin{aligned} \int_a^b g(x) f_1(x)\, f_2(x) \cdots f_n(x)\, dx &= \int_a^b g(x) \exp\left[\sum_{j=1}^n h_j(x)\right] dx \\ &= \int_a^b g(x) \exp\left[n\,\overline{h}(x)\right]\, dx \quad (6.58) \end{aligned}$$

where $h_j(x) = \ln f_j(x)$, and

$$\overline{h}(x) = n^{-1} \sum_{j=1}^n h_j(x)\,.$$

The case we have considered previously is now the special case where $f_j(x) = f(x)$ for all j. Superficially, our integral here does not look that different from the previous ones. The function $h(x)$ considered previously has been replaced by $\overline{h}(x)$. So, we might wonder if all the formulas derived earlier will work if we substitute $\overline{h}(x)$ for $h(x)$ in the earlier formulas. Such a hope will turn out to be too optimistic. However, there will be some important cases where the Laplace method works when the integral has a product of functions.

Crucial to the success of the Laplace method in earlier formulas is that the integrand is approximately shaped as a "bell curve." Figure 6.3 illustrates how the product of five functions can–to a greater or lesser degree–be bell-shaped. In each of four graphs, the five individual functions are plotted as dashed lines and the product of the five functions as an unbroken line. In the graphs shown, the major effect on the shape of the product is determined by the amount of separation of the component functions in the product. In the lower right-hand corner, the separation is zero, and all the functions are identical. As expected, the product of these functions is reasonably bell-shaped. However, in the top left, the functions are widely separated, and the resulting product is much less bell-shaped.

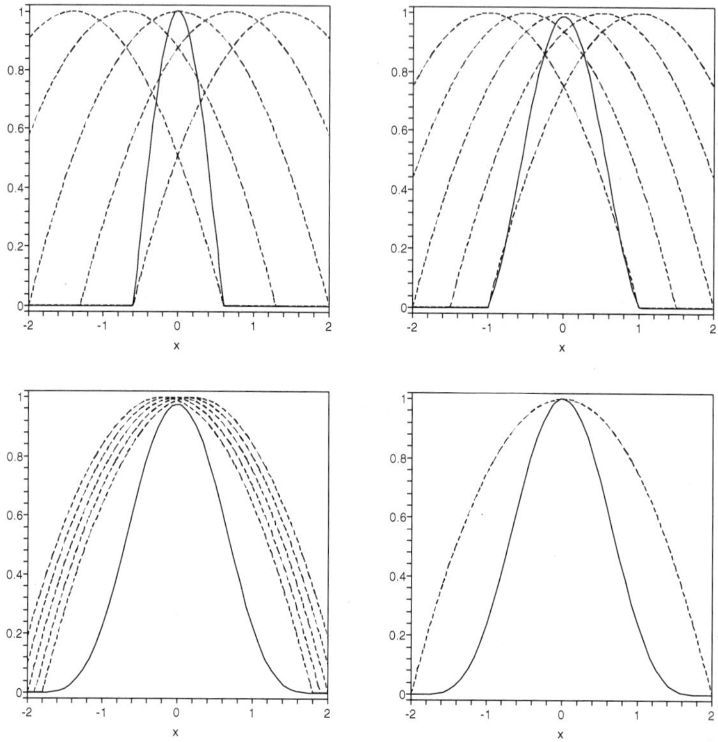

Figure 6.3 *The effect of separation on the bell-shaped property of a product of five functions*

Kass *et al.* (1990) provide sufficient conditions for an asymptotic expansion to order $O(n^{-1})$ for higher dimensional integrals. Like other results described earlier, their method depends upon the ability to partition the integral. Let $x = (x_1, \ldots, x_k)$, and let $\overline{h}(x)$ be maximised at $x = x_0$ which can depend upon n. Let $\mathcal{B}(x_0, \epsilon)$ denote the ball of radius ϵ in $\mathcal{D}$ that is centred at x_0. We partition the integral as

$$\int_{\mathcal{D}}^{[k]} g(x)\, e^{n\overline{h}(x)}\, dx = \underbrace{\int_{\mathcal{B}(x_0,\delta)}^{[k]} g(x)\, e^{n\overline{h}(x)}\, dx}_{A} + \underbrace{\int_{\mathcal{D}-\mathcal{B}(x_0,\delta)}^{[k]} g(x)\, e^{n\overline{h}(x)}\, dx}_{B}$$

where δ is chosen so that the integral B is negligible, and the integrand in A can be closely approximated by a bell-shaped function centred at

x_0. Let

$$H = \left(\frac{\partial^2 \overline{h}}{\partial x_j \, \partial x_m} \right)_{x=x_0} = (\overline{h}_{jm})$$

be the matrix of second partials of $\overline{h}(x)$ at $x = x_0$, and let the jm-th entry of the matrix H^{-1} be denoted by $\overline{h}^{jm}$. Let g_{jm} denote the mixed partial derivative of g with respect to x_j and x_m evaluated at x_0. Similar conventions apply to other derivatives. For example, $\overline{h}_{jms}$ denotes the third mixed partial derivative of $\overline{h}(x)$, and so on. Finally, let

$$Y = (Y_1, Y_2, \ldots, Y_k)$$

denote a random vector having a multivariate normal distribution centred at zero with covariance matrix $-H^{-1}$, and with product moments§

$$\begin{aligned} \sigma^2_{ij} &= E(Y_i \, Y_j) \\ &= -\overline{h}_{ij} \, , \end{aligned}$$

$$\begin{aligned} \sigma^4_{ijms} &= E(Y_i \, Y_j \, Y_m \, Y_s) \\ &= \overline{h}^{ij} \, \overline{h}^{ms} + \overline{h}^{im} \, \overline{h}^{js} + \overline{h}^{is} \, \overline{h}^{jm} \, , \end{aligned}$$

$$\begin{aligned} \sigma^6_{ijmqrs} &= E(Y_i \, Y_j \, Y_m \, Y_q \, Y_r \, Y_s) \\ &= -\overline{h}^{ij} \, \overline{h}^{mq} \, \overline{h}^{rs} - \overline{h}^{ij} \, \overline{h}^{mr} \, \overline{h}^{qs} - \overline{h}^{ij} \, \overline{h}^{ms} \, \overline{h}^{qr} \\ &\quad -\overline{h}^{im} \, \overline{h}^{jq} \, \overline{h}^{rs} - \overline{h}^{im} \, \overline{h}^{jr} \, \overline{h}^{qs} - \overline{h}^{im} \, \overline{h}^{js} \, \overline{h}^{qr} \\ &\quad -\overline{h}^{iq} \, \overline{h}^{jm} \, \overline{h}^{rs} - \overline{h}^{iq} \, \overline{h}^{jr} \, \overline{h}^{ms} - \overline{h}^{iq} \, \overline{h}^{js} \, \overline{h}^{mr} \\ &\quad -\overline{h}^{ir} \, \overline{h}^{jm} \, \overline{h}^{qs} - \overline{h}^{ir} \, \overline{h}^{jq} \, \overline{h}^{ms} - \overline{h}^{ir} \, \overline{h}^{js} \, \overline{h}^{mq} \\ &\quad -\overline{h}^{is} \, \overline{h}^{jm} \, \overline{h}^{qr} - \overline{h}^{is} \, \overline{h}^{jq} \, \overline{h}^{mr} - \overline{h}^{is} \, \overline{h}^{jr} \, \overline{h}^{mq} \, . \end{aligned}$$

Kass *et al.* (1990) proved the following.

Proposition 5. *Let $\mathcal{D}$ be an open subset of $\mathbb{R}^k$. Assume that the following conditions hold.*

1. *The function $g(x)$ is four times continuously differentiable, and the functions $h_j(x)$ are six times continuously differentiable in $\mathcal{D}$.*

§ The reader can prove the formulas for the following moments by successively differentiating the multivariate moment generating function of Y and evaluating at zero. The formula for σ^4_{ijms} has a term for every partition of a set of four items into subsets of size two. Thus there are three terms. Correspondingly, the formula for σ^6_{ijmqrs} has fifteen terms, a term for every partition of six items into three subsets each of size two. Note, for example, that the formula for σ^6_{111111} has fifteen terms, all of which are equal.

2. *The function $\overline{h}(x)$ is maximised at x_0, which may depend upon n.*

3. *There exist positive numbers ϵ, M and η such that, for sufficiently large values of n, we have*

   (a)
$$\left|\frac{\partial^d \overline{h}}{\partial x_{j_1}\cdots\partial x_{j_d}}(x)\right| < M$$
   for all $1 \le d \le 6$ and all $1 \le j_1, \ldots, j_d \le k$, and

   (b)
$$\det(H) > \eta.$$

   (c) For all $0 < \delta < \epsilon$, we have $\mathcal{B}(x_0, \delta) \subset \mathcal{D}$, and

   (d)
$$\sqrt{\det(n\,H)}\cdot\int_{\mathcal{D}-\mathcal{B}(x_0,\delta)}^{[k]} g(x)\,e^{n\overline{h}(x)-n\overline{h}(x_0)}\,dx = O(n^{-2})\,.$$

4. *The integral below exists.*

Then

$$\int_{\mathcal{D}}^{[k]} g(x)\,e^{n\overline{h}(x)}\,dx = \sqrt{\frac{(2\,\pi)^k}{-\det(n\,H)}}\cdot e^{n\overline{h}(x_0)}\cdot\left[a_0 + \frac{a_1}{n} + O\left(\frac{1}{n^2}\right)\right],$$

where

$$a_0 = g\,,$$

$$\begin{aligned} a_1 &= \frac{1}{2}\sum_{ij} g_{ij}\,\sigma^2_{ij} + \frac{1}{6}\sum_{ijms}\overline{h}_{ijm}\,g_s\,\sigma^4_{ijms} \\ &\quad + \frac{g}{24}\sum_{ijms}\overline{h}_{ijms}\,\sigma^4_{ijms} + \frac{g}{72}\sum_{ijmqrs}\overline{h}_{ijm}\,\overline{h}_{qrs}\,\sigma^6_{ijmqrs}. \end{aligned}$$

and all functions and derivatives are evaluated at x_0.¶

Proof. See Kass *et al.* (1990). ∎

Note that the determinant in this asymptotic formula is a multilinear function. So $\det(n\,H) = n^k \det(H)$. Therefore, the order of the leading term is $O(n^{-k/2})$.

¶ In comparing Kass *et al.* (1990) with the formula displayed here, the reader should note that the function $\overline{h}$ in the text here is $-h$ in the notation of Kass *et al.* (1990).

6.11 Applications to statistical inference

In practice, it may be very hard to verify the conditions of Proposition 5, particularly when the functions $h_1, h_2, \ldots$ are generated by some random mechanism. For example, in Bayesian statistics it is often necessary to calculate density functions which are represented as integrals of the form

$$\int_{\Theta}^{[k]} g(\theta) \prod_{j=1}^{n} f(x_j;\, \theta)\, d\theta$$

where Θ is a space of parameters $\theta \in I\!R^k$, and $x_1, \ldots, x_n$ are the observed values of a set of independent identically distributed random variables or vectors with common density function $f(x;\, \theta)$.

Let $X_1, X_2, X_3, \ldots$ be a sequence of independent identically distributed random vectors taking values in some space with common density function $f(x;\, \theta)$, where θ lies in some open subset Θ of $I\!R^k$. Let

$$\overline{\ell}_n(\theta) = n^{-1} \sum_{j=1}^{n} \ln\, f(x_j;\, \theta)\,. \tag{6.59}$$

The function $\overline{\ell}_n(\theta)$ is called the *averaged log-likelihood function* for θ. The averaged log-likelihood plays the same role in our integral that $h(x)$ did earlier. Here we adopt standard Bayesian statistical notation in the context of random sampling. The *maximum likelihood estimate* for θ is the value which maximises the averaged log-likelihood. Thus

$$\widehat{\theta}_n = \arg\max_{\theta \in \Theta} \overline{\ell}_n(\theta)\,. \tag{6.60}$$

In the context of Laplace's method, the estimator $\widehat{\theta}_n$ is also the central value about which the approximation of the integrand is performed. The next result, which links the work of Johnson (1967, 1970) to Proposition 5, is due to Kass *et al.* (1990).

Proposition 6. *We assume the following.*

1. *For any $\theta \neq \theta'$ in Θ, there exists an x such that $f(x;\theta) \neq f(x;\theta')$ (i.e., θ is identifiable).*
2. *For each x, the function $f(x;\theta)$ is a six times continuously differentiable function of θ.*
3. *For all $\theta^\star \in \Theta$, there exist a neighbourhood $\mathcal{N}_1(\theta^\star)$ of $\theta^\star$, a positive integer n, and a random variable Y_1 such that*

$$E_{\theta^\star}(Y_1) < \infty\,,$$

and for all $\theta \in \mathcal{N}_1(\theta^\star)$,

$$\overline{\ell}_n(\theta) - \overline{\ell}_n(\theta^\star) < Y_1 \,.$$

4. *For all $\theta^\star \in \Theta$, there exist a neighbourhood $\mathcal{N}_2(\theta^\star)$ of $\theta^\star$, and a random variable Y_2 such that*

$$E_{\theta^\star}(Y_2) < \infty$$

and for all $\theta \in \mathcal{N}_2(\theta^\star)$, and all $1 \le d \le 6$, and all $1 \le j_1, \ldots, j_d \le k$, we have

$$\left| \frac{\partial^d \ln f(X_1; \theta)}{\partial\theta_{j_1} \cdots \partial\theta_{j_d}} \right| < Y_2 \,.$$

5. *For each $\theta^\star \in \Theta$, define M to be the $k \times k$ matrix whose j, m-th entry is*

$$M_{jm} = \frac{\partial^2}{\partial\theta_j \, \partial\theta_m} E_{\theta^\star} \left[\ln \frac{f(x; \theta^\star)}{f(x; \theta)} \right] .$$

Then $\det(M) > 0$.

6. *For all $\theta^\star \in \Theta$ the maximum likelihood estimate $\widehat{\theta}_n$ is a strongly consistent estimate for $\theta^\star$. In other words,*

$$P_{\theta^\star} \left(\lim_n \widehat{\theta}_n = \theta^\star \right) = 1 \,.$$

Suppose also that $g(\theta)$ is four times continuously differentiable. Then under the assumptions listed above, the integral

$$\int_\Theta^{[k]} g(\theta) \, e^{n\overline{\ell}_n(\theta)} \, d\theta$$

admits an asymptotic expansion as in Proposition 5, with $\overline{h}(x)$ replaced by $\overline{\ell}(\theta)$, and x_0 replaced by $\widehat{\theta}_n$.

Proof. See Kass *et al.* (1990). ∎

6.12 Estimating location parameters

In this section, we shall consider an application of the methods of the previous section to computing efficient equivariant estimates.

Let $X_1, \ldots, X_n$ denote a random sample of independent identically distributed random variables generated from a distribution with density $f(x - \theta)$. We consider the task of estimating θ based upon the sample $X_1 = x_1, \ldots, X_n = x_n$. Although estimation of θ can sometimes be problematic, there is an estimate which has uniformly minimum variance among all location equivariant estimates.|| This is the *Pitman estimator*, which has the form

$$t_n(x_1, \ldots, x_n) = \frac{\displaystyle\int_{-\infty}^{\infty} \theta \prod_{j=1}^{n} f(x_j - \theta)\, d\theta}{\displaystyle\int_{-\infty}^{\infty} \prod_{j=1}^{n} f(x_j - \theta)\, d\theta}. \tag{6.61}$$

Despite having a uniformly smaller variance than other location equivariant estimators, the Pitman estimator has not had the popularity of its rivals such that the maximum likelihood estimate for θ. The reason for this is the difficulty in evaluating the integrals in both numerator and denominator. In many cases, these integrals have to be evaluated numerically.

An obvious alternative is to apply the Laplace expansion to the integrals in both the numerator and denominator. We set

$$\overline{\ell}_n(\theta) = \frac{1}{n} \sum_{j=1}^{n} \ln f(x_j - \theta).$$

Let $\widehat{\theta}_n$ denote the maximum likelihood estimate for θ. We can apply a Laplace approximation to the numerator and denominator of the Pitman estimator. In particular, using formulas (6.13) and (6.14) in both the numerator and the denominator. For the numerator, we replace the function g by the identity function and h by $\overline{\ell}$. In the denominator, we set g to be the constant function one, and again h to be $\overline{\ell}$. Let a_0 and a_1 be the coefficients for the numerator, and b_0 and b_1 the respective coefficients for the denominator. Then

$$\begin{aligned} t_n &= \frac{\displaystyle\int_{-\infty}^{\infty} \theta\, e^{n\,\overline{\ell}_n(\theta)}\, d\theta}{\displaystyle\int_{-\infty}^{\infty} e^{n\,\overline{\ell}_n(\theta)}\, d\theta} \\ &= \frac{e^{n\,\overline{\ell}_n(\widehat{\theta}_n)} \cdot \left[a_0 + a_1\, n^{-1} + O\left(n^{-2}\right)\right]}{e^{n\,\overline{\ell}_n(\widehat{\theta}_n)} \cdot \left[b_0 + b_1\, n^{-1} + O\left(n^{-2}\right)\right]} \\ &= \widehat{\theta}_n + \frac{\overline{\ell}_n'''(\widehat{\theta}_n)}{2\,n\,[\overline{\ell}_n''(\widehat{\theta}_n)]^2} + O(n^{-2}). \end{aligned} \tag{6.62}$$

The simplicity of the term of order n^{-1} compared to the coefficient in

|| An estimator $t(x_1, \ldots, x_n)$ is said to be location equivariant if $t(x_1+a, \ldots x_n+a) = t(x_1, \ldots, x_n) + a$ for all $a, x_1, \ldots, x_n \in \mathbb{R}$.

(6.14) arises because there is substantial cancellation of terms at this order.

This result is not simply of interest as an approximation to the Pitman estimator. It tells us that asymptotically, the Pitman estimator can be regarded as a perturbed maximum likelihood estimate, with the size of the perturbation determined by the second and third derivatives of the log-likelihood function. In particular, if the log-likelihood is symmetric about $\widehat{\theta}_n$, then this term will vanish.

6.13 Asymptotic analysis of Bayes estimators

The Pitman estimator of the previous section is a limiting case of a family of Bayesian methods. Suppose that $\alpha(\theta)$ is a prior probability density function defined on a parameter space Θ, and that Θ is an open subset of $\mathbb{R}^k$. Suppose we wish to estimate some real-valued function of the parameter, say $\beta(\theta)$. Within the Bayes framework, a natural choice for this is the *posterior mean* of $\beta(\theta)$ given a random sample $X_1 = x_1, \ldots, X_n = x_n$. This posterior expectation has integral representation

$$E\left[\beta(\theta) \mid x_1, x_2, \ldots, x_n\right] = \frac{\displaystyle\int_\Theta^{[k]} \beta(\theta) \prod_{j=1}^n f(x_j - \theta)\, \alpha(\theta)\, d\theta}{\displaystyle\int_\Theta^{[k]} \prod_{j=1}^n f(x_j - \theta)\, \alpha(\theta)\, d\theta}. \tag{6.63}$$

Note that the Pitman esimator for location is the special case where $\Theta = \mathbb{R}$, $\beta(\theta) = \theta$, and where $\alpha(\theta) = 1$ is an improper uniform prior on all of $\mathbb{R}$.

By setting $g(\theta) = \alpha(\theta)\,\beta(\theta)$ in the numerator of (6.63), we can apply the Laplace method as earlier. In the denominator, we set $g(\theta) = \alpha(\theta)$. Upon expansion of the integrals, it can be shown that

$$E[\beta(\theta) \mid x_1, \ldots, x_n] = \beta(\widehat{\theta}_n) + a_1\, n^{-1} + O(n^{-2}), \tag{6.64}$$

where, once again, $\widehat{\theta}_n$ is the maximum likelihood estimate for θ, and

$$a_1 = \frac{1}{2} \sum_{ij} \beta_{ij}\, \sigma_{ij}^2 + \sum_{ij} \alpha^{-1}\, \alpha_i\, \beta_j\, \sigma_{ij}^2 + \frac{1}{6} \sum_{ijms} \beta_s\, \overline{\ell}_{ijm}\, \sigma_{ijms}^4. \tag{6.65}$$

Standard conventions from Proposition 5 apply in this formula. For example, $\overline{\ell}_{ijm}$ represents the mixed partial derivative of $\overline{\ell}(\theta)$ with respect to components θ_i, θ_j and θ_m, and so on. All functions are evaluated at $\widehat{\theta}_n$.

6.14 Notes

The classic monograph on the asymptotic expansions of integrals, including the Laplace approximation, is Erdélyi (1956). De Bruijn (1981) also has an excellent summary of the essential results. For a more complete coverage of the subject, see Wong (2001). None of these sources considers statistical applications directly.

For the applications to problems in statistics, the reader may wish to refer to the brief exposition given by Barndorff-Nielsen and Cox (1989), and the papers by Kass *et al.* (1988, 1990).

6.15 Problems

1. In this problem, we evaluate the asymptotic form of the even moments of the standard normal distribution

$$\int_{-\infty}^{\infty} x^{2n}\,\phi(x)\,dx\,, \quad \text{where } \phi(x) = \frac{1}{\sqrt{2\pi}}\,e^{-x^2/2}\,.$$

   (a) Make the substitution $y = x^2/2$ in the integral, and evaluate the integral using the gamma function. Show that for $n = 1, 2, 3, \ldots$, the moments are given by

$$\int_{-\infty}^{\infty} x^{2\,n}\,\frac{1}{\sqrt{2\,\pi}}\,e^{-x^2/2}\,dx = \frac{2^n\,\Gamma\left(n+\frac{1}{2}\right)}{\sqrt{\pi}}\,.$$

   (b) Use Stirling's approximation to determine an asymptotic form for the expression in part (a) as $n \to \infty$.

2. The members of the family of *t-distributions* have densities of the form

$$f(x) = c(\nu)\left(1+\frac{x^2}{\nu}\right)^{-(\nu+1)/2}, \qquad -\infty < x < \infty\,,$$

   where ν is the *degrees of freedom*. As $\nu \to \infty$, the t-density converges to the density of the standard normal distribution. The constant $c(\nu)$ standardises the function to integrate to one.

   (a) The precise formula for the constant of integration is

$$c(\nu) = \frac{1}{\sqrt{\pi\,\nu}}\,\frac{\Gamma\left(\frac{\nu+1}{2}\right)}{\Gamma\left(\frac{\nu}{2}\right)}\,.$$

   For $\nu \to \infty$, apply Stirling's approximation to the gamma functions in this expression to show that

$$c(\nu) \sim \frac{1}{\sqrt{2\,\pi\,e}} \left(\frac{\nu+1}{\nu}\right)^{\frac{\nu}{2}} .$$

(b) We can also write

$$c(\nu)^{-1} = \int_{-\infty}^{\infty} \left(1 + \frac{x^2}{\nu}\right)^{-(\nu+1)/2} dx\,.$$

Change variables in this integral so that $y = x/\sqrt{\nu}$. Then compute the Laplace approximation for this integral as $\nu \to \infty$ to show that

$$\begin{aligned} c(\nu)^{-1} &= \sqrt{\nu} \int_{-\infty}^{\infty} \left(\frac{1}{1+y^2}\right)^{(\nu+1)/2} dy \\ &\sim \sqrt{\frac{2\,\pi\,\nu}{\nu+1}}\,. \end{aligned}$$

Compare this result with that from part (a) of this question. For each case consider the limiting value as $\nu \to \infty$.

3. The function $u(x) = x^{-x}$, $x > 0$ is quite well behaved over the interval $0 < x < \infty$. However, many properties of this function are analytically intractable.

   (a) Determine $\lim_{x \to 0+} u(x)$.
   (b) Determine the maximum value of $u(x)$ on the interval $0 < x < \infty$. Where is the function increasing, and where is it decreasing?
   (c) For each t, determine the maximum value of the function

   $$u_t(x) = x^{-x}\,e^{t\,x}$$

   and the value of x where this maximum is obtained.
   (d) Prove that as $t \to \infty$,

   $$\int_0^{\infty} x^{-x}\,e^{t\,x}\,dx \sim \sqrt{2\,\pi}\,\exp\left(\frac{t-1}{2} + e^{t-1}\right).$$

   To prove this result, you will need to make an appropriate change of variables. Try $s = e^{t-1}$, and $y = x\,e^{1-t}$.

4. In Section 1 of this chapter, we needed a result which is a consequence of Problem 16 at the end of Chapter 1. Prove this result, namely that

$$\left(1 - \frac{y^2}{n}\right)^n = \exp(-y^2)\,[1 + O(n^{-1})]$$

as $n \to \infty$.

5. Verify the formula for coefficient a_0 given in (6.13).

6. Verify the formula for coefficient a_1 given in (6.14).

7. Verify the formulas for the coefficients up to and including order n^{-2} in formula (6.17) for Stirling series.

8. Verify the steps leading to (6.19).

9. Fill in the details leading to formula (6.28).

10. The final steps in the proof of Proposition 2 were omitted. Complete them.

11. Let $X_1, \ldots, X_n$ be independent, identically distributed random variables, having some continuous distribution with distribution function $F(x)$ and density $f(x) = F'(x)$. We suppose that $F(0) = 0$, and that $f(0) > 0$. Furthermore, we suppose that $F(x)$ has a power series representation in some interval $[0, b)$. We define the minimum order statistic

$$X_{(1)} = \min(X_1, X_2, \ldots, X_n).$$

(a) Prove that the density function of $X_{(1)}$ is given by

$$n \cdot f(x) \cdot [1 - F(x)]^{n-1}.$$

(b) Prove that the moment generating function of $X_{(1)}$ has asymptotic representation

$$\int_0^\infty e^{t\,x} \cdot n\, f(x)[1 - F(x)]^{n-1}\, dx \sim 1 + \frac{t}{n\, f(0)} + O\left(\frac{1}{n^2}\right).$$

12. In Section 6.12, we considered the Pitman estimator for the location parameter. One family of distributions with its Pitman estimator in closed form is the Cauchy distribution $\mathcal{C}(\theta, 1)$. For all sample sizes, the Pitman estimator for θ is a rational function of the data for this model. Use Maple or otherwise to prove that the estimator for sample size $n = 3$ is

$$t_3 = \frac{8\,p_1 + 4\,p_1\,p_2 - 3\,p_3 - p_1^3}{24 + 3\,p_2 - p_1^2},$$

where

$$p_j = x_1^j + x_2^j + x_3^j, \quad j = 1, 2, 3.$$

CHAPTER 7

The saddle-point method

7.1 The principle of stationary phase

7.1.1 Complex-valued integrals

Since the Laplace approximation is available for the asymptotic approximation of integrals of real-valued functions, it is quite natural to seek a "Laplace type" approximation for integrals of complex-valued functions which arise through Fourier inversion. Integrals which arise in Fourier inversion formulas may be considered special cases of complex-valued integrals such as

$$\int_a^b \psi(t)\, \xi^n(t)\, dt\,. \tag{7.1}$$

We are interested in the asymptotic behaviour of such integrals as $n \to \infty$.

For real-valued integrals of the Laplace type, we discovered that largest asymptotic contribution to the value of such an integral comes locally from the point t_0 where the integrand is maximised. However, we cannot speak directly about the maximum of a complex-valued function, and must seek an alternative. An early statement of this principle, due to Cauchy, Stokes, Riemann and Kelvin is the *principle of stationary phase.* Consider the asymptotic value of an integral of the form

$$\int_a^b \beta(t)\, e^{i\,n\,\alpha(t)}\, dt \tag{7.2}$$

as $n \to \infty$. We shall suppose that $\alpha(t)$ and $\beta(t)$ are real-valued functions of the real variable t. We can borrow terminology from physics and call $\alpha(t)$ the phase function. The principle of stationary phase states that the largest asymptotic contribution to the value of the integral in (7.2). as $n \to \infty$ comes locally from the neighbourhood of the point t_0 where the phase function $\alpha(t)$ is stationary—a point t_0 where $\alpha'(t_0) = 0$. That this should be the case becomes intuitively clear when we consider the

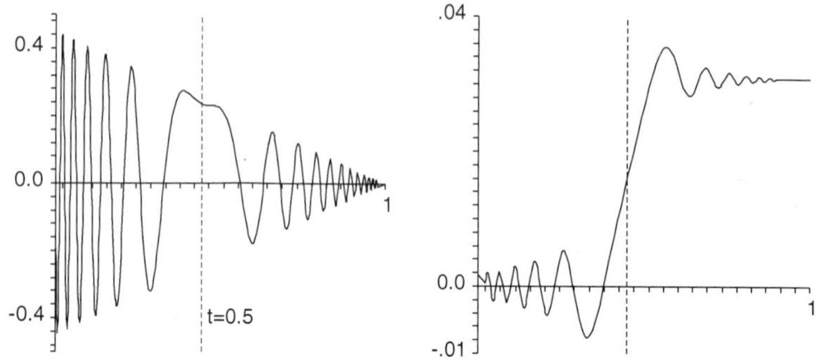

Figure 7.1 *An illustration of the principle of stationary phase*

real and imaginary parts of the integral in (7.2), namely

$$\int_a^b \beta(t) \cos\left[n\,\alpha(t)\right] dt\,, \qquad \int_a^b \beta(t) \sin\left[n\,\alpha(t)\right] dt\,.$$

Suppose that $\beta(t)$ varies smoothly from a to b. Then for large values of n, the sine and cosine factors will oscillate rapidly over small intervals of t where the function $\beta(t)$ is almost constant. Locally, the contribution of such rapidly oscillating components of the integrand is negligible because the positive and negative parts approximately cancel. The frequencies of the oscillations locally around t are proportional to $n\,|\alpha'(t)|$, which is large for large values of n, except at any point t_0 where $\alpha'(t_0) = 0$. At any such value t_0, there is no local approximate cancellation.

7.1.2 An example of stationary phase

To illustrate the principle of stationary phase, let us consider the two plots shown in Figure 7.1. The first plot is the real (cosine) part of the integrand of the integral in (7.2) when

$$n = 200\,, \qquad \beta(t) = (1-t)/2\,, \qquad \text{and } \alpha(t) = \sqrt[3]{t\,(1-t)}$$

over the interval $0 < t < 1$. In this example, the phase function $\alpha(t)$ has a stationary point at $t_0 = 1/2$. The second plot is that of the integral of the first function for t running over the interval from 0 to u, where $0 < u < 1$. It is readily seen that the main asymptotic contribution to the integral for large n is to be found close to $t_0 = 1/2$.

7.2 Perron's saddle-point method

The principle of stationary phase cannot be applied to integrals which are not of the form given in (7.2). Thus the method of the previous section does not immediately apply to general integrals of the form given in (7.1) where the function $\xi(t)$ has both varying phase and varying amplitude. We may also wish to approximate more general contour integrals* of the form

$$\int_C \psi(z)\,\xi^n(z)\,dz = \int_C \psi(z)\,e^{n\,\nu(z)}\,dz \tag{7.3}$$

where C is some oriented contour in the complex plane whose endpoints a and b may be infinite, and $\nu(z) = \log \xi(z)$ is complex-valued. In order to extend the principle of stationary phase to the integral in (7.3), we would like to find a point $z_0 \in C$ such that

- the argument of $\xi(z)$, namely arg $\xi(z) = \Im\,\nu(z)$, is stationary at $z = z_0$, and
- the modulus of $\xi(z)$, namely $|\xi(z)| = \Re\,\nu(z)$, is locally maximised at $z = z_0$.

Note that the second of these two conditions is similar to the requirements for the Laplace approximation of the previous chapter. This gives us two conditions to be satisfied simultaneously along a one-dimensional contour. In general we cannot expect that such a point $z_0 \in C$ will exist. A solution to this problem can be found by exploiting the particular character of contour integration using Cauchy's integral theorem.

Suppose we can find some other contour C^* such that the following hold.

Assumption 1. *The contour C^* has the same endpoints as C, namely a and b. Furthermore, the functions $\psi(z)$ and $\nu(z)$ are analytic throughout the region enclosed by the contours C and C^* with no poles or singularities in the interior.*

This condition and Cauchy's theorem imply that

$$\int_C \psi(z)\,e^{n\,\nu(z)}\,dz = \int_{C^*} \psi(z)\,e^{n\,\nu(z)}\,dz\,. \tag{7.4}$$

Assumption 2. *The contour C^* passes through a point z_0 where*

$$\nu'(z_0) = 0\,.$$

* Here I use the terms "contour" and "contour integral" to denote what some authors call a line integral or curvilinear integral in the complex plane. The term "closed contour" will be used for contours which start and end at the same point.

Oskar Perron (1880–1975)

Oskar Perron is remembered for his work in analysis, algebra, and continued fractions. Important results of interest to statisticians include his saddle-point method and the Perron-Frobenius theorem.

> "There were four facets to his life .... [H]is enthusiasm for mathematics ... his love of the mountains of his surroundings ... his love for his family, which was reciprocated by the care he received at home ... [and] his principles in personal lifestyle, comparable to the exactness shown in his science."
>
> Evelyn Frank, "In Memorial Oskar Perron" *J. Number Th.*, 1982, p. 283.

It follows from this that the real and imaginary components of $\nu'(z)$ vanish at z_0.

Thus

$$\begin{aligned} \nu(z) &= \nu(z_0) + \nu'(z_0)\,(z - z_0) + \frac{1}{2}\,\nu''(z_0)\,(z - z_0)^2 + \cdots \\ &= \nu(z_0) + \frac{1}{2}\,\nu''(z_0)\,(z - z_0)^2 + \cdots . \end{aligned}$$

This will be the justification for using the Laplace method, which approximates $\nu(z)$ locally around z_0 using the displayed complex Taylor expansion. The point z_0 is often called a saddle-point, and gives its name to the saddle-point approximation which we shall derive.

By a δ-neighbourhood of z_0 on the contour we shall mean the set of points on the contour whose arc-length distance along the contour is less than δ from z_0. Let $B(\delta)$ denote this δ-neighbourhood. The arc-length from z_0 to z along the contour is at least as great as the Euclidean distance. Therefore $|z - z_0| < \delta$ for all $z \in B(\delta)$.

Assumption 3. *We further require that the contour $C^\star$ passes through z_0 so that at the point z_0, in such a way that for some δ-neighbourhood $B(\delta)$ the following hold.*

1. *$\Im\,\nu(z)$ is constant on $B(\delta)$.*
2. *$\Re\,\nu(z)$ achieves its unique maximum† in $B(\delta)$ at z_0.*

To apply the Laplace method, we need to ensure that the contribution to the leading term of the Laplace expansion comes from the saddle-point z_0 and nowhere else along the contour. To do this we may assume the following.

Assumption 4. *Outside $B(\delta)$ on the contour $C^\star$, the function $\Re\,\nu(z)$ is bounded away from its maximum $\Re\,\nu(z_0)$. That is,*

$$\sup_{z \in C^\star - B(\delta)} \Re\,\nu(z) < \Re\,\nu(z_0)\,.$$

† However, $\Re\,\nu(z)$ is not maximised with respect to directions away from the contour. Similarly, $\Im\,\nu(z)$ is not constant in directions that depart from the contour. This point is considered more fully in the next section, where we shall consider the geometric interpretation of this third assumption. At this stage it is not clear why a contour satisfying the condition is preferable to any other. We shall defer such issues until later.

Now suppose we parametrise the contour $C^\star$ by some coordinate s such that

$$C^\star = \{z(s) \ : \ -\infty \le s \le \infty\}$$

and such $z_0 = z(0)$ is the point of stationarity determined above. Let $z_0' = z'(0)$, and so on. Also, let

$$\begin{aligned} \nu_0 &= \nu(z_0), \\ \nu_0' &= \nu'(z_0) \qquad (= 0), \\ \nu_0'' &= \nu''(z_0), \end{aligned}$$

and so on. Then

$$\nu[\,z(s)\,] = \nu_0 + \frac{\nu_0'' \cdot [\,z_0'\,]^2}{2}\, s^2 + \cdots . \tag{7.5}$$

Assumption 3 and formula (7.5) imply that

$$\Im\{\,\nu_0'' \cdot [\,z_0'\,]^2\,\} = 0 . \tag{7.6}$$

In other words, $\nu_0'' \cdot [\,z_0'\,]^2$ is real-valued. Assumption 3 and (7.5) also imply that $\nu_0'' \cdot [\,z_0'\,]^2 \le 0$. Henceforth, let us interpret Assumption 3 to imply the slightly stronger condition that

$$\nu_0'' \cdot [\,z_0'\,]^2 < 0 . \tag{7.7}$$

From here on we may proceed in the spirit of Laplace. We have

$$\begin{aligned} \int_{C^\star} \psi(z)\, e^{n\,\nu(z)}\, dz &= \int_{-\infty}^{\infty} \psi[z(s)]\, z'(s)\, e^{n\,\nu[z(s)]}\, ds \\ &\sim \psi_0\, z_0'\, \exp(n\,\nu_0) \int_{-\infty}^{\infty} \exp\left[\frac{n\,\nu_0''\,(z_0')^2}{2}\, s^2\right] ds \end{aligned}$$

as $n \to \infty$. The displayed integrand is a complex version of the "bell-shaped" approximant used in the Laplace method of the previous chapter. We can evaluate this integral to obtain the approximation

$$\int_{C^\star} \psi(z)\, e^{n\,\nu(z)}\, dz \sim \psi_0\, z_0'\, e^{n\,\nu_0} \sqrt{\frac{2\,\pi}{-n\,\nu_0''\,(z_0')^2}}$$

as $n \to \infty$.

Next, we can write

$$\frac{z_0'}{\sqrt{-\nu_0''\,(z_0')^2}} = \frac{|z_0'|\, e^{i\,\arg z_0'}}{\sqrt{|(z_0')^2\,\nu_0''|}} = e^{i\,\arg z_0'}\, \frac{1}{\sqrt{|\nu_0''|}},$$

where arg(.) is the argument function. Putting the pieces together we obtain the following saddle-point approximation for the original contour integral.

Proposition 1. *Under Assumptions 1–4 and assuming that the integral below is finite for some n, we have*

$$\int_C \psi(z)\, e^{n\,\nu(z)}\, dz \;\sim\; \psi_0\, e^{i \arg z_0'} \exp(n\,\nu_0) \sqrt{\frac{2\,\pi}{n\,|\nu_0''|}} \tag{7.8}$$

as $n \to \infty$.

Note that the final saddle-point approximation does not depend on the choice of the parametrisation of the contour $C^\star$ by the coordinate s. While z_0' is dependent upon the parametrisation, its direction—tangent to the contour—is not.

A way to sharpen the result is to include extra terms in the expansion. This leads to an asymptotic expansion for the integral which is similar in form to that derived in Proposition 3 of the previous chapter. This is not surprising, because the saddle-point approximation of this chapter is essentially the Laplace method applied to a modified contour integral. Therefore, the method for obtaining the terms of the expansions is essentially the same as that of the previous chapter. Assumption 4 above needs to be strengthened to control the size of the tail of the contour integral away from the saddle-point.

Assumption 5. *Suppose that for every positive integer m and* $\delta > 0$,

$$\int_{z \notin B(\delta)} \psi(z)\, e^{n\,\nu(z)}\, dz = e^{n\,\nu_0}\, O(n^{-m}) \tag{7.9}$$

as $n \to \infty$.

Proposition 2. *Suppose that Assumptions 1–3 hold. We replace Assumption 4 by Assumption 5. Provided the contour integral below is finite, the integral has asymptotic expansion*

$$\int_C \psi(z)\, e^{n\,\nu(z)}\, dz \;\sim\; \frac{\exp(n\,\nu_0)}{\sqrt{n}} \left(a_0 + \frac{a_1}{n} + \frac{a_2}{n^2} + \cdots\right) \tag{7.10}$$

where the coefficient a_0 *is determined by Proposition 1 above.*

All other coefficients are determined from the Taylor expansions of $\psi[z(s)]$ and $\nu[z(s)]$. The formulas for the coefficients a_j in (7.10) may be obtained from Proposition 3 of the previous chapter by replacing the function g and h by $\psi[z(\cdot)]\, z'(\cdot)$ and $\nu[z(\cdot)]$, respectively. For a proof of Proposition 2, see the discussion of this expansion in de Bruijn (1981).

7.3 Harmonic functions and saddle-point geometry

The point z_0 is a point of stationarity of $\nu(z)$. However, unlike the real line, the point z_0 never represents a local extremum of $\nu(z)$, unless $\nu(z)$ is constant throughout the region. For example, if we plot the surface generated by the real part $v = \Re\,\nu(z)$, namely

$$\mathcal{S}_{\text{real}} = \{\, (x,\, y,\, v) \in \mathbf{R}^3 \;:\; v = \Re\,\nu(x + i\,y) \,\}$$

we find that the surface so generated is "peakless," in the sense that there are no local maxima or minima. Such surfaces are *harmonic functions*, which satisfy Laplace's differential equation

$$\frac{\partial^2 v}{\partial x^2} + \frac{\partial^2 v}{\partial y^2} = 0$$

at all points on the surface. Similar remarks hold for the surface

$$\mathcal{S}_{\text{imag}} = \{\, (x,\, y,\, w) \in \mathbf{R}^3 \;:\; w = \Im\,\nu(x + i\,y) \,\}$$

generated from the imaginary part of $w = \Im\,\nu(z)$. Figure 7.2 illustrates this geometrical property with the hyperbolic cosine function. The surfaces generated by the real and imaginary parts of the hyperbolic cosine are displayed in the left and right plots, respectively. On both of these surfaces, we see that saddle-points are the only flat points where the gradient vanishes in all directions. For this reason, the point z_0 at which $\nu'(z_0) = 0$ is often called a *saddle-point* of ν.

Another property of the two harmonic functions displayed in Figure 7.2 is that they are conjugate harmonic functions. Two such functions are said to be conjugate if they satisfy the Cauchy-Riemann equations, namely that

$$\frac{\partial v}{\partial x} = \frac{\partial w}{\partial y}, \qquad \frac{\partial v}{\partial y} = -\frac{\partial w}{\partial x}.$$

Geometrically, this can be seen in the family of level curves of the contour plots for each function. For example, Figure 7.3 shows what happens when the contours of the hyperbolic cosine, displayed in Figure 7.2, are superimposed on each other, and correctly scaled so that axes are commensurate. In Figure 7.3, it can be seen that the superimposed contours are orthogonal to each other. This is a direct consequence of the conjugacy property defined by the Cauchy-Riemann equations.

A saddle-point z_0 of a function $\nu(z)$ is said to be of order m for $m = 1, 2, 3, \ldots$ if

$$\nu_0' = \nu_0'' = \cdots = \nu_0^{(m)} = 0$$

and $\nu_0^{(m+1)} \neq 0$. For the geometrical discussion which follows below, we

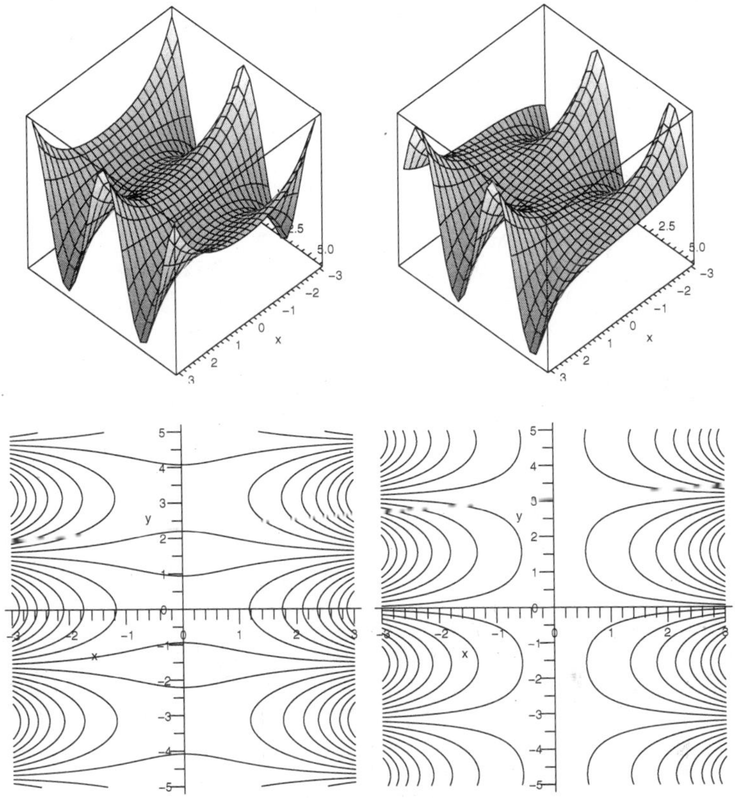

Figure 7.2 *Peakless surfaces generated by plotting the real and imaginary parts of the hyperbolic cosine function. On the left above is the surface plot of the real part* $v = \Re \cosh(x + i\,y)$ *and directly below is the corresponding contour plot of the same function. On the right is the surface plot of the imaginary part* $w = \Im \cosh(x + i\,y)$ *and the corresponding contour plot.*

shall restrict our attention to saddle-points of order $m = 1$, which are most commonly encountered.

Let us imagine that we are standing at a saddle-point on the surface, and that we wish to descend from the saddle-point as quickly as possible. In two diametrically opposite directions from the saddle-point, the surface drops off, however slowly at first. Perpendicular to these directions, the surface ascends above the saddle-point. See Figure 7.4 which shows the behaviour of a surface in the neighbourhood of a typical saddle-point. Plotted here are the tangent vectors to the paths of steepest descent and ascent from the saddle-point. Between the paths of steepest descent

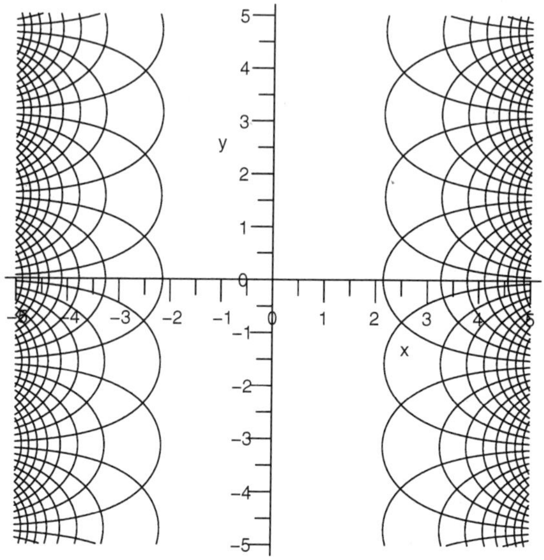

Figure 7.3 *Superimposed contours of the real and imaginary parts of the hyperbolic cosine function. See also Figure 7.2. The real and imaginary parts are conjugate functions, which implies that the level curves are orthogonal families. In this plot, the axes are constrained to be commensurate, so that the orthogonality is evident upon superposition.*

and steepest ascent are another pair of paths where the surface neither rises nor falls and the elevation remains constant. These paths are level curves of the surface. The tangent vectors to these level curves through the saddle-point are perpendicular to each other, and make 45° angles with the tangent vectors to the paths of steepest descent and ascent. These vectors are plotted as dashed lines in the figure.

Now let us consider how this picture changes if we switch from one such harmonic function to its conjugate. A geometrical consequence of the Cauchy-Riemann equations is that conjugate harmonic functions have the same saddle-points. So the conjugate diagram to that shown in Figure 7.4 also has a saddle-point at its centre. However, the level curves of conjugate functions are orthogonal to each other. Therefore the directions of the dashed line tangent vectors in Figure 7.4 now correspond to paths of steepest ascent and descent. Similarly, the directions of the unbroken line tangent vectors now correspond to level curves. In other words, our diagram undergoes a 45° rotation.

Let us now return to the Laplace method which was the basis for the

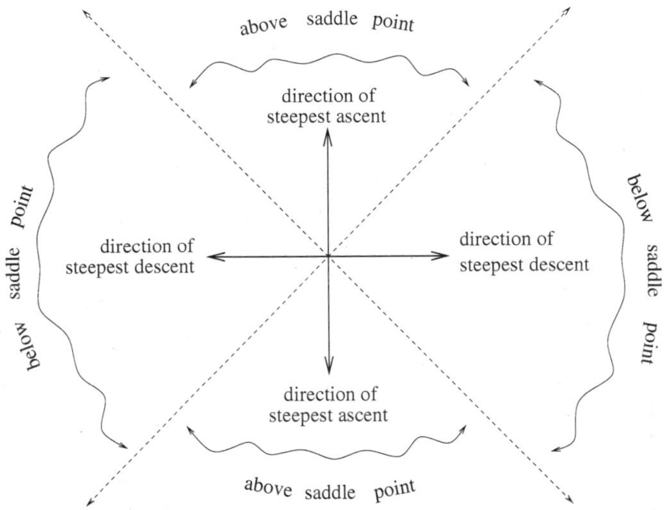

Figure 7.4 *A typical geometry around a saddle-point. Directions of steepest ascent and descent are plotted as solid line vectors, while directions of constant elevation are plotted as dashed line vectors.*

saddle-point approximation in Proposition 1. The contour $C^\star$ used in that approximation was chosen to pass through the saddle-point z_0 of the function $\nu(z)$. We now consider a geometrical interpretation of Assumption 3, which determined the direction of $C^\star$ through z_0.

We note first that

$$|\xi(z)| = |e^{\nu(z)}| = e^{\Re\,\nu(z)}\,.$$

So Assumption 3, which requires that $\Re\,\nu(z)$ is maximised at $z = z_0$ can be interpreted to mean that $|\xi(z)|$ is maximised at $z = z_0$. Assumption 3 also requires that $\Im\,\nu(z)$ is locally constant around $z = z_0$. So the contour $C^\star$ is tangent to a level curve of $w = \Im\,\nu(x + i\,y)$. However, the level curves of $w = \Im\,\nu(x + i\,y)$ through the saddle-point are paths of steepest descent for the function $v = \Re\,\nu(x + i\,y)$.

These two conditions determine the behaviour of the modulus $|\xi(z)|$ at $z = z_0$ along the contour $C^\star$. Among all contours passing through z_0, the contour for which the $|\xi(z)|$ is most tightly concentrated at z_0 are the contours satisfying Assumption 3. In order for a Laplace approximation to be accurate, it is natural to require that the function $|\xi(z)|$. The saddle-point approximation can be regarded as a local form of the method of steepest descent which uses the contour of steepest descent of $w = \Re\,\nu(x + i\,y)$ through the saddle-point. The original form of the method of steepest descent was introduced by Debye (1909) to obtain asymptotic expansions of contour integrals. In practice the precise form of this contour may be difficult to determine. The saddle-point method, developed by Perron (1917) avoids the tricky problem of constructing the path of steepest descent. It requires only that the steepest descent property holds locally in the sense that

$$\Re[\nu_0''\,(z_0')^2] < 0\,, \qquad \Im[\nu_0''\,(z_0')^2] = 0$$

which is a path of locally steepest descent.

7.4 Daniels' saddle-point approximation

Let us return to the evaluation of the Fourier inversion integral. Let $X_1, X_2, \ldots, X_n$ be independent, identically distributed continuous random variables with common characteristic function $\chi(t)$, which is assumed to be analytic in some neighbourhood of the origin. Let $M(t)$ be the moment generating function of X_j, and $K(t)$ its cumulant generating function. Let $\mu = E(X_j)$ and $\sigma^2 = \mathrm{Var}(X_j) > 0$.

We can obtain the density function for the sample average

$$\overline{X}_n = \frac{X_1 + \cdots + X_n}{n}$$

using a Fourier inversion integral, namely

$$f(\overline{x}) = \frac{1}{2\,\pi} \int_{-\infty}^{\infty} e^{-i\,t\,\overline{x}}\,\chi^n(t/n)\,dt\,, \tag{7.11}$$

which has a form suitable for saddle-point approximation.‡ We can write $\chi(t) = e^{K(i\,t)}$. Then

$$f(\overline{x}) = \frac{1}{2\,\pi} \int_{-\infty}^{\infty} e^{-i\,t\,\overline{x}} \exp\left[n\,K\left(\frac{i\,t}{n}\right)\right]\,dt\,. \tag{7.12}$$

Making the substitution $z = n^{-1}\,i\,t$, the integral formula becomes

$$\begin{aligned} f(\overline{x}) &= \frac{n}{2\,\pi\,i} \int_{-i\,\infty}^{i\,\infty} e^{-n\,z\,\overline{x}} \exp\left[n\,K\,(z)\right]\,dz \\ &= \frac{n}{2\,\pi\,i} \int_{-i\,\infty}^{i\,\infty} \exp\left\{n\left[K\,(z) - z\,\overline{x}\right]\right\}\,dz\,. \end{aligned} \tag{7.13}$$

We can recognise that this integral is in the form required for Proposition 1 with $\nu(z) = K(z) - z\,\overline{x}$ and $\psi(z) = 1$. The contour C of the integral is the imaginary axis $-i\,\infty < z < i\infty$.

‡ It would be more consistent to write $\overline{x}_n$ here. However, that gets messy in subsequent formulas.

The saddle-point of $K(z) - z\,\overline{x}$ is determined by solving

$$K'(z_0) = \overline{x}\,. \tag{7.14}$$

From the form of this equation, we might expect to find a solution z_0 lying on the real axis. This turns out to be the case. In fact for large values of n, the saddle-point z_0 will lie on the real axis close to the origin provided that $\overline{x}$ lies in the centre of the distribution of $\overline{X}_n$. To see this, note first that by the law of large numbers, $\overline{X}_n \to \mu$ with probability one. Furthermore, for t in some neighbourhood of the origin on the real axis

$$K'(t) = \mu + \sigma^2\,t + O(t^2) \qquad \text{as } t \to 0\,.$$

Thus with probability one for sufficiently large n there is some neighbourhood $(-\epsilon,\,\epsilon)$ such that $K'(\epsilon) > \overline{x}$ and $K'(-\epsilon) < \overline{x}$. Asymptotically $K'(t_0) = \overline{x}$ has a unique solution t_0 in the interval $-\epsilon < t < \epsilon$ and $K''(t_0) > 0$.

We can perturb the contour of the inversion integral in (7.13) to pass through this saddle-point, so that

$$f(\overline{x}) = \frac{n}{2\,\pi\,i}\int_{t_0-i\,\infty}^{t_0+i\,\infty} \exp\left\{n\left[K\left(z\right) - z\,\overline{x}\right]\right\}\,dz\,. \tag{7.15}$$

In this integral the contour of integration is understood to be the line parallel to the imaginary axis passing through the point t_0. This contour can be parametrised by a real coordinate s to that $z(s) = t_0 + i\,s$, for $-\infty < s < \infty$.

We now turn to Assumption 3 of Proposition 1, or more precisely the assumptions on the second derivative in (7.6) and (7.7). Note that along the contour

$$\left\{\frac{d^2}{ds^2}\,\Im\left[K(t_0 + i\,s) - (t_0 + i\,s)\,\overline{x}\right]\right\}_{s=0} = 0\,, \tag{7.16}$$

and

$$\left\{\frac{d^2}{ds^2}\,\Re\left[K(t_0 + i\,s) - (t_0 + i\,s)\,\overline{x}\right]\right\}_{s=0} < 0\,. \tag{7.17}$$

See Problem 1. In addition, for $z = t_0 + i\,s$, the modulus of the integrand

$$\left|e^{-z\,\overline{x}}\,M(z)\right| = e^{\Re\,\{K(z)-z\,\overline{x}\}} \tag{7.18}$$

is uniquely maximised at the saddlepoint t_0. This follows from the inequality

$$\begin{aligned}\left|e^{-z\,\overline{x}}\,M(z)\right| &= \left|e^{-(t_0+i\,s)\,\overline{x}}\right|\cdot\left|E(e^{(t_0+i\,s)\,X_j})\right| \\ &= e^{-t_0\,\overline{x}}\left|E(e^{(t_0+i\,s)\,X_j})\right|\end{aligned}$$

$$
\begin{aligned}
&= e^{-t_0 \overline{x}} \left| E(e^{t_0 X_j} e^{i s X_j}) \right| \\
&\le e^{-t_0 \overline{x}} E(e^{t_0 X_j} |e^{i s X_j}|) \\
&= e^{-t_0 \overline{x}} M(t_0) .
\end{aligned}
$$

Therefore the contour $z(s) = t_0 + i s$ is tangent to the path of steepest descent from the saddlepoint t_0, where the modulus of the integrand achieves its global maximum. Thus Assumption 3 holds.

Assumption 4 can also be verified. The Riemann-Lebesgue lemma implies that

$$M(t_0 + i s) \to 0$$

as $s \to \pm\infty$. It follows from this that the modulus of the integrand must go to zero at infinity in both directions along the contour. See Problem 2.

Therefore, we may apply Proposition 1 to determine that

Proposition 3. *The density of $\overline{X}_n$ is asymptotically*

$$f(\overline{x}) \sim \sqrt{\frac{n}{2\pi K''(t_0)}} \exp\{n [K(t_0) - t_0 \overline{x}]\} . \tag{7.19}$$

as $n \to \infty$.

Daniels (1954) showed that for many distributions, the *relative* error in the approximation to $f(\overline{x})$ is $O(n^{-1})$ uniformly in $\overline{x}$. That is, if $g(\overline{x})$ represents the saddle-point approximation to $f(\overline{x})$ as in Proposition 3, then for a large class of distributions

$$\left| \frac{f(\overline{x})}{g(\overline{x})} - 1 \right| \le \frac{A}{n}$$

for some constant A which does not depend upon n or $\overline{x}$.

Daniels also obtained an asymptotic expansion for $f(\overline{x})$. Using Proposition 2 we get the following result.

Proposition 4. *The density of $\overline{X}_n$ has aymptotic expansion*

$$f(\overline{x}) \sim \sqrt{\frac{n}{2\pi K''(t_0)}} \exp\{n [K(t_0) - t_0 \overline{x}]\} \left(1 + \frac{a_1}{n} + \frac{a_2}{n^2} + \cdots\right) , \tag{7.20}$$

where, in particular,

$$a_1 = \frac{K^{(iv)}(t_0)}{8 [K''(t_0)]^2} - \frac{5 [K'''(t_0)]^2}{24 [K''(t_0)]^3} . \tag{7.21}$$

The reader should note that the coefficient a_1 is approximately

$$\frac{\rho_4}{8} - \frac{5\,\rho_3^2}{24},$$

where ρ_n is the nth standardised cumulant. The main difference between this quantity and a_1 is that the cumulant generating function is evaluated at the saddle-point rather than at the origin. Asymptotically these are close, because the saddle-point converges to zero as $n \to \infty$. We can distinguish the true standardised cumulants such as ρ_n from those obtained by differentiating the cumulant generating function at t_0. Let us call the latter the standardised tilted cumulants of the distribution.

Assumption 5 needs to be verified to prove that the formal expansion in (7.20) is asymptotic. Outside a δ-neighbourhood of t_0, the upper and lower tails of the contour integral are $O(r^n)$, where $0 < r < 1$. Thus the contribution from these components of the integral is negligible.

If we look at the asymptotic series in Proposition 4, we observe that it does not depend heavily upon $\overline{x}$. The reason for this is that when $\overline{x}$ is in the middle of its distribution around μ, the saddle-point t_0 will be close to zero. So the various standardised tilted cumulants that arise in the expansion will be close to their untilted counterparts. Thus the effect of the adjustment arising from the asymptotic expansion is most strongly seen in the tails of the distribution of $\overline{X}_n$.

This suggests that the asymptotic series behaves like an integrating factor in the centre of the distribution. With this in mind, an obvious alternative is to try to normalise the saddle-point approximation directly by setting

$$f(\overline{x}) \sim c_n \sqrt{\frac{n}{2\,\pi\,K''(t_0)}}\,\exp\{n\,[K(t_0) - t_0\,\overline{x}]\} \tag{7.22}$$

where the constant c_n is chosen to make the saddle-point approximation integrate exactly to one. The asymptotic effect of this will be much the same as adding terms of order $O(n^{-1})$ and beyond to the approximation. For a given sample size n the two methods are not equivalent. One may recommend itself over the other usually for computational reasons.

7.5 Towards the Barndorff-Nielsen formula

7.5.1 Normal distribution

In this section, we shall consider a number of examples of the saddle-point approximation for $\overline{X}_n$, culminating in the Bandorff-Nielsen for-

Henry Daniels (1912–2000)

President of the Royal Statistical Society from 1974–1975, Henry Daniels combined exceptional abilities in applied mathematics with a pragmatic understanding of statistics. The full importance of his 1954 paper, which introduced the saddle-point method into statistics, was more fully understood as the connection between saddle-point adjustments and exponential tilting was better known. The saddle-point approximation is now also known as the tilted Edgeworth expansion. This latter interpretation, motivated by statistical inference, has brought the saddle-point method into the mainstream of statistical asymptotics.

Henry Daniels is also remembered for his love of the English concertina, on which he played chamber music for friends and colleagues. He had an engineer's eye for mechanical devices, with his expertise ranging from concertinas to mechanical watches.

mula for the approximate conditional distribution of the maximum likelihood estimator given a maximal ancillary statistic.

Let X_j be $\mathcal{N}(\mu, \sigma^2)$. Then X_j has cumulant generating function

$$K(t) = \mu\, t + \frac{\sigma^2\, t^2}{2} \tag{7.23}$$

The saddle-point equation $K'(t_0) = \overline{x}$ reduces to $\mu + \sigma^2\, t_0 = \overline{x}$ which is solved by

$$t_0 = \frac{\overline{x} - \mu}{\sigma^2} \qquad \text{so that} \qquad K''(t_0) = \sigma^2\,. \tag{7.24}$$

Since $K^{(n)}(t) = 0$ for $n \geq 3$, the coefficients $a_1, a_2, \ldots$ in (7.20) vanish. For the normal distribution, the saddle-point approximation for the density of $\overline{X}_n$ is exact with no error.

7.5.2 Exponential distribution

When X_j has distribution $\mathcal{E}(\lambda)$ then the cumulant generating function has the form

$$K(t) = \ln(\lambda) - \ln(\lambda - t)\,. \tag{7.25}$$

So the saddle-point equation $K'(t_0) = \overline{x}$ reduces to $\lambda - t_0 = \overline{x}^{-1}$, which is solved by

$$t_0 = \lambda - \overline{x}^{-1}\,. \tag{7.26}$$

Therefore $K''(t_0) = \overline{x}^2$. Plugging these into (7.19) we get

$$f(\overline{x}) \;\sim\; \sqrt{\frac{n}{2\,\pi}}\, e^n\, \lambda^n\, \overline{x}^{n-1}\, e^{-n\,\lambda\,\overline{x}}\,. \tag{7.27}$$

This is not the true density of $\overline{X}_n$. The true density is

$$f(\overline{x}) = \frac{(n\,\lambda)^n}{(n-1)!}\, \overline{x}^{n-1}\, e^{-n\,\lambda\,\overline{x}} \tag{7.28}$$

which is the density of a $\mathcal{G}(n, n\,\lambda)$ distribution. The discrepancy between the true density and the saddle-point approximation (7.27) is in the constant of integration. To compare the two, we can use Stirling's approximation. We have

$$\begin{aligned} \frac{(n\,\lambda)^n}{(n-1)!} \quad &= \quad \frac{n^{n+1}\,\lambda^n}{n!} \\ &\sim \quad \frac{n^{n+1}\,\lambda^n}{\sqrt{2\,\pi\, n}\, n^n\, e^{-n}} \\ &= \quad \sqrt{\frac{n}{2\,\pi}}\, e^n\, \lambda^n \end{aligned}$$

which is the constant appearing in (7.27).

7.5.3 *Gamma distribution*

The saddle-point approximation for the sum of gamma distributed random variables is similar to the exponential case above, of which it is a generalisation. Suppose X_j, $1 \leq j \leq n$ have distribution $\mathcal{G}(\alpha, \lambda)$. The saddle-point density of $\overline{X}_n$ is

$$f(\overline{x}) \sim \sqrt{\frac{n\,\alpha}{2\,\pi}}\, e^{n\,\alpha} \left(\frac{\lambda}{\alpha}\right)^{n\,\alpha} \overline{x}^{n\,\alpha-1}\, e^{-n\,\lambda\,\overline{x}} . \tag{7.29}$$

See Problem 3, which also asks the reader to show that this approximation is exact except for the constant of integration. The true distribution of $\overline{X}_n$ is $\mathcal{G}(n\,\alpha, n\,\lambda)$. The constant appearing in the saddle-point approximation is related to the correct constant of integration through Stirling's approximation—a result similar to that obtained for the exponential distribution above.

7.5.4 *Uniform distribution*

Suppose X_j is $\mathcal{U}(-1, 1)$. In this case, the density of $\overline{X}_n$ is a piecewise polynomial of degree $n-1$ with knots at $j/(n-1)$ for $j = 1 - n, \ldots, 0, \ldots n-1$.

The moment generating function of X_j is

$$M(t) = \frac{1}{2}\int_{-1}^{1} e^{t\,x}\,dx = \begin{cases} \frac{e^t - e^{-t}}{2\,t} = \frac{\sinh t}{t} & t \neq 0 \\ 1 & t = 0 . \end{cases} \tag{7.30}$$

So, for $t \neq 0$,

$$K(t) = \ln \sinh t - \ln t , \qquad K'(t) = \coth t - t^{-1} . \tag{7.31}$$

Therefore, the saddle-point equation reduces to

$$\coth t_0 - t_0^{-1} = \overline{x} , \tag{7.32}$$

so that, at the saddle-point, we obtain

$$K''(t_0) = t_0^{-2} - \operatorname{csch}^2 t_0 . \tag{7.33}$$

Unfortunately, we cannot find a closed form expression for the saddle-point t_0 in this case. It is easy to solve the saddle-point equation numerically. However, if the saddle-point approximation is to be evaluated over an interval of values of $\overline{x}$, this is computationally expensive. Alternatively, we could find an analytic approximation to t_0 using an approximation to the saddle-point equation itself. For example, when $\overline{x}$ is close to one, then t_0 is large. So we may approximate the equation by $1 - t_0^{-1} = \overline{x}$ which leads to the approximation

$$t_0 \sim \frac{1}{1-\overline{x}}.$$

We may also make the approximations

$$K(t_0) \sim \ln\left[\frac{(1-\overline{x})\,\exp((1-\overline{x})^{-1}}{2}\right], \qquad K''(t_0) \sim (1-\overline{x})^2.$$

Thus the saddle-point approximation to the density of $\overline{X}_n$ is roughly

$$f(\overline{x}) \sim \sqrt{\frac{n}{2\pi}}\left(\frac{e}{2}\right)^n (1-\overline{x})^{n-1}, \qquad \overline{x} \to 1.$$

The exact density close to one differs only in the normalising constant. Close to one, the density has the form

$$f(\overline{x}) = \frac{n^n}{2^n\,(n-1)!}\,(1-\overline{x})^{n-1} \qquad \overline{x} > 1 - 2\,n^{-1}.$$

For $\overline{x} \to 0$, we can use the local approximation $t_0 \sim 3\,\overline{x}$, which is obtained by a Taylor expansion of $K'(t)$ about zero. In practice, the approximation $t_0 \sim 3\,\overline{x}$ is sufficient for all but the most extreme part of the tails when n is large, because $\overline{X}_n = O_p(1/\sqrt{n})$ as $n \to \infty$.

To fit the centre and the tails simultaneously, a numerical approximation to the saddle-point function can be obtained from the Maple command[§]

$$> t0 := x \to solve\,((e^t + e^{-t}) \cdot t - e^t + e^{-t} = x \cdot t \cdot (e^t - e^{-t}),\, t)$$

where $\overline{x}$ is represented by the variable x above. Figure 7.5 displays the density of $\overline{X}_n$ as a solid line and the saddle-point approximation as a broken line using this approximation for the cases $n = 1, 2, 3, 4$. The exact density for the mean of $n > 1$ independent $\mathcal{U}(-1, 1)$ random variables is easily computed using the formula[¶]

$$f(\overline{x}) = \frac{n^n}{2^n\,(n-1)!}\sum_{j=0}^{n}(-1)^j\binom{n}{j}\left\langle 1 - \overline{x} - \frac{2\,j}{n}\right\rangle^{n-1} \tag{7.34}$$

where

$$\langle y \rangle = \begin{cases} y & \text{if } y \geq 0 \\ 0 & \text{if } y < 0. \end{cases}$$

[§] I recommend not solving $\coth t - t^{-1} = \overline{x}$ numerically without clearing the denominators first. Numerical methods which interate from $t = 0$ may have difficulty with $\coth t$ and t^{-1} which are not defined at zero.

[¶] This is the version of the formula provided by Daniels. Note that on the interval from -1 to $+1$ the last term in the sum is always zero. We leave this term in for the appearance of symmetry.

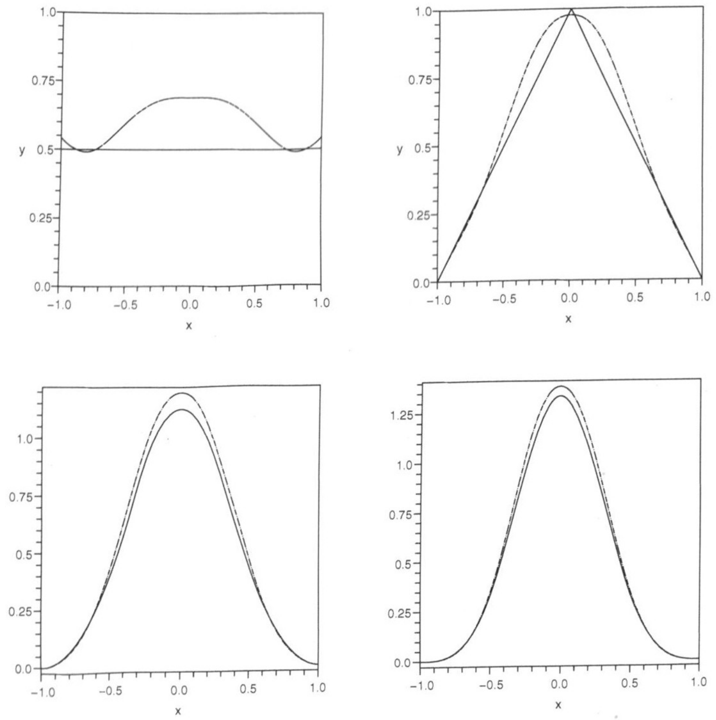

Figure 7.5 *The saddle-point approximation to the sum of n independent* $\mathcal{U}(-1, 1)$ *random variables for* $n = 1, 2, 3, 4$. *The solid line is the true density and the broken line is the saddle-point approximation*

Comparing the true densities and the saddle-point approximations in this example, we note the high accuracy of the saddle-point approximation in the tails of the distribution. The accuracy in the middle of the distribution may look disappointing. However, we should note that, close to the centre, the saddle-point approximation is not substantially different from the central limit approximation. Note also that the saddle-point densities have not been standardised to integrate to one.

7.5.5 The Barndorff-Nielsen formula

Consider a random sample $X_1, \ldots, X_n$ from a continuous distribution belonging to the exponential family form, namely with density

$$f(x; \theta) = \exp\left[\theta\, s(x) - K(\theta)\right] f_0(x) .$$

The statistic $S = \sum s(X_j)$ is complete and sufficient for θ, and the maximum likelihood estimator for θ is obtained by solving $S = n\,K'(\widehat{\theta}_n)$. Next, let us consider the saddle-point approximation for the distribution of S when θ is the true value of the parameter. The following can be verified.

- The cumulant generating function for S is $n\,[\,K(\theta+t) - K(\theta)\,]$.
- The equation for the saddle-point is
$$n\,K'(\theta + t_0) = S\,,$$
which is solved by $\theta + t_0 = \widehat{\theta}_n$, or equivalently, $t_0 = \widehat{\theta}_n - \theta$.
- The observed information for θ is
$$i_n(\widehat{\theta}_n) = n\,K''(\widehat{\theta}_n)\,.$$
See Section 5.3.
- The likelihood function has the form
$$\begin{aligned} L_n(\theta;\, x_1, \ldots, x_n) &= L_n(\theta;\, s) \\ &= \exp\left[\theta\, s - n\,K(\theta)\right]\,. \end{aligned}$$

Therefore, the saddle-point approximation for the density of S is

$$p(s;\,\theta) \sim \sqrt{\frac{1}{2\,n\,\pi\,K''(\widehat{\theta}_n)}}\,\exp\left\{n\left[K(\widehat{\theta}_n) - K(\theta)\right] - (\widehat{\theta}_n - \theta)\,s\right\}.$$

Replacing $n\,K''(\widehat{\theta}_n)$ by $i_n(\widehat{\theta}_n)$, writing the exponential in terms of the likelihood ratio $\widetilde{L}_n = L_n(\theta)/L_n(\widehat{\theta}_n)$, to use Barndorff-Nielsen's notation here, this becomes

$$p(s;\,\theta) \sim (2\,\pi)^{-1/2} \cdot i_n^{-1/2}(\widehat{\theta}_n) \cdot \widetilde{L}_n\,, \tag{7.35}$$

where $\widetilde{L}_n = \widetilde{L}_n(\theta)$.

Next, we reparametrise the model by introducing a new parameter $\tau = E_\theta(S)$. Then s becomes the maximum likelihood estimator for τ, so that we may write $s = \widehat{\tau}_n$. In addition, the reader can check that the observed information for τ is the reciprocal of the observed information for θ. Let us abuse terminology and write the observed information for τ as $i_n(\widehat{\tau})$. Let us also write the density function of $\widehat{\tau}_n$ as $p^*(\widehat{\tau}_n;\,\tau)$. This density has saddle-point approximation

$$p^*(\widehat{\tau}_n;\,\tau) \sim (2\,\pi)^{-1/2} \cdot i_n^{1/2}(\widehat{\tau}_n) \cdot \widetilde{L}_n\,, \tag{7.36}$$

where now $\widetilde{L}_n = L_n(\tau)/L_n(\widehat{\tau}_n)$. This is a version of Barndorff-Nielsen's celebrated formula for the distribution of $\widehat{\tau}_n$. While the derivation given here depends upon the particular parametrisation of the model by τ, the Barndorff-Nielsen formula in (7.36) is parametrisation equivariant as the true density is. If we transform the parameter τ in (7.36) to some other parameter, such as our original θ, and apply standard change of variable techniques to transform the density—and its approximation—we can check that the resulting formula can also be organised into the form given by (7.36). This follows from the fact that $\widetilde{L}_n(\theta) = \widetilde{L}_n(\tau)$ and

$$\sqrt{i_n(\widehat{\tau}_n)} = \sqrt{i_n(\widehat{\theta}_n)} \cdot \left| \frac{d\tau}{d\theta} \right|$$

when $\tau = \tau(\theta)$ and $\widehat{\tau}_n = \tau(\widehat{\theta}_n)$. Therefore

$$\begin{aligned} p^*(\widehat{\theta}_n;\, \theta) &= p^*(\widehat{\tau}_n; \tau) \cdot \left| \frac{d\tau}{d\theta} \right| \\ &\sim (2\,\pi)^{-1/2} \cdot i_n^{1/2}(\widehat{\tau}_n) \cdot \widetilde{L}_n \cdot \left| \frac{d\tau}{d\theta} \right| \\ &= (2\,\pi)^{-1/2} \cdot i_n^{1/2}(\widehat{\theta}_n) \cdot \widetilde{L}_n \cdot \left| \frac{d\theta}{d\tau} \right| \cdot \left| \frac{d\tau}{d\theta} \right| \\ &= (2\,\pi)^{-1/2} \cdot i_n^{1/2}(\widehat{\theta}_n) \cdot \widetilde{L}_n\,. \end{aligned}$$

Thus the formula in (7.36) applies more generally, beyond the parametrisation $\tau = E_\theta(S)$.

To illustrate this formula, consider a random sample of size n from $\mathcal{N}(\theta,\, 1)$. Then $\widehat{\theta}_n = \overline{x}_n$, and

$$\begin{aligned} \widetilde{L}_n &= \exp\left\{ -\frac{1}{2} \left[\sum_{j=1}^n (x_j - \theta)^2 - \sum_{j=1}^n (x_j - \overline{x}_n)^2 \right] \right\} \\ &= \exp\left[-\frac{n}{2}\, (\overline{x}_n - \theta)^2 \right] \\ &= \exp\left[-\frac{n}{2}\, (\widehat{\theta}_n - \theta)^2 \right]. \end{aligned}$$

We also have $i_n(\widehat{\theta}) = n$. Plugging these into the Barndorff-Nielsen formula we get

$$p^*(\widehat{\theta}_n;\, \theta) \sim \sqrt{\frac{n}{2\,\pi}}\, \exp\left[-\frac{n}{2}\, (\widehat{\theta}_n - \theta)^2 \right]$$

which is the exact formula for $\widehat{\theta}_n$ in this case. That the formula works for the normal distribution may not be a surprise.

A more interesting test case is that of a random sample from an exponential distribution with mean θ^{-1}, for which the maximum likelihood estimator is $\widehat{\theta} = n / \sum x_j$. In this case the exact distribution of $\widehat{\theta}_n$ is that of the reciprocal of a Gamma distributed random variable. It is left to the reader to show that

$$p^*(\widehat{\theta}_n;\,\theta) \sim \sqrt{\frac{n}{2\,\pi}}\cdot e^n\cdot\theta^n\cdot\widehat{\theta}_n^{\,-n-1}\,\exp(-n\,\theta/\widehat{\theta}_n)\,,$$

which is the exact formula except for the constant of integration. Once again, we can recognise that the constant that appears in this formula is related to the exact constant via Stirling's approximation to $n!$.

In the general case of formula (7.36), adjustments can also be made to the constant to ensure that the approximation exactly integrates to one. If this is done, it is often the case that

$$p^*(\widehat{\theta}_n;\,\theta) = c_n\cdot i_n^{1/2}(\widehat{\theta}_n)\cdot\widetilde{L}_n\cdot\left[1+O(n^{-3/2})\right]\,,$$

when c_n is chosen to renormalise the saddle-point approximation.

For k-dimensional full exponential families, the formula is similar, except that $i_n(\widehat{\theta}_n)$ must be replaced by $\det i_n(\widehat{\theta}_n)$, the determinant of the observed information matrix, and $(2\,\pi)^{-1/2}$ becomes $(2\,\pi)^{-k/2}$. The formula can also be fine-tuned by rescaling so that it integrates exactly to one.

Outside the exponential family, the Barndorff-Nielsen formula does not provide an approximation for the marginal distribution of the maximum likelihood estimator. The properties of the exponential family are essential here. However, Barndorff-Nielsen (1983) noted that more generally the formula remains a valid approximation to the conditional distribution of $\widehat{\theta}_n$ given an approximate maximal ancillary statistic‖ for θ. In the particular case where the model is a full exponential family, any such maximal ancillary will be independent of $\widehat{\theta}_n$ by Basu's theorem, making the conditional and marginal distributions of $\widehat{\theta}_n$ the same. So for models with vector parameter θ and maximal ancillary $\Omega = \omega(X_1,\ldots,X_n)$, Barndorff-Nielsen's formula becomes

$$p^*(\widehat{\theta}_n;\,\theta\,|\,\Omega) = c_n\cdot\sqrt{\det\,i_n(\widehat{\theta}_n)}\cdot\widetilde{L}_n\cdot\left[1+O(n^{-3/2})\right]\,. \qquad (7.37)$$

See Barndorff-Nielsen (1980) and Barndorff-Nielsen (1983). This celebrated formula was given considerable attention when it first appeared in the literature in this form. The discussion which follows in the remainder of this section may be of interest to statisticians who are familiar with

‖ A statistic is said to be ancillary if its distribution is not dependent on the value of the parameter. That is, if its distribution is functionally free of the parameter. Such a statistic will be a maximal ancillary if every other ancillary statistic can be written as a function of the statistic. If θ is a location parameter then a maximal location invariant statistic is the sort of maximal ancillary intended here. The maximal ancillary may be approximate in the sense that its asymptotic distribution is independent of θ.

the issues involved in arguments involving conditioning. Other readers can skip this discussion with no loss in subsequent sections and chapters.

The Barndorff-Nielsen formula uses two ideas—the conditionality principle from the foundations of inference and the saddle-point approximation from asymptotic analysis. It is worth separating out the foundational aspects of the formula from its analytic aspects in considering its importance.

The conditionality principle was introduced by Fisher (1934), who considered transformation models involving location and scale parameters, where it is possible to condition on a maximal invariant.** Based on the examples given in his paper, Fisher concluded that

> The process of taking account of the distribution of our estimate in samples of the particular configuration observed has therefore recovered the *whole* of the information available... .[I]n general the theoretical process illustrated here uses the information *exhaustively*... . The reduction of the data is sacrificed to its *complete* interpretation.

The italics are mine. Here, the word "configuration" is Fisher's term for a maximal invariant. Fisher's arguments are theoretically sound, but his optimism for conditioning must be treated with some care. In particular, the claim that any procedure is optimal or uses all information can often be rigorously justified, but is rarely independent of the precise interpretation of the optimality involved. In general models, the principle that one can—and should—condition on an ancillary statistic cannot be justified by the same decision-theoretic arguments which can be used when the ancillary is a maximal invariant. Even if such foundational questions are set aside, there remain additional problems. A maximal ancillary statistic may not exist. The principle that one may—or should—condition on an approximate ancillary is also problematic, if one accepts, as I have argued in earlier chapters, that an asymptotic argument is not sufficient in itself to justify the use of a particular method. If conditioning on a statistic is warranted by small sample considerations, then the use of Barndorff-Nielsen's formula is appropriate if that statistic is asymptotically a maximal ancillary. Determining whether one should condition an inference on a statistic requires an understanding of the context of the problem. It is not possible to make sweeping generalisations. In particular, Cox and Reid (1987) state the following.

** In a transformation model, a maximal invariant statistic is ancillary. However, in general an ancillary statistic cannot always be represented as invariant. Transformation models, which possess invariant statistics, will have a maximal invariant statistic. A model which has ancillary statistics will not necessarily have a maximal ancillary.

> Conditioning plays at least two roles in the sampling theory of statistical inference; one is to induce relevance of the probability calculations to the particular data under analysis, and the other to eliminate or reduce the effect of nuisance parameters.

The reader is cautioned not to confuse these two roles.

The other issue to be considered is the validity of the formula as an asymptotic approximation. The formula was originally obtained as a generalisation from special cases. Bardorff-Nielsen (1980, 1983) demonstrated that the formula in (7.37) holds exactly for transformation models. See Problem 11, which illustrates this fact for location models involving the group of translations. For an exponential transformation model—that is, a transformation model which is also of exponential family form—the maximum likelihood estimator is complete and sufficient, and therefore independent of any ancillary statistic. So when applied to an exponential transformation model, the Barndorff-Nielsen formula exactly reproduces the marginal distribution of $\widehat{\theta}_n$. Both the normal and the exponential models, considered earlier, are exponential transformation models. Beyond the class of transformation models, and the exponential family, the formula is also applicable to various curved exponential families. See Butler (2007) for more information. Also motivated by conditional inference, Cox and Reid (1987) have provided a general framework for approximations which are closely related to the Barndorff-Nielsen formula for models involving nuisance parameters.

7.6 Saddle-point method for distribution functions

In statistical applications, it is often more important to compute distribution functions than density functions. For example, in hypothesis testing, the distribution function of a statistic is more useful than the density function, because the former allows us to compute critical values. There are two ways to approximate the distribution function of $\overline{X}_n$ using the saddle-point technique.

The first way is to integrate the saddle-point approximation given in Proposition 3. Suppose $\overline{X}_n$ has distribution function $F(y)$. Then under reasonable regularity conditions

$$F(y) \sim \int_{-\infty}^{y} \sqrt{\frac{n}{2\,\pi\,K''(t)}}\,\exp\left\{n\left[K(t) - t\,x\right]\right\}\,dx\,, \tag{7.38}$$

where $t = t(x)$ is such that $K'(t) = x$. It is helpful to perform a transformation of variables on this integral by integrating with respect to the saddle-point variable t instead of x. Let us assume that we can solve the saddle-point equation to write t as a strictly increasing function of x. Suppose also that $t \to -\infty$ as $x \to -\infty$. Using

$$K''(t)\,dt = dx$$

we can write the integral as

$$F(y) \sim \int_{-\infty}^{t(y)} \sqrt{\frac{n\,K''(t)}{2\,\pi}}\,\exp\{n\,[K(t) - t\,K'(t)]\}\,dt \qquad (7.39)$$

where $K'[t(y)] = y$. The integral given in (7.39) may be easier to compute than that in (7.38) because it does not explicitly use the saddle-point function in the integrand. An additional transformation of this integral is possible. The expression $K(t) - t\,K'(t)$ has a maximum at $t = 0$, which suggests that we should define

$$e^{n\,[K(t) - t\,K'(t)]} = e^{-n\,q^2/2}$$

and integrate with respect to

$$q = \text{sgn}(t)\,\sqrt{2\,[t\,K'(t) - K(t)]}\,, \qquad (7.40)$$

where $\text{sgn}(t)$ denotes the sign of t_0. Using transformation of variables again,

$$F(y) \sim \int_{-\infty}^{q(y)} \sqrt{\frac{n}{2\,\pi\,K''(t)}}\,\left(\frac{q}{t}\right)\,e^{-n\,q^2/2}\,dq\,. \qquad (7.41)$$

Generally, the integrals in (7.38), (7.39) and (7.41) will all be difficult to evaluate. So another approximation is needed. The integrals can be approximated using integration by parts. Substituting into the result developed in Problem 4 at the end of the chapter with $x = q$, $u = q(y)$, and

$$a(u) = \sqrt{\frac{1}{K''[t(y)]}}\,\left[\frac{q(y)}{t(y)}\right]$$

we get the next result.

Proposition 5. *Let*

$$r = \sqrt{n}\,q\,, \qquad \text{and} \qquad v = t\,\sqrt{n\,K''(t)}\,.$$

As $n \to \infty$,

$$F(y) \sim \Phi[r(y)] + \left[\frac{1}{r(y)} - \frac{1}{v(y)}\right]\,\phi[r(y)]\,, \qquad (7.42)$$

where ϕ *and* Φ *are the standard normal density and distribution function, respectively. The error is* $O(n^{-1})$ *uniformly in* y.

Formula (7.42) has attracted attention from statisticians because of the appearance of some well known statistics in this formula. If the density function $f(x)$ is extended to an exponential family of the canonical form

$$f(x;\,t) = e^{t\,x-K(t)}\,f(x)$$

where t is the natural parameter, then the saddle-point equation $K'(t) = \overline{x}$ becomes the likelihood equation for the maximum likelihood estimate of t based on a sample from $f(x;\,t)$ of size n. Therefore, t_0 is the maximum likelihood estimate for t based upon the sufficient statistic $\overline{x}$. The statistic r is seen to be the signed likelihood ratio test statistic for testing the hypothesis that the natural parameter is zero. The statistic v is seen to be a Wald statistic for testing the same hypothesis. Both have an asymptotic normal distribution under standard regularity assumptions. When the density $f(x)$ is normal, the canonical exponential family is the normal location model, with the mean as its natural parameter. In this case, the Wald statistic and the signed likelihood ratio statistic are identical. So for the normal location model, the second term in (7.42) vanishes, and the first term is seen to be exact.

The formula for $F(y)$ given in Proposition 5 is usually known as the Lugannani-Rice formula, and is named after the Robert Lugannani and Stephen Rice who derived an asymptotic expansion for $F(y)$, of which (7.42) displays the first two terms. This asymptotic expansion can be obtained in several ways. The original approach due to Lugannani and Rice involved an expansion of the Fourier inversion integral for $F(y)$. It is also possible to derive the asymptotic expansion by integrating the formula in Proposition 4 term by term.

7.7 Saddle-point method for discrete variables

7.7.1 The formula for integer-valued random variables

When $X_1, X_2, \ldots, X_n$ are integer-valued random variables, then the average $\overline{X}_n$ will be discrete, and take values of the set $j\,n^{-1}$, where j is an integer. The saddle-point method used earlier for continuous random variables has an extension to this case via the Fourier inversion formula for such random variables. Let X_j have cumulant generating function K as earlier. Then the probability mass function of $\overline{X}_n$ is

$$p(\overline{x}) = \frac{1}{2\,\pi\,i}\int_{-i\,\pi}^{i\,\pi} \exp\{n\,[K(z) - z\,\overline{x}]\}\,dz\,.$$

This integral has much the same form as the continuous case, with the exception that the contour of integration is of finite length. Once again, we can perturb the contour, so that it passes through the saddle-point $t_0 = t_0(\overline{x})$, so that

$$p(\overline{x}) = \frac{1}{2\,\pi\,i}\int_{t_0-i\,\pi}^{t_0+i\,\pi} \exp\{n\,[K(z) - z\,\overline{x}]\}\,dz\,. \tag{7.43}$$

and proceed with a Laplace expansion about t. We find that

$$p(\overline{x}) = \frac{1}{\sqrt{2\,n\,\pi\,K''(t_0)}}\exp\{n\,[K(t_0) - t_0\,\overline{x}]\}\,\left[1 + O(n^{-1})\right]\,, \tag{7.44}$$

for $\overline{x} = k\,n^{-1}$. The extension of this formula to a full asymptotic series is parallel to the continuous case as well.

7.7.2 An example with the binomial distribution

Suppose X_j is $\mathcal{B}(N,\,\theta)$, so that $n\,\overline{X}_n$ has binomial distribution $\mathcal{B}(N\,n,\,\theta)$. It is useful to compare the exact distribution of $\overline{X}_n$ with its saddle-point approximation. Then

$$K(t) = N\,\ln[1 + \theta\,(e^t - 1)] \qquad \text{and} \qquad K'(t) = \frac{N\,\theta\,e^t}{1 + \theta\,(e^t - 1)}\,.$$

The saddle-point equation $K'(t_0) = \overline{x}$ is solved by

$$t_0 = \ln\left[\frac{\overline{x}\,(1-\theta)}{(N - \overline{x})\,\theta}\right]\,.$$

The saddle-point approximation for the mass function of $\overline{X}_n$ is then found to be

$$p(\overline{x}) = \frac{N^{n\,N+1/2}}{(2\,\pi\,n)^{1/2}}\cdot\frac{\theta^{n\,\overline{x}}\,(1-\theta)^{n\,(N-\overline{x})}}{\overline{x}^{n\,\overline{x}+1/2}\,(N-\overline{x})^{n\,(N-\overline{x})+1/2}}\,[1 + O(n^{-1})]\,. \tag{7.45}$$

This formula is equivalent to replacing the binomial coefficient in the exact formula by its Stirling approximation. See Problem 5 at the end of the chapter for the derivation and interpretation of this expression.

7.8 Ratios of sums of random variables

The saddle-point method can be applied to ratios $R = X_1/X_2$ where X_1 and X_2 are independent continuous random variables. Suppose that X_2 is strictly positive with finite mean. Then the density of R can be obtained by the Fourier inversion formula

$$f(r) = \frac{1}{2\,\pi\,i}\int_{-\infty}^{\infty} \chi_1(t)\,\chi_2'(-r\,t)\,dt \tag{7.46}$$

where $\chi_1(t)$ and $\chi_2(t)$ are the characteristic functions of X_1 and X_2, respectively. This inversion formula can be derived[††] by using the Fourier inversion formula for the distribution function of $X_1 - r\,X_2$ and the identity

$$P(R \le r) = P(X_1 - r\,X_2 \le 0)\,.$$

When X_1 (respectively, X_2) is the sum of n_1 (respectively, n_2) independent identically distributed random variables, then we may write

$$\chi_1(t) = e^{n_1\,K_1(i\,t)} \qquad \text{and} \qquad \chi_2'(t) = n_2\,i\,K_2'(i\,t)\,e^{n_2\,K_2(i\,t)},$$

where $K_1(t)$ and $K_2(t)$ are the respective cumulant generating functions of any random variable in the numerator and denominator. Therefore,

$$\begin{aligned} f(r) &= \frac{n_2}{2\,\pi}\int_{-\infty}^{\infty} e^{n_1\,K_1(i\,t)+n_2\,K_2(-i\,r\,t)}\,K_2'(-r\,t)\,dt \\ &= \frac{n_2}{2\,\pi\,i}\int_{-i\,\infty}^{i\,\infty} e^{n_1\,K_1(z)+n_2\,K_2(-r\,z)}\,K_2'(-r\,z)\,dz\,. \end{aligned}$$

We wish to apply the saddle-point method to this integral to obtain its asymptotic form as $n = \min(n_1,\,n_2)$ goes to infinity. The contour along the imaginary axis can be perturbed to pass through the saddle-point t_0 so that

$$f(r) = \frac{n_2}{2\,\pi\,i}\int_{t_0-i\,\infty}^{t_0+i\,\infty} e^{n_1\,K_1(z)+n_2\,K_2(-r\,z)}\,K_2'(-r\,z)\,dz\,, \qquad (7.47)$$

where

$$n_1\,K_1'(t_0) - n_2\,r\,K_2'(-r\,t_0) = 0\,. \qquad (7.48)$$

Expanding the exponent about the saddle-point, and evaluating the series from the Laplace expansion, we find that the dominant term is

$$f(r) \sim \frac{n_2\,K_2'(-r\,t_0)\,\exp[n_1\,K_1(t_0) + n_2\,K_2(-r\,t_0)]}{\sqrt{2\,\pi\,[n_1\,K_1''(t_0) + n_2\,r^2\,K_2''(-r\,t_0)]}}\,, \qquad (7.49)$$

where the relative error is $O(n^{-1})$.

Gurland (1948) also obtained an inversion formula for

$$R = \frac{a_1\,X_1 + a_2\,X_2 + \cdots + a_n\,X_n}{b_1\,X_1 + b_2\,X_2 + \cdots + b_n\,X_n}\,.$$

In practice, cases where the numerator and denominator are not independent are more important than the independence case studied above. However, the ratio of dependent continuous variables may not be continuous, and the regularity conditions necessary to obtain a Fourier inversion formula are more difficult to obtain. Cramér (1946a, p. 317) obtained the following result. Suppose X_1 and X_2 have joint characteristic

[††] For the regularity assumptions and a detailed derivation, see Gurland (1948).

function $\chi(t_1, t_2)$, and X_2 is strictly positive. Let $R = X_1/X_2$. Then

$$f(r) = \frac{1}{2\pi i} \int_{-\infty}^{\infty} \left[\frac{\partial \chi(t_1, t_2)}{t_2}\right]_{t_2 = -t_1 r} dt_1$$

provided that the integral is uniformly convergent with respect to x.

7.9 Distributions of M-estimators

7.9.1 The general case

The method used in the previous section can be also be used to approximate the distributions of M-estimators. Let $Y_1, Y_2, \ldots, Y_n$ be independent identically distributed random variables whose common distribution is determined by some real-valued unknown parameter θ. Let $h(\theta, y)$ be a real-valued function such that

$$E_\theta \left[h(\theta, Y) \right] = 0$$

for all θ. An M-estimator for θ is a solution $\widetilde{\theta}_n$ to the equation

$$\sum_{j=1}^{n} h(\widetilde{\theta}_n, Y_j) = 0\,. \tag{7.50}$$

Such M-estimators include the maximum likelihood estimator $\widehat{\theta}_n$ as a special case where h is the score function.

Suppose that $h(\theta, y)$ is continuous and strictly decreasing in θ for all values of y. Suppose also that $h(\theta, y) \to \mp\infty$ as $\theta \to \pm\infty$. Then there will exist a unique solution $\widetilde{\theta}$ to the M-estimating equation (7.50) and

$$P_\theta(\widetilde{\theta} \le a) = P_\theta\left[\sum_{j=1}^{n} h(a, Y_j) \le 0\right]\,. \tag{7.51}$$

We can apply the saddle-point method on the right-hand side of this identity. Define $X_j = h(a, Y_j)$ so that

$$P_\theta(\widetilde{\theta}_n \le a) = P_\theta(\overline{X}_n \le 0)\,. \tag{7.52}$$

The distribution function of $\overline{X}_n$ can be approximated by the methods of Section 7.6.

7.9.2 An example

As an example of this method, consider the maximum likelihood estimate for the shape parameter α of a $\mathcal{G}(\alpha, 1)$ distribution with density

$$f(y;\, \alpha) = \begin{cases} \dfrac{y^{\alpha-1}\, e^{-y}}{\Gamma(\alpha)} & y > 0 \\ 0 & y \le 0\,. \end{cases}$$

For a random sample $y_1, \ldots, y_n$, the maximum likelihood estimate $\widehat{\alpha}_n$ is the solution to the equation

$$n^{-1} \sum_{j=1}^{n} \ln\, y_i - \Psi(\widehat{\alpha}_n) = 0\,, \tag{7.53}$$

where $\Psi(\alpha)$ is the logarithmic derivative of the gamma function‡‡ defined by

$$\Psi(\alpha) = \frac{\Gamma'(\alpha)}{\Gamma(\alpha)}\,.$$

The cumulant generating function $K(t)$ of $X_j = \ln\, Y_j$ and the saddle-point equation are

$$K(t) = \ln\, \Gamma(t + \alpha) - \ln\, \Gamma(\alpha)\,, \qquad \Psi(t_0 + \alpha) = \overline{x}\,.$$

We apply the Lugannani-Rice formula in (7.42) to determine

$$P_\alpha(\widehat{\alpha}_n \le a) = P_\alpha[\overline{X}_n \le \Psi(a)]\,,$$

where the functions $t(y)$, $r(y)$ and $v(y)$ must be evaluated at $y = \Psi(a)$. It can be checked that

$$\begin{aligned} t[\Psi(a)] &= a - \alpha\,, \\ r[\Psi(a)] &= \operatorname{sgn}(a-\alpha)\, \sqrt{2\,n\, [(a-\alpha)\, \Psi(a) + \ln\, \Gamma(\alpha) - \ln\, \Gamma(a)]}\,, \\ v[\Psi(a)] &= (a - \alpha)\, \sqrt{n\, \Psi(1,\, a)}\,. \end{aligned}$$

Here, $\Psi(1, a) = \Psi'(a)$. Programming this in Maple code is straightforward. For example, the statements

$$> \; \Phi := x \rightarrow \frac{1}{2} + \frac{1}{2} \cdot \operatorname{erf}\left(\frac{x}{\sqrt{2}}\right)$$

$$> \; \phi := x \rightarrow \frac{1}{\sqrt{2 \cdot \pi}} \cdot \exp\left(-\frac{x^2}{2}\right)$$

$$> \; t := (a,\, \alpha) \rightarrow a - \alpha$$

‡‡ We are recycling notation here. This is not to be confused with Mills' ratio from Chapter 2.

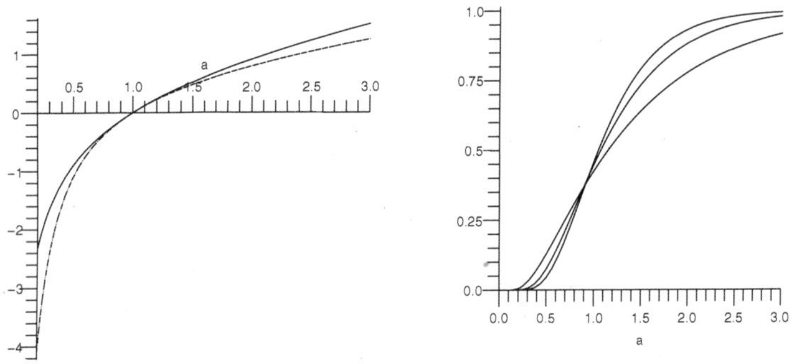

Figure 7.6 *Left: plot of test statistics r and v as functions of a with $n = 1$ and $\alpha = 1$. Right: Saddle-point approximation to the distribution function of $\widehat{\alpha}_n$ for $n = 1, 2, 3$ when $\alpha = 1$.*

$$> \ v := (a,\, \alpha,\, n) \to (a - \alpha) \cdot \sqrt{n \cdot \Psi(1,\, a)}$$

$$> \ r := (a,\, \alpha,\, n) \to$$
$$\text{signum}(a - \alpha) \cdot \sqrt{2 \cdot n \cdot (\,(a - \alpha) \cdot \Psi(a) + \ln(\Gamma(\alpha)) - \ln(\Gamma(a))\,)}$$

$$> F := (a,\, \alpha,\, n) \to$$
$$\Phi(\, r(a,\, \alpha,\, n)\,) + \left(\frac{1}{r(a,\, \alpha,\, n)} - \frac{1}{v(a,\, \alpha, n)}\right) \cdot \phi(\, r(a,\, \alpha,\, n)\,)$$

define the function F which is the saddle-point approximation to the distribution function of $\widehat{\alpha}_n$. The plot on the left of Figure 7.6 shows the test statistics $r(a, 1, 1)$ (unbroken line) and $v(a, 1, 1)$ (dotted line) as functions of a. Note that the two functions vanish at $a = 1$. This is where there is no evidence against the hypothesis $\alpha = 1$. The two functions are tangent here, a reflection of the fact that the two test statistics are asymptotically equivalent as $n \to \infty$. On the right of Figure 7.6 the saddle-point approximation to the distribution function of $\widehat{\alpha}_n$ is displayed for sample sizes $n = 1, 2, 3$, when the true value of the parameter α is set to one.

7.10 The Edgeworth expansion

The determination of the saddle-point t_0 in (7.20) of Proposition 4, or in (7.22) is often difficult. From the asymptotic theory we know that it is close to the origin of the complex plane for sufficiently large n. However, it may be that there is no simple expression for the solution to the equation for the saddle-point.

So it is natural to explore approximations to the saddle-point formula itself. The simplest way to approximate the formula in (7.20) is to replace $K(t)$ by a quadratic approximation. We have

$$K(t) = \mu\, t + \sigma^2\, t^2/2 + O(t^3)$$

and $t_0 = (\overline{x} - \mu)/\sigma^2 + o_p(n^{-1/2})$. So

$$\begin{aligned} K(t_0) - t_0\, \overline{x} &= -\frac{(\overline{x} - \mu)^2}{2\,\sigma^2} + O_p(n^{-3/2}) \text{ and} \\ K''(t_0) &= \sigma^2 + O_p(n^{-1/2}). \end{aligned}$$

Therefore from (7.20), we get

$$\begin{aligned} f(\overline{x}) &\sim \sqrt{\frac{n}{2\,\pi\,\sigma^2}}\, \exp\left\{-\frac{n\,(\overline{x} - \mu)^2}{2\,\sigma^2}\right\} \\ &= \frac{1}{\sigma}\,\phi\left(\frac{\overline{x} - \mu}{\sigma}\right). \end{aligned}$$

This is the normal approximation for $\overline{X}_n$, which can also be derived by a local central limit theorem. We have obtained this result by being quite casual with terms of higher order.

The *Edgeworth approximation* is a sharpening of the central limit approximation to include terms involving the cumulants of the distribution which appear in these higher order terms. Suppose $X_1, \ldots, X_n$ are independent and identically distributed continuous random variables with mean zero and variance one. Define $S_n^* = \sqrt{n}\,\overline{X}_n$, or equivalently, $S_n^* = \sum_{j=1}^n X_j/\sqrt{n}$. By the central limit theorem, S_n^* converges in distribution to standard normal as $n \to \infty$. Furthermore, $E(S_n^*) = 0$ and $\text{Var}(S_n^*) = 1$. The mth cumulant of S_n^* is $\kappa_m\, n^{1-m/2}$, where κ_m is the mth cumulant of X_1. Let $f^*(x)$ be the density function of S_n^*.

We apply formula (2.30) that was derived in Chapter 2. Expanding about the standard normal distribution, we have

$$\begin{aligned} f^*(x) &\sim \exp\left[\sum_{m=3}^{\infty} \frac{\kappa_m}{n^{m/2-1}\, m!}\,(-\partial)^m\right] \phi(x) \\ &= \exp\left[-\frac{\kappa_3}{6\,\sqrt{n}}\,\partial^3 + \frac{\kappa_4}{24\,n}\,\partial^4 - \cdots\right] \phi(x). \end{aligned}$$

For large n, the exponential can be approximated to order $O(n^{-1})$ by

$$f^*(x) \sim \left[1 - \frac{\kappa_3}{6\,\sqrt{n}}\,\partial^3 + \frac{\kappa_4}{24\,n}\,\partial^4 + \frac{\kappa_3^2}{72\,n}\,\partial^6 + \cdots\right] \phi(x). \qquad (7.54)$$

This is the Edgeworth expansion for the density of S_n^*. However, it is

Francis Edgeworth (1845–1926)

Francis Ysidro Edgeworth is best known to statisticians for the expansion that bears his name. A large portion of his research was in economics, where he sought a mathematical framework for utilitarian theory. His work is famous for making no concessions to the reader.

> "In his lectures, his writings, and his conversation Edgeworth leaped from peak to peak and was often difficult to follow.... To the discerning, however, he had always a message which was worth pondering.... To a courtly grace, derived perhaps from his Spanish mother, he added the Irish characteristics of humour, imagination, and generosity."
>
> *The Times* obituary of Edgeworth, February 1926.

often written in a different form, using Hermite polynomials. Define the mth Hermite polynomial $H_m(x)$, where $m \ge 0$ by

$$(-1)^m \, \partial^m \, \phi(x) = H_m(x) \, \phi(x) \, . \tag{7.55}$$

This is a polynomial of degree m for all m. For example, $H_0(x) = 1$, $H_1(x) = x$, $H_2(x) = x^2 - 1$, $H_3(x) = x^3 - 3\,x$. Also

$$H_4(x) = x^4 - 6\,x^2 + 3 \tag{7.56}$$

$$H_5(x) = x^5 - 10\,x^3 + 15\,x \tag{7.57}$$

$$H_6(x) = x^6 - 15\,x^4 + 45\,x^2 - 15 \, . \tag{7.58}$$

Other Hermite polynomials can be computed recursively using the equation

$$H_{m+1}(x) = x\, H_m(x) - H'_m(x) \, . \tag{7.59}$$

This is Problem 6 at the end of the chapter. Alternatively, the Hermite polynomials can be computed from the generating function

$$\sum_{m=0}^{\infty} \frac{H_m(x)}{m!} \, t^m = \exp(x\,t - t^2/2) \tag{7.60}$$

which is Problem 7.

Now let us return to the formula for $f^*(x)$. We can also write the Edgeworth expansion for $f^*(x)$ using Hermite polynomials as

$$f^*(x) \;\sim\; \phi(x) \left[1 + \frac{\kappa_3}{6\,\sqrt{n}} \, H_3(x) + \frac{\kappa_4}{24\,n} \, H_4(x) + \frac{\kappa_3^2}{72\,n} \, H_6(x) + \cdots \right] . \tag{7.61}$$

Upon truncation, the error term is $O_p(n^{-3/2})$, so that

$$f^*(x) = \phi(x) \left[1 + \frac{\kappa_3}{6\sqrt{n}} H_3(x) + \frac{\kappa_4}{24n} H_4(x) + \frac{\kappa_3^2}{72n} H_6(x) + O\left(\frac{1}{n\sqrt{n}}\right)\right] .$$

Note that the fifth order Hermite polynomial is missing among the early terms of the series. This is because the expansion of the exponential is in increasing powers of $1/\sqrt{n}$ rather than increasing orders of the Hermite functions. To expand out additional terms, it is only necessary to collect terms together which have common powers of n.[§§] A series for the distribution function of S_n^* can be obtained by integrating term by term. Note that for $m \ge 1$,

$$\int_{-\infty}^{x} \phi(t) \, H_m(t) \, dt = -\phi(x) \, H_{m-1}(x) \, . \tag{7.62}$$

[§§] The alternative way of grouping terms, which orders them in increasing orders of $H_m(x)$ leads to the Gram-Charlier A series. Unlike the Edgeworth ordering of the series, which is truly asymptotic in $1/\sqrt{n}$, the Gram-Charlier A series does not generally have satisfactory properties in the remainder terms.

See Problem 8. Using this result, we can integrate the Edgeworth expansion for the density to get

$$F^*(x) \sim \Phi(x) - \phi(x)\left[\frac{\kappa_3}{6\sqrt{n}}H_2(x) + \frac{\kappa_4}{24n}H_3(x) + \frac{\kappa_3^2}{72n}H_5(x) + \cdots\right] \tag{7.63}$$

where, once again, the remainder term upon truncation is $O(n^{-3/2})$.

The Edgeworth expansion is often easier to calculate than the saddle-point expansion. This convenience comes with a price. A principal virtue of the saddle-point approximation is that the relative error in the approximation is uniformly small. Unfortunately, this fails for the Edgeworth approximation. While the approximation is good when $\overline{x}$ is close to μ, the approximation can be very poor in the tails. For this reason, the Edgeworth expansion should not be used uncritically to approximate significance levels for tests. However, the (unmodified) expansion is of particular importance for studying the centre of the distribution, as the following application illustrates.

7.11 Mean, median and mode

Using the Edgeworth expansion, we can study the relative locations of the mean, median and mode for distributions which are close to normal. The following approximations are due to Haldane (1942).

We study the distribution up to order $O(1/\sqrt{n})$. From (7.61), an approximation to the mode is given by solving

$$\begin{aligned} 0 &= \frac{d}{dx}f^*(x) \\ &= \phi'(x)\left[1 + \frac{\kappa_3}{6\sqrt{n}}H_3(x)\right] + \phi(x)\frac{\kappa_3}{6\sqrt{n}}H_3'(x) + O(n^-1) \\ &= \phi(x)\left[-x - \frac{\kappa_3}{6\sqrt{n}}(x^4 - 6x^2 + 3)\right] + O(n^{-1}) \end{aligned}$$

This will be solved for $x = O(n^{-1/2})$, so that $x^4 - 6x^2 + 3 = 3 + O(n^{-1})$. So the equation is solved by $x = -\kappa_3/(2\sqrt{n}) + O(n^{-1})$, which is the mode.

To determine the median in terms of κ_3, we use (7.63), by solving

$$\begin{aligned} 1/2 &= F^*(x) \\ &= \Phi(x) - \phi(x)\frac{\kappa_3(x^2-1)}{6\sqrt{n}} + O(n^{-1}) \\ &= 1/2 + x\,\phi(x) + \phi(x)\frac{\kappa_3}{6\sqrt{n}} + O(n^{-1}). \end{aligned}$$

In the last step, we have used the fact that $x = O(n^{-1/2})$. This reduces to

$$0 = x + \frac{\kappa_3}{6\sqrt{n}} + O(n^{-1})$$

which is solved by $x = -\kappa_3/(6\sqrt{n}) + O(n^{-1})$.

Finally, we note that the mean of the distribution $f^*(x)$ was originally set to zero, with no error in this case. Therefore

$$\text{Mode} = -\frac{\kappa_3}{2\sqrt{n}} + O(n^{-1}) \tag{7.64}$$

$$\text{Median} = -\frac{\kappa_3}{6\sqrt{n}} + O(n^{-1}) \tag{7.65}$$

$$\text{Mean} = 0\,. \tag{7.66}$$

From these formulas, some rough conclusions can be drawn about the relative locations of these three measures of centrality. For positively skewed distributions, the three measures will usually be ordered as

$$\text{Mode} < \text{Median} < \text{Mean}$$

such that

$$3 \times (\text{Median} - \text{Mode}) \approx 2 \times (\text{Mean} - \text{Mode})$$

For negatively skewed distributions, the ordering will be reversed with similar proportions as above.

7.12 Hayman's saddle-point approximation

In some problems, it is possible to obtain the probability generating function for the distribution of a discrete random variable without having a simple expression for its probability mass function. Even when an explicit formula for the probability mass function is available, it may not be a simple matter to determine the asymptotic tail behaviour of the random variable from an examination of the form of the function. The method that we shall consider in this section can be used to construct asymptotic expansions for probability mass function in the tail of the distribution.

Suppose X is a random variable taking values on the nonnegative integers, with probability generating function

$$A(t) = \pi_0 + \pi_1\, t + \pi_2\, t^2 + \pi_3\, t^3 + \cdots\,,$$

where $\pi_j = P(X = j)$. We wish to find an asymptotic formula for π_n as $n \to \infty$. The coefficient π_n can be obtained from $A(t)$ via the formula

$$\pi_n = \frac{A^{(n)}(0)}{n!}\,.$$

This is quite useful when n is small, but is difficult to work with when n is large. Alternatively, π_n can be obtained using the Cauchy residue theorem as

$$\pi_n = \frac{1}{2\,\pi\,i} \oint_C A(z)\, z^{-n-1}\, dz\,, \tag{7.67}$$

where C is a simple closed contour around $z = 0$, such that C and its interior lie within the domain of the complex plane where $A(z)$ is analytic.

Next, suppose that $A(z)$ is analytic in the entire complex plane.¶¶ Consider the contour C which is the circle of radius $t > 0$ about the origin. Using polar coordinates in equation (7.67) along the contour gives us the integral formula

$$\begin{aligned} \pi_n &= \frac{1}{2\,\pi\,t^n} \int_{-\pi}^{\pi} A(t\,e^{i\,\theta})\, e^{-n\,i\,\theta}\, d\theta \\ &= \frac{1}{2\,\pi\,t^n} \int_{-\pi}^{\pi} \exp\left[\ln\, A(t\,e^{i\,\theta}) - n\,i\,\theta\right]\, d\theta\,. \end{aligned} \tag{7.68}$$

The presence of an n in the exponent of the integrand suggests the use of the Laplace method. To do this, we shall need to maximise the entire exponent because $n\,i\,\theta$ has no quadratic component in θ. Expanding in θ we get

$$\ln\, A(t\,e^{i\,\theta}) - n\,i\,\theta = \ln\, A(t) + i\,\theta\,[a(t) - n] - \frac{1}{2}\,\theta^2\, b(t) + \cdots,$$

where

$$a(t) = t\frac{d}{dt}\, \ln\, A(t)\,, \qquad b(t) = t\,\frac{d}{dt}\, a(t)\,.$$

Suppose we can choose t so that the linear term in θ vanishes. Let us call this value t_n, with its defining saddle-point equation $a(t_n) = n$. Then the exponent of the integrand has a local maximum at $\theta = 0$, which corresponds to the point where the contour intersects the real axis. Then we may use the Laplace method to obtain

$$\begin{aligned} \pi_n &= \frac{1}{2\,\pi\,t^n} \int_{-\pi}^{\pi} A(t_n) \exp\left[-\frac{1}{2}\,\theta^2\, b(t_n) + o(\theta^2)\right]\, d\theta \\ &\sim \frac{A(t_n)}{t_n^n\,\sqrt{2\,\pi\,b(t_n)}}\,. \end{aligned}$$

Note that the standard saddle-point assumptions must be satisfied. In particular, $b(t_n) \to \infty$. A full set of sufficient conditions has been provided by Hayman (1956) as follows.

¶¶ This is a very strong condition. Many nonnegative integer-valued random variables in common use do not satisfy this condition. For these random variables, we shall need the method of Darboux which we consider later.

Definition 1. *A generating function* $A(t)$ *is said to be* Hayman admissible *if it satisfies the following three conditions.*

1. *There is some* $t^* > 0$ *such that* $A(t) > 0$ *for* $t > t^*$. *(When* $A(t)$ *is a probability generating function, then this condition is automatically satisfied.)*
2. $\lim_{t\to\infty} b(t) = \infty$.
3. *For* $t > t^*$, *there exists a function* $\delta(t)$ *such that*

$$0 < \delta(t) < \pi\,, \qquad \text{for all } t > t^*\,,$$

$$A(t\,e^{i\,\theta}) \;\sim\; A(t)\,e^{i\,\theta\,a(t)-\theta^2\,b(t)/2}$$

as $t \to \infty$ *uniformly in* $|\theta| \le \delta(t)$, *and*

$$A(t\,e^{i\,\theta}) = o\left(\frac{A(t)}{\sqrt{b(t)}}\right)$$

as $t \to \infty$ *uniformly in* $\delta(t) \le |\theta| \le \pi$.

The major obstacle to the implementation of Hayman's method is the verification that the relevant generating function is admissible according to Definition 1. Hayman (1956) provided the following result, which can be used to verify that a variety of generating functions are admissible.

Proposition 6. *Let* $A(z)$ *and* $B(z)$ *be Hayman admissible functions, analytic in the entire complex plane. Let* $p(z)$ *be a polynomial in* z. *Then the following functions are also Hayman admissible.*

1. $A(z)\,B(z)$;
2. $\exp[A(z)]$;
3. $A(z) + p(z)$;
4. $p(z)\,A(z)$ *provided the leading coefficient of* $p(z)$ *is positive;*
5. $p[A(z)]$ *provided the leading coefficient of* $p(z)$ *is positive;*
6. $\exp[p(z)]$ *provided the coefficients* c_n *of*

$$\exp[p(z)] = \sum_{n=0}^{\infty} c_n\,z^n$$

are positive for sufficiently large n.

For a proof, see Hayman (1956). For Hayman admissible generating functions, we have the next asymptotic result.

Proposition 7. *Let X be a nonnegative integer-valued random variable whose probability generating function $A(t)$ has analytic extension $A(z)$ into the entire complex plane. If $A(z)$ is Hayman admissible, then as $n \to \infty$, there exists a unique solution t_n in the interval $t > t^*$ to the equation $a(t_n) = n$ such that*

$$P(X = n) \sim \frac{A(t_n)}{t_n^n \sqrt{2\pi\, b(t_n)}}. \tag{7.69}$$

For a detailed proof, see Wong (2001).

The method outlined above can also be used to compute an asymptotic formula for $\tau_n = P(X > n)$. The condition that $A(t)$ be analytic in the entire complex plane is a very strong assumption, and implies that

$$\begin{aligned} P(X > n) &= P(X = n+1) + P(X > n+1) \\ &\sim P(X = n+1). \end{aligned}$$

That is, the $(n+1)^{\text{st}}$ term in the sum over tail probabilities dominates the remaining terms. However, the asymptotic formula for the tail probability can be computed directly. The generating function

$$A_1(t) = \tau_0 + \tau_1 t + \tau_2 t^2 + \tau_3 t^3 + \cdots$$

can be computed from $A(t)$ using the formula

$$\begin{aligned} A_1(t) &= \frac{1 - A(t)}{1 - t} \\ &\sim \frac{A(t)}{t} \text{ as } t \to \infty. \end{aligned}$$

The singularity of $A_1(t)$ at $t = 1$ can be removed by setting $A_1(1) = E(X)$. Define

$$\begin{aligned} a_1(t) &= t \frac{d}{dt} \ln \frac{1 - A(t)}{1 - t} \\ &\sim a(t) - 1. \end{aligned}$$

In addition,

$$\begin{aligned} b_1(t) &= t \frac{d}{dt} a_1(t) \\ &\sim b(t). \end{aligned}$$

This can be tightened up to yield

$$a_1(t) = a(t) - 1 + o(1) \tag{7.70}$$

as $t \to \infty$. So there exists a solution t_{1n} to the equation $a_1(t_{1n}) = n$ in the interval between t_n and t_{n+2}, where

$$a(t_{1n}) = a(t_{n+1}) + o(1)\,, \tag{7.71}$$

To prove the statements for t_{1n}, note that asymptotically

$$\begin{aligned} a_1(t_n) - n &= -1 + o(1) < 0 \text{ and} \\ a_1(t_{n+2}) - n &= 1 + o(1) > 0\,. \end{aligned}$$

Since $a_1(t)$ is continuous there must lie a root of the equation $a_1(t) = n$ between t_n and t_{n+2}. Statement (7.71) follows immediately from (7.70).

Proposition 8. *Suppose a nonnegative integer-valued random variable X has probability generating function $A(z)$ that is analytic in the entire complex plane and Hayman admissible. Define*

$$A_1(z) = \begin{cases} \frac{1-A(z)}{1-z} & z \neq 1 \\ \\ E(X) & z = 1\,. \end{cases}$$

Then $A_1(z)$ is analytic in the entire plane and is also Hayman admissible.

Proof. As noted above, the singularity in $A_1(z)$ at $z = 1$ is removable by setting $A(1) = E(X)$. To check Hayman admissibility, we note that condition 1 is straightforward, and that condition 2 follows from $b_1(t) \sim b(t)$. For condition 3, note that for θ and δ as specified for the condition on $A(z)$,

$$\begin{aligned} A_1(t\,e^{i\,\theta}) &\sim \frac{A(t\,e^{i\,\theta})}{t\,e^{i\,\theta}} \\ &\sim \frac{A(t)\,e^{i\,\theta\,a(t)-\theta^2\,b(t)/2}}{t\,e^{i\,\theta}} \\ &= \frac{A(t)}{t}\,e^{i\,\theta\,[a(t)-1]-\theta^2\,b(t)/2} \\ &\sim A_1(t)\,e^{i\,\theta\,a_1(t)-\theta^2\,b_1(t)/2}\,. \end{aligned}$$

The fourth condition also follows from elementary manipulations. ∎

Finally, the following holds.

Proposition 9. *If $A_1(z)$ is analytic in the entire complex plane, and Hayman admissible, then*

$$P(X > n) \sim \frac{A_1(t_{1n})}{t_{1n}^n\,\sqrt{2\,\pi\,b_1(t_{1n})}}\,, \tag{7.72}$$

where $a_1(t_{1n}) = n$.

7.13 The method of Darboux

The assumption that $A(t)$ is analytic throughout the complex plane is necessary for Hayman's method. Note that the saddle-point equation $a(t_n) = n$ is solved by increasingly large values t_n which correspond to contours of increasing radius for the Cauchy residues. However, if $A(t)$ has singularities away from the origin, then these contours will eventually expand to include these points in their interiors. Thus the original formula (7.67) will cease to be valid.

Many probability generating functions have singularities. Not only is the methodology above invalid, but the asymptotics for the probabilities π_n become qualitatively different in character. Typically, such singularities occur when π_n goes to zero at an exponential rate or slower—a common property of many probability distributions. For such cases, the method of Darboux can often be used to provide an asymptotic expansion of π_n. Although this method is not a saddle-point method, it is natural to consider it here as a solution to the problem considered earlier.

Suppose X is a nonnegative integer-valued random variable with probability generating function

$$A(z) = \pi_0 + \pi_1 z + \pi_2 z^2 + \pi_3 z^3 + \cdots$$

where $\pi_n = P(X = n)$. In general, $A(z)$ is analytic in the disc $|z| < 1$ but may have singularities outside this disc. Let us assume that there is some point $t \geq 1$ where A has a singularity, and that $A(z)$ is analytic strictly at all points of the disc $|z| \leq t$ except the point $z = t$. Suppose we can write

$$A(z) = (1 - q\,z)^\alpha\, P(q\,z) \tag{7.73}$$

where $q = t^{-1}$ and $P(z)$ is analytic at $z = 1$. In this expression, α is not constrained to be an integer. When this is the case, we define $(1 - q\,z)$ in the region where $|z| < q^{-1}$ through its principal branch. Now let us expand the function $P(z)$ about the point $z = 1$. Suppose that this expansion has the form

$$P(z) = p_0 + p_1\,(1 - z) + p_2\,(1 - z)^2 + \cdots . \tag{7.74}$$

We can approximate $P(z)$ by the polynomial $P_m(z) = \sum_{j=0}^{m} p_j\,(1 - z)^j$. So combining this polynomial approximation with (7.73), we can define the mth Darboux approximant to $A(z)$ to be

$$A_m(z) \quad = \quad (1 - q\,z)^\alpha \sum_{j=0}^{m} p_j\,(1 - q\,z)^j$$

$$= \sum_{j=0}^{m} p_j \, (1 - q\, z)^{\alpha+j} \, . \tag{7.75}$$

The mth Darboux approximant $A_m(z)$ is analytic in the disk where $|z| < q^{-1}$. So we can expand $A_m(z)$ about $z = 0$ to obtain

$$A_m(z) = \sum_{n=0}^{\infty} \frac{A_m^{(n)}(0)}{n!} \, z^n \, . \tag{7.76}$$

The coefficients in (7.76) can easily be determined to be

$$\frac{A_m^{(n)}(0)}{n!} = (-q)^n \sum_{k=0}^{m} p_k \binom{\alpha + k}{n} . \tag{7.77}$$

See Problem 9. Then the following result can be proved.

Proposition 10. *Under the regularity conditions described above,*

$$P(X = n) = q^n \left[(-1)^n \sum_{k=0}^{m} p_k \binom{\alpha + k}{n} + o(n^{-\alpha-m-1}) \right] .$$

for all $m = 0, 1, 2, \cdots$.

See Wong (2001) for a proof. Note that the series displayed above is not an asymptotic series as defined in Section 2.4 of Chapter 2. However, it can be regarded as asymptotic in a generalised sense that the remainder term is of asymptotic type. For example, when $m = 1$ and $\alpha = -1$, the equation reduces to $\pi_n = q^{-n} \, [p_0 + o(1)]$. So when $\alpha = 1$, we see that the tail of the distribution is asymptotically equivalent to a geometric tail. When $\alpha = -1$, setting m to successively higher values, we see that $\pi_n = q^{-n} \, [p_0 + o(n^{-m})]$ for all m.

7.14 Applications to common distributions

7.14.1 Poisson distribution

When X has a Poisson distribution with parameter λ, then the probability generating function $A(t)$, and the functions $a(t)$ and $b(t)$ have the form

$$\begin{aligned} A(t) &= \exp[\lambda\, (t - 1)] \, , \\ a(t) &= \lambda\, t \, , \text{ and} \\ b(t) &= \lambda\, t \, . \end{aligned}$$

The probability generating function $A(t)$ has an analytic continuation into the entire complex plane. We can use property 6 of Proposition 6 to verify that $A(z)$ is Hayman admissible. The saddle-point equation $a(t_n) = n$ is solved by $t_n = n/\lambda$. It can be checked that Hayman's saddle-point approximation reduces to

$$P(X = n) \sim \frac{\lambda^n \exp(n - \lambda)}{n^n \sqrt{2\pi n}} \tag{7.78}$$

which is the exact formula for $P(X = n)$ with the exception that $n!$ has been replaced by its Stirling approximation. The reader will recall that a similar phenomenon occurs with Daniels' saddle-point approximation to the gamma distribution.

To approximate $P(X > n)$, we determine that the generating function for the sequence $\tau_n = P(X > n)$ is

$$A_1(t) = \frac{1 - \exp[\lambda\,(t-1)]}{1-t},$$

which has an analytic extension into the entire complex plane. and is Hayman admissible. The equation $a_1(t_{1n}) = n$ cannot be solved in simple closed form. It is helpful to use the asymptotic formulas

$$t_{1n} = \frac{n+1}{\lambda} + O(n^{-1})$$

and

$$b_1(t) = \lambda\, t + o(t)\,.$$

See Problem 10. Using the asymptotic formulas

$$A_1(t) \sim t^{-1}\,, \qquad t_1 \sim (n+1)/\lambda\,, \qquad \text{and} \qquad b_1(t) \sim \lambda\, t\,,$$

and Proposition 9, we find that

$$\begin{aligned} P(X > n) &\sim \frac{\lambda^{n+1}\, e^{-\lambda}}{\sqrt{2\pi\,(n+1)}\,(n+1)^{n+1}\, e^{-(n+1)}} \\ &\sim P(X = n+1)\,, \end{aligned}$$

which should come as no surprise. Sharper results can be obtained by using the exact formula for $A_1(t)$ and better approximations to other quantities. See Problem 10. Once again, the last step in the approximation above is obtained from Stirling's approximation to $(n+1)!$. The final result is in keeping with a fact that the leading term of the tail sum is asymptotically dominant for the Poisson distribution.

7.14.2 Binomial distribution

The right tail of a $\mathcal{B}(N, p)$ random variable X is asymptotically degenerate in the sense that $P(X = n) = 0$ when $n > N$. So it is instructive to see what happens with Hayman's method on this example. The observations below also hold for any nonnegative integer valued random variable with finite support.

The probability generating function $A(t)$ of X is a polynomial which has an analytic extension into the entire complex plane. However, it is easy to check that $A(z)$ is not Hayman admissible. As $t \to \infty$,

$$A(t) \sim p^N t^N, \quad a(t) \sim N, \quad \text{and } b(t) \to 0.$$

So condition 2 of Definition 1 fails.

7.14.3 Neyman type A distribution

The Neyman type A distribution is produced by a sum of independent identically distributed Poisson random variables, where the number of terms in the sum is itself Poisson distributed and independent of the sequence of terms of the sum. We can write the probability generating function for such a distribution in the form

$$A(t) = \exp\left\{\lambda\left[e^{\mu(t-1)} - 1\right]\right\}.$$

This can be seen to be of the form $A(t) = A_\lambda[A_\mu(t)]$, where A_λ and A_μ are Poisson probability generating functions. The parameter μ is often called the *clumping index* of the distribution, as the probability function can be multimodal, with the number of modes increasing for larger values of μ.

The function $A(z)$ is defined for all complex values of z, but we need to check that $A(t)$ is Hayman admissible.

- Using our results from Section 7.14.2, we deduce that
$$A_\mu(z) = \exp[\mu(z-1)]$$
is Hayman admissible.
- Next, substituting $p(z) = \lambda(z-1)$ in property 5 of Proposition 6, we see that
$$\lambda\left[e^{\mu(z-1)} - 1\right]$$
is admissible.
- The admissibility of $A(t)$ then follows from property 2 of the same proposition.

Expressions for $a(t)$ and $b(t)$ are given by

$$a(t) = \lambda\,\mu\,t\,e^{\mu\,(t-1)} \qquad \text{and} \qquad b(t) = a(t)\,(1+\mu\,t)\,.$$

Although a simple expression for t_n is not available, the solution can be formally as

$$t_n = \frac{1}{\mu}\,W\left(\frac{n\,e^{\mu}}{\lambda}\right)$$

where W is the Lambert W function. In Maple, the Lambert W function is called using the command *LambertW*.

Now we can write

$$A(t) = e^{-\lambda}\,\exp\left[\frac{a(t)}{\mu\,t}\right]\,.$$

Therefore, using $a(t_n) = n$ we deduce that

$$A(t_n) = e^{-\lambda}\,\exp\left[\frac{n}{W(n\,e^{\mu}/\lambda)}\right] \qquad \text{and} \qquad b(t_n) = n\,\left[1 + W(n\,e^{\mu}/\lambda)\right]\,.$$

So if X has a Neyman type A distribution, then

$$P(X = n) \sim \frac{\mu^n\,e^{-\lambda}}{[W(n\,e^{\mu}/\lambda)]^n\,\sqrt{2\,\pi\,n\,[1 + W(n\,e^{\mu}/\lambda)]}}\,\exp\left[\frac{n}{W(n\,e^{\mu}/\lambda)}\right]$$

as $n \to \infty$. In this formula, the values of W and t_n increase quite slowly in n. This is in keeping with the well-known fact that the Neyman-type A distribution has long tails. Bounds on the rate at which $t_n \to \infty$ can be obtained from inequalities. For example, when $\lambda = \mu = 1$, then $a(t) \geq e^{t-1}$ for $t \geq 1$. So $t_n < 1 + \ln\,n$. It is left as an exercise for the reader to show that

$$1 + \ln\,n - \ln\,(1 + \ln\,n) \leq t_n \leq 1 + \ln\,n$$

for all values of $n \geq 1$. The lower bound is a better approximation, although neither bound is satisfactory for the purposes of asymptotics.

7.14.4 Negative binomial (Pascal) distribution

In a sequence of Bernoulli trials with success probability p, let X be the number of trials until r successes have occurred, for $r = 1, 2, \ldots$. If π_n is the probability that exactly n failures occur before the rth success, then

$$\pi_n = \binom{-r}{n}\,p^r\,(p-1)^n \qquad n = 0, 1, 2, \cdots\,.$$

The distribution of the number of failures is known as the negative binomial or Pascal distribution. When $r = 1$, this reduces to $\pi_n = p\,(1-p)^n$, which is the geometric distribution.

The probability generating function for the negative binomial has the form

$$A(z) = \left(\frac{p}{1 - q\,z}\right)^r, \qquad \text{where } q = 1 - p\,.$$

This function is not analytic through the complex plane, and has a pole at $z = q^{-1}$ of order r. We can apply the method of Darboux to this case, with $\alpha = -r$ and $P \equiv p^r$. Therefore $p_0 = p^r$ and $p_k = 0$ for $k \geq 1$. Applying Proposition 10 and using $q = 1 - p$, we find that

$$\begin{aligned} \pi_n &= q^n \left[(-1)^n\, p_0 \binom{-r}{n} + o(n^{-m})\right] \\ &= \binom{-r}{n} p^r\,(p-1)^n + o(n^{-m}) \end{aligned}$$

for all $m \geq 0$. In fact, the error term is zero and the approximation is exact.

7.14.5 Return to equilibrium in a random walk

Consider a simple random walk $S_n = Y_1 + \cdots + Y_n$, where $S_0 = 0$,

$$Y_j = \begin{cases} +1 & \text{with probability } \alpha \\ -1 & \text{with probability } \beta = 1 - \alpha\,, \end{cases}$$

and Y_j, $j \geq 1$ are independent. We shall assume that $0 < \alpha < 1$. The random walk S_n is said to return to equilibrium at time k if $S_k = 0$. Clearly such a return to equilibrium can only occur when k is even. So we may write $k = 2\,n$. Let $\pi_n = P(S_{2\,n} = 0)$. The generating function for the sequence π_n is

$$\begin{aligned} A(t) &= \pi_0 + \pi_1\,t + \pi_2\,t^2 + \cdots \\ &= 1 + t\,P(S_2 = 0) + t^2\,P(S_4 = 0) + \cdots\,. \end{aligned}$$

The exact value of π_n may be determined by a binomial probability to be

$$\pi_n = (-1)^n \binom{-\frac{1}{2}}{n} (4\,\alpha\,\beta)^n$$

The generating function is given in explicit form by

$$A(t) = \frac{1}{\sqrt{1 - 4\,\alpha\,\beta\,t}}\,.$$

See Feller (1968, p. 273). In this case, the generating function $A(t)$ is not a probability generating function. However, the coefficients are positive, and the method of Darboux applies in the same way. Choosing $p_0 = 1$,

$p_k = 0$ for $k \geq 1$, and $q = 4\,\alpha\,\beta$ in Proposition 10, we determine that for all $m \geq 0$,

$$\pi_n = (4\,\alpha\,\beta)^n \left[(-1)^n \binom{-\frac{1}{2}}{n} + o(n^{m-1/2})\right].$$

Once again, the error term is seen to be zero. In both this example, and the Pascal distribution above, the approximation is exact because the factor $P(z)$ is a polynomial, of degree zero in particular.

7.15 Problems

1. Prove (7.16) and (7.17).

2. In Section 7.4, the derivation of Daniels' saddle-point approximation required that Assumption 4 be verified. To prove this, you will need the Riemann-Lebesgue lemma, which states that the Fourier transform of an integrable function vanishes at infinity.

   (a) First, argue that $e^{t_0\,x}\,f(x)$ is an integrable function of x.

   (b) Apply the Riemann-Lebesgue Lemma to the integral

   $$\int_{-\infty}^{\infty} e^{t_0\,x}\,e^{i\,s\,x}\,f(x)\,dx$$

   and show that $M(t_0 + i\,s) \to 0$ as $s \to \pm\infty$.

   (c) Conclude from this fact that Assumption 4 holds.

3. The following problem relates to the derivation of formula (7.29) in Section 7.5.3.

   (a) Obtain the moment generating function and the cumulant generating function for the Gamma density $\mathcal{G}(\alpha,\,\lambda)$.

   (b) Derive the saddle-point approximation for the Gamma density given in (7.29).

   (c) Find the exact density for $\overline{X}_n$ when $X_1,\,\ldots,\,X_n$ is a random sample from $\mathcal{G}(\alpha,\,\lambda)$.

   (d) Compare the true density for $\overline{X}_n$ with the saddle-point approximation above, and prove that, with the exception of the constant of integration, the saddle-point formula is exact.

   (e) Use Stirling's approximation to show that the saddle-point approximation for $\overline{X}_n$ is asymptotically equivalent to the true density for large n.

4. Consider an integral of the form

$$I_n(u) = \int_{-\infty}^{u} a(x)\, \phi(x;\, n^{-1})\, dx$$

where $\phi(x;\, n^{-1})$ is the normal density function with mean zero and variance n^{-1} evaluated at x. Assume that $a(x)$ is a smooth function such that $a(0) \neq 0$ and such that $\exp(-x^2)\, a(x)$ goes to zero as x goes to $\pm\infty$. We wish to obtain an asymptotic expression for $I_n(u)$ as $n \to \infty$.

(a) Show that

$$I_n(u) = a(0)\, \Phi(\sqrt{n}\, u) + \int_{-\infty}^{u} [\, a(x) - a(0)\,] \cdot \sqrt{n}\, \phi(\sqrt{n}\, x)\, dx\,,$$

where ϕ and Φ are the standard normal density and distribution functions, respectively.

(b) Show that for fixed $u > 0$,

$$I_n(u) = a(0)\, [1 + O(n^{-1})]$$

(c) Use integration by parts on the expression in part (a) to show that

$$I_n(u) = a(0)\, \Phi(\sqrt{n}\, u) - \frac{a(u) - a(0)}{\sqrt{n}\, u}\, \phi(\sqrt{n}\, u) + R_n(u)\,,$$

for suitable remainder term $R_n(u)$. Hint: $\phi' = -x \cdot \phi$.

(d) Suppose in addition that we are given

$$I_n(\infty) \sim 1 \qquad \text{as } n \to \infty\,.$$

Use these results above to obtain the formal expansion

$$I_n(u) \sim \Phi(\sqrt{n}\, u) + \left[\frac{1}{\sqrt{n}\, u} - \frac{a(u)}{\sqrt{n}\, u} \right] \phi(\sqrt{n}\, u)\,.$$

The error in the expansion on the right-hand side can be shown to be $O(n^{-1})$ uniformly in u. See Temme (1982).

5. Suppose $X_1, \ldots, X_n$ is a random sample from $\mathcal{B}(N,\, \theta)$. See Section 7.7.2. We consider the steps to derive the saddle-point approximation to the probability mass function of $\overline{X}_n$, taking values on the set of integer multiples of n^{-1}.

(a) Show that the cumulant generating function for $\mathcal{B}(N,\, \theta)$ is

$$K(t) = N \ln \left[1 + \theta\, (e^t - 1) \right]\,.$$

(b) Solve the equation $K'(t_0) = \overline{x}$ to obtain the saddle-point

$$t_0 = \ln \left[\frac{\overline{x}\,(1-\theta)}{(N-\overline{x})\,\theta} \right] .$$

(c) Derive the formula for the saddle-point approximation given in (7.45).

6. Prove the recursion for Hermite polynomials

$$H_{m+1}(x) = x\,H_m(x) - H'_m(x) , \qquad (7.79)$$

as in equation (7.59).

7. The generating function $A(x,\, t)$ of a sequence $g_n(x)$ of functions of a real variable x can be defined in a way that is similar to the generating function of a sequence, namely

$$A(x,\, t) = g_0(x) + g_1(x)\,t + g_2(x)\,t^2 + \cdots .$$

Let $H_m(x)$ denote the mth Hermite polynomial. Prove that the sequence $H_m(x)\,/\,m!$ has generating function

$$\sum_{m=0}^{\infty} \frac{H_m(x)}{m!}\,t^m = \exp(x\,t - t^2/2) .$$

8. Let $\phi(x)$ denote the standard normal probability density function. Prove that the Hermite polynomials satisfy the integral equation

$$\int_{-\infty}^{x} \phi(t)\,H_m(t)\,dt = -\phi(x)\,H_{m-1}(x) .$$

9. Verify equation (7.77).

10. Complete the derivation of the saddle-point approximation for $P(X > n)$ when X has a Poisson distribution as in Section 7.14.1. In particular, you should check that

$$t_{1n} = \frac{n+1}{\lambda} + O(n^{-1})$$

and

$$b_1(t) = \lambda\,t + o(t) .$$

as required.

Continue the investigation of the asymptotic behaviour of $P(X > n)$ by improving the result given in Section 7.14.1. In particular, write

$$P(X > n) \sim c(\lambda, n) \cdot P(X = n + 1)$$

where $c(\lambda, n) \sim 1$. Find a formula for $c(\lambda, n)$.

11. Suppose $X_1, \ldots, X_n$ is a random sample from continuous distribution on the real line with density function $f(x-\theta)$. It is required to estimate the location parameter θ. Let us assume that the maximum likelihood estimator $\widehat{\theta}_n$ exists and is unique with probability one.

(a) Define the random vector $\Omega = \omega(X_1, \ldots, X_n)$ by

$$\begin{aligned} \omega(x_1, x_2, \ldots, x_n) &= (\omega_1, \omega_2, \ldots, \omega_n) \\ &= (x_1 - \widehat{\theta}_n, x_2 - \widehat{\theta}_n, \ldots, x_{n-1} - \widehat{\theta}_n)\,. \end{aligned}$$

Prove that Ω is a maximal location invariant statistic. In other words, show that any real valued function $h(x_1, \ldots, x_n)$ which is invariant in the sense that

$$h(x_1 + y, \ldots, x_n + y) = h(x_1, \ldots, x_n) \qquad \forall\, y, x_1, \ldots, x_n$$

can be written in the form

$$h(x_1, \ldots, x_n) = h_1(x_1 - \widehat{\theta}_n, \ldots, x_n - \widehat{\theta}_n)$$

for some function h_1.

(b) Prove that the conditional density of $\widehat{\theta}_n$ has the form

$$\begin{aligned} p(\widehat{\theta}_n; \theta \,|\, \Omega) &= c(\omega_1, \ldots, \omega_n) \cdot \prod_{j=1}^{n} f(x_j - \theta) \\ &:= c(\omega_1, \ldots, \omega_n) \cdot \prod_{j=1}^{n} f(\widehat{\theta}_n + \omega_j - \theta)\,. \end{aligned}$$

(c) Show that

$$L_n(\widehat{\theta}_n) = \prod_{j=1}^{n} f(\omega_j)\,.$$

(d) Prove that Barndorff-Nielsen's formula also has the form of the expression in part (b) above.

(e) Finally, prove that when the constant c_n Barndorff-Nielsen's p^* formula is chosen to make the expression integrate to one, then the p^* formula is exact for location models.

CHAPTER 8

Summation of series

8.1 Advanced tests for series convergence

8.1.1 Comparison tests

Let us begin by restricting attention to nonnegative series (*i.e.*, series whose terms are all nonnegative). We remind the reader that convergence is a *tail property* of a series. So the investigation of the convergence of a series

$$a_0 + a_1 + a_2 + a_3 + \cdots$$

depends only on the properties of

$$a_m + a_{m+1} + a_{m+2} + a_{m+3} + \cdots$$

for any positive integer m. This is particularly important for series where some test for convergence may be applied, but the conditions of the test only hold for terms that are sufficiently far out in the tail.

The *boundedness criterion* for series convergence is the following. A nonegative series

$$\sum_{j=0}^{\infty} a_j = a_0 + a_1 + a_2 + \cdots$$

is convergent to some value s if and only if the partial sums

$$s_n = \sum_{j=0}^{n} a_j \quad n \geq 0,$$

are bounded above. That is, there is some real number u such that $s_n \leq u$ for all n. In this case $u \geq s$, and s is the smallest such upper bound.

From this criterion, several important results can be proved, including the comparison test. Tests of this sort can be used to reduce the problem of convergence of one series to convergence for another.

Proposition 1. *Let* $\sum a_j$ *and* $\sum b_j$ *be two series.*

1. *Comparison test. Suppose that* $0 \leq a_j \leq b_j$ *for all* j. *If* $\sum b_j$ *converges, then* a_j *converges.*
2. *Limit comparison test. Suppose that* $0 < a_j, b_j$ *for all* j, *and that* $\lim_{j\to\infty} a_j/b_j = c$, *where* $0 < c < \infty$. *Then* $\sum b_j$ *converges if and only if* $\sum a_j$ *converges.*

Proof. To prove statement 1, we use the boundedness criterion. Assume that $0 \leq a_j \leq b_j$ for all j, and suppose $\sum b_j$ converges to b. Then the partial sums of $\sum b_j$ are bounded above by some value u, by the boundedness criterion. Since $0 \leq a_j \leq b_j$, the partial sums of $\sum a_j$ are also bounded by u. So, by the boundedness criterion, the series $\sum a_j$ converges.

To prove statement 2, suppose $\sum b_j$ converges, and that $\lim a_j/b_j = c$. Since the ratio a_j/b_j has limit c, it must eventually be less than or equal to $c+1$, for sufficiently large j. Equivalently, there is some positive integer m such that

$$a_j \leq (c+1)\, b_j$$

for all $j \geq m$. However,

$$\sum_{j=m}^{\infty} (c+1)\, b_j = (c+1) \sum_{j=m}^{\infty} b_j\,.$$

Since the right-hand side of this equality is convergent, it follows that the left-hand side is convergent as well. So $\sum_{j=m}^{\infty} a_j$ converges by statement 1 of the comparison test. From this fact, we can conclude that $\sum_{j=0}^{\infty} a_j$ also converges. The proof of the converse follows by reversing the roles of $\sum a_j$ and $\sum b_j$, and using c^{-1} as the limiting ratio. ∎

8.1.2 Cauchy's condensation theorem

Our next result is known as *Cauchy's condensation theorem.*

Proposition 2. *Suppose the terms of the series* $\sum a_j$ *are positive and decreasing monotonically to zero. Then*

$$\sum_{j=0}^{\infty} a_j = a_0 + a_1 + a_2 + a_3 + \cdots$$

converges if and only if

$$\sum_{j=0}^{\infty} 2^j\, a_{2^j} = a_1 + 2\, a_2 + 4\, a_4 + 8\, a_8 + \cdots$$

converges.

Proof. Since the terms of the series are monotone decreasing,

$$\begin{aligned} a_1 + a_2 + a_3 + a_4 + \cdots &= a_1 + (a_2 + a_3) + (a_4 + a_5 + a_6 + a_7) + \cdots \\ &\le a_1 + (a_2 + a_2) + (a_4 + a_4 + a_4 + a_4) + \cdots \\ &= a_1 + 2\,a_2 + 4\,a_4 + \cdots . \qquad (8.1) \end{aligned}$$

where the inequality is understood to be a term-by-term comparison between the series. Furthermore,

$$\begin{aligned} a_1 + a_2 + a_3 + a_4 + \cdots &= a_1 + a_2 + (a_3 + a_4) + (a_5 + a_6 + a_7 + a_8) \\ &\ge a_1 + a_2 + 2\,a_4 + 4\,a_8 + \cdots \\ &= a_1 + \frac{1}{2}(2\,a_2 + 4\,a_4 + 8\,a_8 + \cdots) . \qquad (8.2) \end{aligned}$$

In inequality (8.1), we apply the comparison test from Proposition 1. If the right-hand side converges, then the left-hand side converges. In the second inequality of (8.2) we apply the comparison test around the other way. We conclude that if the left-hand side converges, then the right-hand side must converge. ∎

Note that it really does not matter whether the original series $\sum a_j$ starts at $j = 0$ or $j = 1$.

8.1.3 Bertrand's series

The Cauchy condensation theorem provides a powerful tool for checking the convergence of series. The following example illustrates its use.

Consider the series

$$\sum_{j=2}^{\infty} \frac{1}{j^p\,(\ln j)^q} = \frac{1}{2^p\,(\ln 2)^q} + \frac{1}{3^p\,(\ln 3)^q} + \frac{1}{4^p\,(\ln 4)^q} + \cdots , \qquad (8.3)$$

where $p,\ q \ge 0$.* The series given in (8.3) is known as *Bertrand's series*, and the case

$$\sum_{j=1}^{\infty} \frac{1}{j^p} = 1 + \frac{1}{2^p} + \frac{1}{3^p} + \frac{1}{4^p} + \cdots , \qquad (8.4)$$

* Clearly for some values of p and q, the series can start at $j = 1$. To apply Cauchy condensation to this series, we can start the series at $j = 1$ or $j = 0$ with some arbitrarily assigned terms which make the sequence of terms positive and decreasing.

where $q = 0$ (and j starts at one) is called a *p-series*. By the Cauchy condensation theorem, the convergence of Bertrand's series is equivalent to that of

$$\sum_{j=3}^{\infty} \frac{2^j}{2^{j\,p}\,(\ln\,2^j)^q} = (\ln\,2)^{-q} \sum_{j=3}^{\infty} \frac{1}{j^q\,2^{j\,(p-1)}}\,. \tag{8.5}$$

Suppose $p > 1$. Using the comparison test from Proposition 1, we can compare the "condensed" series in (8.5) to a geometric series $\sum 2^{-j\,(p-1)}$ to see that (8.5) converges in this case. On the other hand, when $p < 1$, the terms of the condensed series (8.5) go to infinity. Thus the series does not converge.

The case $p = 1$ is left for consideration. For this case, our condensed series (8.5) has the form $\sum j^{-q}$, if we ignore the factor $(\ln\,2)^{-q}$ which does not influence convergence. This is a p-series, which we could have obtained by assigning $p' = q$ and $q' = 0$ in (8.3). So we have already seen that this converges when $q > 1$, and diverges when $q < 1$. The case $p = 1$ and $q = 1$ is the only case left to consider. When $p = 1$ and $q = 1$, the condensed series has the form $p' = 1$ and $q' = 0$. Applying the Cauchy condensation yet again we see that the convergence of this series is equivalent to $p'' = 0$ and $q'' = 0$, which is divergent.

Summarising, we see that our original series converges if $p > 1$, or if both $p = 1$ and $q > 1$. Otherwise the series diverges. ■

8.1.4 *The family of Kummer tests*

By a *Kummer test* for convergence, we shall mean a test of the following type. Let $b_0, b_1, b_2, \ldots$ be a sequence of positive constants. For any given series $\sum a_j$ whose terms are positive, let

$$\rho(j) = b_j \frac{a_j}{a_{j+1}} - b_{j+1} \tag{8.6}$$

and let

$$\rho = \lim_{j\to\infty} \rho(j)$$

provided this limit exists. If $\rho > 0$, we conclude that the series $\sum a_j$ converges. If $\rho < 0$, we conclude that the series diverges. If $\rho = 0$, the test reaches no conclusion. In particular, the following choices for the sequence b_j, $j \geq 0$ are popular:

- $b_j = 1$, which is the *ratio test*;
- $b_j = j$, which is *Raabe's test*;
- $b_j = j\,\ln\,j$, which is *Bertrand's test*.

Proposition 3. *Suppose* $\liminf_{j\to\infty} \rho(j) > 0$. *Then the series* $\sum a_j$ *converges.*

Proof. Suppose that $\liminf_{j\to\infty} \rho(j) > 0$. Then there exists a real number $c > 0$, and a positive integer n such that

$$b_j \frac{a_j}{a_{j+1}} - b_{j+1} \geq c \tag{8.7}$$

whenever $j \geq n$. This is equivalent to the statement that $a_j\, b_j - a_{j+1}\, b_{j+1} \geq c\, a_{j+1}$ for all $j \geq n$. Summing both sides of this inequality over j, we get

$$\sum_{j=n}^{n+m-1} (a_j\, b_j - a_{j+1}\, b_{j+1}) \geq c \sum_{j=n}^{n+m-1} a_{j+1}\,. \tag{8.8}$$

The summation on the left-hand side telescopes. So the inequality reduces to

$$a_n\, b_n - a_{n+m-1}\, b_{n+m-1} \geq c \sum_{j=n}^{n+m-1} a_{j+1} = c\,(s_{n+m} - s_n)\,. \tag{8.9}$$

In addition, we have

$$a_n\, b_n \geq a_n\, b_n - a_{n+m-1}\, b_{n+m-1}\,.$$

Combining this last inequality with (8.9), we get

$$s_{n+m} - s_n \leq c^{-1}\, a_n\, b_n\,.$$

But this inequality can be written in the form

$$s_{n+m} \leq s_n + c^{-1}\, a_n\, b_n\,, \quad \text{for all } m\,. \tag{8.10}$$

Since the right-hand side of (8.10) does not involve m, we see that the partial sums s_{n+m} are bounded above. By the boundedness criterion for convergence, the series $\sum a_j$ converges. ■

Proposition 4. *Suppose that* $\limsup_{j\to\infty} \rho(j) < 0$. *Suppose additionally that the series* $\sum b_j^{-1}$ *diverges. Then the series* $\sum a_j$ *diverges.*

Proof. Since $\limsup \rho(j) < 0$, there must exist a positive integer n such that $\rho(j) \leq 0$ for all $j \geq n$. This is equivalent to the statement

$$a_n\, b_n \leq a_{n+1}\, b_{n+1} \leq a_{n+2}\, b_{n+2} \leq \cdots\,.$$

So $a_j \geq (a_n\, b_n)\, b_j^{-1}$ for all $j \geq n$. We sum over all j such that $j \geq n$. Since $\sum b_j^{-1}$ diverges, we can use the comparison test of Proposition 1 to conclude that $\sum a_j$ diverges. ■

Proposition 5. *The ratio test, Raabe's test and Bertrand's test satisfy the conclusions of Propositions 3 and 4:*

- *If* $\liminf \rho(j) > 0$*, then* $\sum a_j$ *converges.*
- *If* $\limsup \rho(j) < 0$*, then* $\sum a_j$ *diverges.*

Proof. It is sufficient to check that $\sum b_j^{-1}$ diverges for each test, as the other assumptions are easily verified. For the ratio test, $b_j = 1$, and the divergence is immediate. For Raabe's test, $b_j = j$. Thus $\sum b_j^{-1}$ is an harmonic series–a special case of Bertrand's series with $p = 1$ and $= 0$. The divergence follows from the arguments given above for Bertrand's series. For Bertrand's test, with $b_j = j\boldsymbol{m} \ln j$, the divergence also follows from a special case of Bertrand's series, with $p = 1$ and $q = 1$. ∎

Henceforth, we will let $\rho_1(j)$, $\rho_2(j)$ and $\rho_3(j)$ denote the values of $\rho(j)$ for the ratio test, Raabe's test, and Bertrand's test, respectively.

The ratio test is certainly the best known of these convergence tests, and needs little introduction here. It is usually the primary tool for the investigation of convergence. Although it is easy to implement, it is often inconclusive. For example, the p-series

$$1 + \frac{1}{2^p} + \frac{1}{3^p} + \frac{1}{4^p} + \cdots$$

converges for $p > 1$ and diverges for $p \leq 1$. Unfortunately, this cannot be determined by the ratio test because $\lim \rho_1(j) = 0$ for all p. Raabe's test, using $\rho_2(j)$ can be regarded as a refinement of the ratio test, for those cases where $\lim \rho_1(j) = 0$. Applying Raabe's test, for example, to the p-series above, we find that

$$\begin{aligned} \rho_2(j) &= j\,\frac{j^{-p}}{(j+1)^{-p}} - (j+1) \\ &= (j+1)\left[\left(1 + \frac{1}{j}\right)^{p-1} - 1\right]. \end{aligned}$$

Thus $\lim \rho_2(j) = p - 1$. This is much more satisfactory, because it tells us that $\sum j^{-p}$ converges whenever $p > 1$, and diverges when $p < 1$. Unfortunately, Raabe's test is inconclusive for $p = 1$. To decide this case, we can apply Bertrand's test. Problem 1 asks the reader to show that the harmonic series–the p-series with $p = 1$–diverges using Bertrand's test.

Of course, in order to use Raabe's and Bertrand's tests here, we must already know the convergence properties of p-series, as these facts are built into Proposition 5. In fact, all three of the tests discussed in Proposition 5 are rather cleverly constructed forms of the limit comparison test. For example, the ratio test is really a limit comparison test of a series with a geometric series. Similarly, Raabe's test is a limit comparison with a p-series, and Bertrand's test is a limit comparison with Bertrand's series. Recognising this, we can make arbitrarily sharp tests, beyond Bertrand's test, by constructing series $\sum b_j^{-1}$ which diverge, but just barely. Beyond a certain point, the exercise becomes contrived. Unfortunately, there is no "slowest" diverging series. So we cannot expect any single Kummer test to be the last word on the issue of series convergence.

Bertrand's test is sharper than Raabe's test, and Raabe's test is sharper than the ratio test. Furthermore, these tests can be made even more powerful with Cauchy's condensation theorem. Nevertheless, the ratio test is not made obsolete by Raabe's test, nor Raabe's test by Bertrand's test. In most cases, it is best to use the simplest test first before trying a more complicated test. If we use a delicate test too quickly, we may miss the big picture about convergence. We are not only interested in convergence per se, but also in the rate at which convergence occurs. Thus it is best to look for "main effects" in convergence before studying more delicate issues.

8.2 Convergence of random series

Random series often have a different character from deterministic series. Nevertheless, it is possible to reduce convergence problems involving some random series to the study of moments. A useful result in this respect is the next proposition.

Proposition 6. *Let X_j, $j \geq 0$ be an infinite sequence of independent random variables. Then the following statements are equivalent.*

1. *The series $\sum X_j$ converges in probability.*
2. *The series $\sum X_j$ converges with probability one.*

The proof of this result can be found in many textbooks in probability, including Breiman (1968, Section 3.4). See also Durrett (1996, Section 1.8). This proposition asserts that for sums of independent random variables, a form of weak convergence, namely convergence in probability, is equivalent to a form of strong convergence, namely convergence almost surely. Weak convergence is often easier to prove because it can be obtained using moment conditions and Chebyshev's inequality. For example, an immediate application of this proposition is through the following corollary.

Proposition 7. *Suppose X_j, $j \geq 0$ is a sequence of independent random variables, with respective means and variances μ_j and σ_j^2. Suppose also that the series $\sum \mu_j$ and $\sum \sigma_j^2$ converge to μ and σ^2, respectively. Then there is some random variable X with mean μ and variance σ^2 such that*

$$P\left(\sum_{j=0}^{\infty} X_j = X\right) = 1 .$$

Proof. This follows from Proposition 6 and standard convergence results for random variables. It can be checked that the sequence of partial sums

$$S_n = \sum_{j=0}^{n} X_j$$

is Cauchy in mean square (that is, in L^2). So there is some X such that $S_n \to X$ in mean square. Convergence of S_n to X in mean square to X implies that S_n converges in probability to X. Applying Proposition 6, we can conclude that $S_n \to X$ with probability one. Convergence of S_n to X in mean square also implies that

$$E(S_n) = \sum_{j=0}^{n} \mu_j \to E(X) , \quad \mathrm{Var}(S_n) = \sum_{j=0}^{n} \sigma_j^2 \to \mathbf{Var}(X) .$$

The required conclusion follows. ∎

For sums of dependent random variables the convergence results are not as straightforward, but can be obtained using tools such as the *martingale convergence theorem*. See Breiman (1968, Section 5.4).

8.3 Applications in probability and statistics

8.3.1 Fluctuations of random variables

Our first example studies the fluctuations in a sequence of random variables. Suppose that $X_0, X_1, X_2, \ldots$ form a sequence of independent $\mathcal{N}(0, 1)$ random variables, and $c_0, c_1, c_2, \ldots$ is a sequence of positive constants. How does the sequence X_j/c_j, $j \geq 0$ behave as $j \to \infty$? Clearly, if $c_j \to \infty$ sufficiently quickly, then

$$\lim_{j\to\infty} \frac{X_j}{c_j} = 0\,.$$

On the other hand, if $c_j = c$ for all j, then X_j/c_j is an unbounded sequence that fluctuates between $-\infty$ and ∞. That is,

$$\limsup_{j\to\infty} \frac{|X_j|}{c_j} = +\infty\,.$$

Where is the cutoff point between these two types of behaviour?

Suppose $\epsilon > 0$ is arbitrary. Let us call a value of j for which

$$\frac{|X_j|}{c_j} > \epsilon$$

an ϵ-fluctuation. Thus we may refine our question by asking how many ϵ-fluctuations should we expect as $j \to \infty$? In particular, will the number of such fluctuations be finite or infinite? The *Borel-Cantelli Lemma*† is the essential tool for answering this question. Applying this lemma, we see that

$$P\left(|X_j| > \epsilon\, c_j \text{ infinitely often}\right) = \begin{cases} 1 & \text{if } \sum P\left(|X_j| > \epsilon c_j\right) \text{ diverges}; \\ 0 & \text{if } \sum P\left(|X_j| > \epsilon c_j\right) \text{ converges}. \end{cases}$$

Thus the Borel-Cantelli lemma reduces the problem to the determining the convergence or divergence of a series. When the constants c_j, $j \geq 0$ are bounded, the number of ϵ- fluctuations in the sequence will be infinite. So let us restrict attention to the case where $c_j \to \infty$. As $x \to \infty$, we have

$$P(X > x) \sim \frac{\phi(x)}{x}\,,$$

from Section 2.4.2. Thus

$$P\left(|X_j| > \epsilon\, c_j\right) \sim 2\,\epsilon^{-1}\, c_j^{-1}\, \phi(\epsilon\, c_j)\,, \tag{8.11}$$

as $n \to \infty$. So, using the limit comparison test of Proposition 1, it suffices to check the convergence of the series

$$\sum_{j=0}^{\infty} \frac{\phi(\epsilon\, c_j)}{c_j} = \sum_{j=0}^{\infty} \frac{1}{c_j\,\sqrt{2\,\pi}} \exp\left(-\frac{\epsilon^2\, c_j^2}{2}\right)\,. \tag{8.12}$$

† Readers can refer to Breiman (1968) or to Billingsley (1995) for an explanation and proof of the Borel-Cantelli lemma.

Suppose we let $c_j = \sqrt{\ln j}$. Then

$$\sum_{j=3}^{\infty} \frac{1}{c_j\sqrt{2\pi}} \exp\left(-\frac{\epsilon^2 c_j^2}{2}\right) = \frac{1}{\sqrt{2\pi}} \sum_{j=3}^{\infty} \frac{1}{j^\delta \sqrt{\ln j}}, \tag{8.13}$$

where $\delta = \epsilon^2/2$. The series on the right-hand side can be recognised as a special case of Bertrand's series with $p = \delta$ and $q = 1/2$. In Section 8.1.3, we found that the series on the right-hand side converges when $\delta > 1$ and diverges when $\delta \le 1$. So by the Borel-Cantelli lemma

$$P\left(\frac{|X_j|}{\sqrt{\ln j}} > \epsilon \text{ infinitely often}\right) = \begin{cases} 1 & \text{when } 0 \le \epsilon \le \sqrt{2}\,; \\ 0 & \text{when } \sqrt{2} < \epsilon\,. \end{cases}$$

This is equivalent to

$$P\left(\limsup_{j\to\infty} \frac{|X_j|}{\sqrt{2\ln j}} = 1\right), \tag{8.14}$$

which is a sharp answer.

8.3.2 Law of the iterated logarithm

Our next example illustrates the relationship between series convergence and one of the most famous achievements in probability theory, namely the law of the iterated logarithm (LIL). The LIL can be approached by considering the "boundary" between two other celebrated results in probability: the law of large numbers (LLN) and the central limit theorem (CLT). Let X_j, $j \ge 1$ be independent and identically distributed random variables with mean zero and variance one. Let a_j, $j \ge 1$ be an increasing sequence of positive constants. What happens to

$$\frac{X_1 + X_2 + X_3 + \cdots + X_n}{a_n}$$

as $n \to \infty$? From the LLN, we know that

$$\lim_{n\to\infty} \frac{|X_1 + X_2 + \cdots + X_n|}{n} = 0 \tag{8.15}$$

with probability one. From the CLT, we can prove that

$$\limsup_{n\to\infty} \frac{|X_1 + X_2 + \cdots + X_n|}{\sqrt{n}} = +\infty\,, \tag{8.16}$$

also with probability one. So it is natural to seek the boundary between these two types of behaviours. The celebrated answer to this question is the LIL, due to Khintchine (1924) and Kolmogorov (1929), who found that

$$\limsup_{n\to\infty} \frac{|X_1 + X_2 + \cdots + X_n|}{\sqrt{2\,n \ln \ln n}} = 1 \tag{8.17}$$

with probability one. This result is superficially similar to that in our previous example. The factor $\sqrt{n}$ in the denominator stabilises the variance of the sum. The important difference between the right-hand sides in (8.14) and (8.17) is that $\sqrt{\ln n}$ has been replaced by $\sqrt{\ln \ln n}$. This is in keeping with the fact that the partial sums are not independent, and do not fluctuate as erratically as independent normal random variables.

It was Feller who found the precise relationship between the fluctuations of the partial sums and the convergence of series. He showed that when a_n is an increasing sequence, we have

$$P\left(|X_1 + X_2 + \cdots + X_n| > a_n \text{ infinitely often}\right) = \begin{cases} 1 \\ 0 \end{cases} \tag{8.18}$$

as

$$\sum_{n=1}^{\infty} \frac{a_n}{n\sqrt{n}} \exp\left(-\frac{a_n^2}{2\,n}\right) \begin{cases} = \infty \\ < \infty \end{cases} \tag{8.19}$$

respectively. So if we choose

$$a_n = (1 \pm \epsilon)\sqrt{2\,n \ln \ln n}\,,$$

in (8.18), then the convergence of series (8.19) is equivalent to the convergence of

$$\sum \frac{1}{n\,(\ln n)^{(1\pm\epsilon)^2}\,(\ln \ln n)^{1/2}}\,.$$

If we apply Cauchy condensation to this series, we see that the convergence is equivalent to the series

$$\sum \frac{1}{n^{(1\pm\epsilon)^2}\,(\ln n)^{1/2}}\,.$$

Once again, this is Bertrand's series, with $p = (1 \pm \epsilon)^2$ and $q = 1/2$. For $0 < \epsilon < 1$, this converges when $p = 1 + \epsilon$ and diverges when $p = 1 - \epsilon$. This is precisely the LIL, seen as a special case of Feller's result.

8.3.3 Estimation for nonhomogeneous Poisson variates

Consider a sequence X_j, $j \geq 1$ of independent Poisson random variables, where the j random variable has mean $\lambda\,\alpha_j$, where $\alpha_j > 0$ for all j. We shall assume that the parameters α_j are given, and that we wish to estimate λ. The maximum likelihood estimator for λ based upon the first n variates is

$$\widehat{\lambda}_n = \frac{\sum_{j=1}^n X_j}{\sum_{j=1}^n \alpha_j}$$

which is a minimum variance unbiased estimator for λ. Let us consider the conditions under which $\widehat{\lambda}_n$ estimates λ consistently as $n \to \infty$. For this particular model, the consistency of $\widehat{\lambda}_n$ is equivalent to the condition that $\mathrm{Var}(\widehat{\lambda}_n)$ goes to zero as $n \to \infty$. It is easily checked that

$$\widehat{\lambda}_n \text{ is } \begin{cases} \text{consistent} & \text{if } \sum \alpha_j \text{ diverges} \\ \\ \text{inconsistent} & \text{if } \sum \alpha_j \text{ converges}. \end{cases} \tag{8.20}$$

The convergence of $\sum \alpha_j$ controls more than the consistency of $\widehat{\lambda}_n$. In fact, if $\sum \alpha_j$ converges, then there is no consistent estimator of λ at all. To prove this, note that when $\sum \alpha_j$ converges, so does

$$\sum_{j=1}^{\infty} \mathrm{Var}(X_j) = \sum_{j=1}^{\infty} (\lambda\, \alpha_j)\,.$$

By Proposition 7 it follows that $\sum X_j$ converges. Suppose we set

$$X = \sum_{j=1}^{\infty} X_j\,.$$

Since the terms of this sum—Poisson random variables—are integers, it follows that all but finitely many terms are zero. It can be checked that X is a Poisson random variable with mean

$$E(X) = \lambda \sum_{j=1}^{\infty} \alpha_j\,.$$

Note that X is a sufficient statistic for λ in the model for the full data set $X_1, X_2, \cdots$. To see this, we can observe that the conditional distribution of the infinite vector

$$P\left(X_j = x_j,\ \forall j \;\middle|\; \sum X_j = \sum x_j\right) = \frac{\left(\sum_{j=1}^{\infty} x_j\right)!}{\prod_{j=1}^{\infty} x_j\,!} \prod_{k=1}^{\infty} \left(\frac{\alpha_k}{\sum_{j=1}^{\infty} \alpha_j}\right)^{x_k}$$

is functionally independent of λ. Note that the infinite products in this expression involve only finitely many factors which differ from one. Since X is sufficient for λ, it follows that any consistent sequence of estimators for λ can be chosen to be functions of X. However, this is impossible, since X has a nondegenerate Poisson distribution.

The other case to be considered is where $\sum \alpha_j$ diverges. In this case, $\widehat{\lambda}_n$

is consistent as an estimator of λ. Inferences about λ are typically made using the statistic

$$\sqrt{\sum_{j=1}^{\infty} \alpha_j} \left(\widehat{\lambda}_n - \lambda\right)$$

which is asymptotically $\mathcal{N}(0, \lambda)$ as $n \to \infty$.

8.4 Euler-Maclaurin sum formula

In the case of a convergent series $\sum a_j$ summing to s, it is often useful to obtain a *generalised asymptotic series* for the tail of the series of the form

$$s_n - s \sim \varphi_1(n) + \varphi_2(n) + \varphi_3(n) + \cdots, \quad \text{as } n \to \infty, \tag{8.21}$$

where $\varphi_{k+1} = o(\varphi_k)$ for all k. Special cases that might arise in practice include the following.

- Cases where the tail of the series goes slowly to zero at roughly the same rate as a p-series. For example, we may be able to write

$$s_n - s \sim \frac{b_1}{n^p} + \frac{b_2}{n^{p+1}} + \frac{b_3}{n^{p+2}} + \cdots, \quad \text{as } n \to \infty.$$

- Cases where the tail of the series goes to zero at roughly the rate of a geometric series. Here, we may have expansions of the form

$$s_n - s \sim c_1 \rho_1^n + c_2 \rho_2^n + c_3 \rho_3^n + \cdots, \quad \text{as } n \to \infty,$$

  where $1 > \rho_1 > \rho_2 > \cdots > 0$.

Such examples are not exhaustive, of course. In practice, the generalised asymptotic expansion for $s_n - s$ may be quite different from these cases.

A basic tool for constructing such asymptotic expansions in n is the *Euler-Maclaurin summation formula.* While our main applications for this formula will be to slowly converging series, the formula will also be valid for diverging series as well. Suppose there is some infinitely differentiable function $f(x)$ such that $a_j = f(j)$, so that

$$s_n = f(1) + f(2) + \cdots + f(n)..$$

Let m be some fixed positive integer. The Euler-Maclaurin formula for s_n has the form

$$\begin{aligned} s_n &= \int_1^n f(x)\,dx + c + \frac{1}{2} f(n) + \frac{B_2}{2!} f'(n) + \frac{B_4}{4!} f'''(n) + \cdots \\ &\quad + \frac{B_{2m}}{(2m)!} f^{(2m-1)}(n) + O\left(\int_n^\infty \left|f^{(2m)}(x)\right| dx\right), \end{aligned} \tag{8.22}$$

Leonhard Euler (1707–1783)

One of the most prolific mathematicians of all time. Among contributions too numerous to mention, Leonhard Euler pioneered many modern tools for series summation.

> "Concerning the summation of very slowly converging series, in the past year I have lectured to our Academy on a special method of which I have given the sums of very many series sufficiently accurately and with very little effort."
>
> "I have very little desire for anything to be detracted from the fame of the celebrated Mr Maclaurin since he probably came upon the same theorem for summing series before me, and consequently deserves to be named as its first discoverer."
>
> Leonhard Euler to James Stirling written between 1736 and 1738. See H. Loeffel, Leonhard Euler (1707-1783), *Mitt. Verein. Schweiz. Versicherungsmath.* 1, 1984, pp. 19–32.

k	B_k	B_k^*	k	B_k	B_k^*
0	1		6	$\frac{1}{42}$	$\frac{691}{2730}$
1	$-\frac{1}{2}$	$\frac{1}{6}$	7	0	$\frac{7}{6}$
2	$\frac{1}{6}$	$\frac{1}{30}$	8	$-\frac{1}{30}$	$\frac{3617}{510}$
3	0	$\frac{1}{42}$	9	0	$\frac{43867}{798}$
4	$-\frac{1}{30}$	$\frac{1}{30}$	10	$\frac{5}{66}$	$\frac{174611}{330}$
5	0	$\frac{5}{66}$			

Table 8.1 *The first few Bernoulli numbers represented in both the older (Whittaker and Watson) and more recent (Ibramowitz and Stegun) notations*

as $n \to \infty$. In (8.22), c is some suitable constant which does not depend upon the value of n. Here, the constants B_1, B_2, and so on, are the respective Bernoulli numbers using the Ibramowitz and Stegun notation. The values of the Bernoulli numbers B_k of low order are given in the displayed table, along with the values according to the Whittaker and Watson notation B_k^* which we used in Chapter 2.‡

The Euler-Maclaurin formula works very effectively when $f(x)$ is a polynomial. In this case, the derivatives $f^{(2m}(x)$ vanish for sufficiently large m, and the order remainder term is zero. For example, we can employ the formula to produce the standard sum formulas for the powers of the integers. To obtain the usual formula for the sum of the first n positive integers, we set $f(x) = x$ for all x. It is easily checked that (8.22) reduces to

$$1 + 2 + \cdots + n = \frac{n\,(n+1)}{2}\,.$$

In a similar way, we can set $f(x) = x^2$. Then (8.22) becomes

$$1^2 + 2^2 + 3^2 + \cdots + n^2 = \frac{n\,(n+1)\,(2n+1)}{6}\,.$$

See Problem 8. The Euler-Maclaurin formula is also particularly useful for series which slowly converge or slowly diverge. In terms of the function $f(x)$, a natural property for this is

$$f^{(2k+1)}(x) = o\left[f^{(2k-1)}(x)\right] \quad \text{as } x \to \infty\,,$$

‡ See the comments footnoted to Problem 20 of Chapter 2.

which ensures that the formula is a generalised asymptotic series as in (8.21), and that the remainder order term is insignificant for large n.

In Maple, the Euler-Maclaurin formula can be invoked by an expression of the form

$> \; eulermac(a(j),\, j = 1..n)$

which provides the Euler-Maclaurin expansion of $s_n = a(1) + \cdots + a(n)$. For example, the command

$> \; eulermac(j,\, j = 1..n)$

produces the output

$$\frac{1}{2}\,n^2 + \frac{1}{2}\,n$$

which is the standard summation formula for the arithmetic series. A slowly diverging series such as $\sum j^{-1}$ can be obtained from

$> \; assume(n, posint)$

$> \; eulermac\left(\frac{1}{j},\, j = 1..n\right)$

yielding an expression that reduces to

$$\ln\, n + \gamma + \frac{1}{2\,n} + \frac{1}{12\,n^2} + \frac{1}{120\,n^4} + \frac{1}{252\,n^6} + O\left(\frac{1}{n^8}\right). \tag{8.23}$$

Here, $\gamma = 0.577\ldots$ is Euler's constant. If the *assume* command is omitted, then the integral expression in the Euler-Maclaurin formula is left unevaluated, because Maple will otherwise allow for the contingency that n is negative or complex. This Euler-Maclaurin formula for this slowly diverging harmonic series can be compared with a slowly converging series such as $\sum j^{-2}$. Invoking the commands

$> \; assume(n, posint)$

$> \; eulermac\left(\frac{1}{j^2},\, j = 1..n\right)$

upon simplification results in

$$\frac{\pi^2}{6} - \frac{1}{n} + \frac{1}{2\,n^2} + \frac{1}{6\,n^3} + \frac{1}{30\,n^5} + \frac{1}{42\,n^7} + O\left(\frac{1}{n^9}\right). \tag{8.24}$$

The constant $\pi^2/6$ is the constant c in (8.22) for this particular series. Here, c cannot be obtained by an evaluation of the series for small n, as above. To find this constant, the full infinite series must be evaluated. It was Euler who first managed to sum this infinite series by means of an algebraic trick that factors a Taylor series as if it were a polynomial.§

While the Euler-Maclaurin formula is a powerful tool for evaluating the *asymptotic form* of a series, it is not a panacea for the *evaluation* of series. As we saw above in formula (8.22), there is an unknown constant c which must be determined. Unfortunately, the determination of this constant may be formally equivalent to the evaluation of the series. Nevertheless, the Euler-Maclaurin formula often gives us insights into the asymptotic nature–if not the asymptotic value–of the tail of a slowly converging series. This is particularly useful when we turn to the problem of extrapolating the partial sums of a slowly converging series to accelerate convergence.

8.5 Applications of the Euler-Maclaurin formula

8.5.1 Distributions of records

Suppose $X_1, X_2, \ldots, X_n$ is a random sample from some continuous distribution with support on the positive real numbers. A random variable X_j is said to be a *record* at time j if

$$X_j > \max(X_1, X_2, \ldots, X_{j-1}).$$

In other words, X_j is a record if it is the largest random variable up to time j. What can we say about the distribution of X_n?

Let $I_j = 1$ if X_j is a record, and set $I_j = 0$ otherwise. Also, for each j, let R_j denote the event that $I_j = 1$, so that $P(R_j) = E(I_j)$. Then the number of records up to time n is

$$T_n = I_1 + \cdots + I_n.$$

Since the rank of X_j among the variates $X_1, X_2, \ldots X_j$ is independent of

§ Euler wrote the function $x^{-1} \sin x$ in two forms, namely

$$\frac{\sin x}{x} = 1 - \frac{x^2}{3!} + \frac{x^4}{5!} - \cdots$$

which is its Taylor series about zero, and

$$\frac{\sin x}{x} = \left(1 - \frac{x^2}{\pi^2}\right)\left(1 - \frac{x^2}{4\,\pi^2}\right)\left(1 - \frac{x^2}{9\,\pi^2}\right)\cdots.$$

The second representation was based on the mathematical fiction that $x^{-1} \sin x$ is a polynomial of infinite degree with roots $\pm\pi$, $\pm(2\,\pi)$, $\pm(3\,\pi)$, and so on. When the second expression is expanded in x and the coefficients of the two series in x^2 are equated the formula emerges (and much more).

the rankings of $X_1, \ldots, X_{j-1}$, it follows that I_j, $j \geq 1$ are independent random variables. In addition, the event R_j occurs if X_j is greater than any X_1, X_{j-1}, which has probability j^{-1}. Therefore,

$$E(T_n) = 1+\frac{1}{2}+\cdots+\frac{1}{n}, \quad \text{Var}(T_n) = 1+\frac{1}{2}\left(1-\frac{1}{2}\right)+\cdots+\frac{1}{n}\left(1-\frac{1}{n}\right).$$

Asymptotic expressions for the mean and variance of T_n can be obtained from the Euler-Maclaurin formulas derived in (8.23) and (8.24), yielding

$$E(T_n) \sim \ln n+\gamma+\frac{1}{2\,n}+\frac{1}{12\,n^2}+\frac{1}{120\,n^4}+\frac{1}{252\,n^6}+\cdots,$$

and

$$\text{Var}(T_n) \sim \ln n+\left(\gamma-\frac{\pi^2}{6}\right)+\frac{3}{2\,n}-\frac{5}{12\,n^2}-\frac{1}{6\,n^3}+\frac{1}{120\,n^4}-\frac{1}{30\,n^5}+\cdots.$$

To find the limiting form for the distribution of T_n, we need to standardise T_n as

$$\begin{aligned} T_n^* &= \frac{T_n - E(T_n)}{\sqrt{\text{Var}(T_n)}} \\ &\sim \frac{T_n - (\ln n + \gamma)}{\sqrt{\ln n + \gamma - \frac{\pi^2}{6}}}. \end{aligned}$$

Then T_n^* is asymptotically $\mathcal{N}(0, 1)$ as $n \to \infty$.

A more satisfying approximation might be a Poisson approximation or binomial approximation. For example, let U_n be a Poisson random variable with mean $\lambda_n = E(T_n)$. Then the distribution of U_n approximates T_n for large n. Alternatively, for relatively small values of n, a binomial random variable with matched mean and variance can be used.

8.5.2 Expectations of record values

The calculations of the previous example can be extended in the following way. Again, suppose that $X_1, X_2, \ldots$ are independent positive random variables with some common continuous distribution. Define

$$M_j = \max(X_1, X_2, \ldots, X_j) \text{ for } j \geq 1.$$

In addition, define

$$\mu_1 = E(M_1), \quad \mu_j = E(M_j - M_{j-1} \mid M_j > M_{j-1}) \text{ for } j \geq 2.$$

As above, we have $P(M_j > M_{j-1}) = j^{-1}$. Therefore, we may write

$$E(M_n) = \mu_1 + \frac{\mu_2}{2} + \frac{\mu_3}{3} + \cdots + \frac{\mu_n}{n}. \tag{8.25}$$

Now suppose that the random variables X_j have a common exponential distribution with mean μ. A straightforward consequence of the memoryless property of exponential random variables is that $\mu_j = \mu$ for all j. So for this case,

$$\begin{aligned} E(M_n) &= \mu\left(1 + \frac{1}{2} + \frac{1}{3} + \cdots + \frac{1}{n}\right) \\ &\sim \mu\left(\ln n + \gamma + \frac{1}{2n} + \cdots\right). \end{aligned}$$

8.6 Accelerating series convergence

8.6.1 Introduction

Some of the series that we have considered earlier converge slowly. In such cases, it is hard to determine the sum of a series by simply calculating a partial sum. It is not unusual to have to sum thousands of terms before the partial sums get close to the sum of the infinite series for practical purposes. While this is well within the range of modern computing, we often encounter numerical problems with series after summing a large number of terms. For example, the terms themselves may be of much smaller magnitude than the tail of the series, leading to a problem of underflow in the calculations. Even when the computer can handle the summation, it may be necessary to calculate the sum of the series many times over. This arises in computing the values of a function which is determined by a series expansion.

Can we do better than to simply evaluate a partial sum of the series in such cases? If the terms of the series are completely arbitrary in form, it is doubtful whether we can do much better. This exploits the only truly general property of a convergent series–namely that its tail is asymptotically negligible–by replacing this tail by zero. However, most series that we encounter in practice have considerable regularity in their terms. For example, we can often determine the rate at which the n-th term goes to zero as n goes to infinity. By a series acceleration method, we shall mean a transformation taking certain series

$$a_1 + a_2 + a_3 + a_4 + \cdots$$

to some other series

$$a'_1 + a'_2 + a'_3 + a'_4 + \cdots$$

where the primed series is chosen

- to have a sum which equals or approximates the sum of the unprimed series, and
- has partial sums which converge at a faster order than the partial sums of the unprimed series.

We can observe that this is possible in a number of simple ways. For example, consider the two series

$$1 - \frac{1}{2} + \frac{1}{3} - \frac{1}{4} + \cdots$$

and

$$\frac{1}{1 \times 2} + \frac{1}{3 \times 4} + \frac{1}{5 \times 6} + \cdots .$$

The first series is a standard expansion for $\ln 2$ as an alternating series. The second series is a transformation of the first series by grouping paired differences in the first series. While this standard "acceleration" of the original alternating series has made the errors smaller, it has not improved the rate at which the errors go to zero as a function of n.

The methods that we shall consider below will take us considerably beyond this simple type of acceleration: we shall be considering methods that improve the order of the error. Of course, there is no free lunch. Any method that works well on a regularly varying series may fail if applied in an inappropriate context. Indeed, for some series, our acceleration methods may introduce some unpredictability into the behaviour of the partial sums. We shall examine some of these issues below.

8.6.2 Kummer acceleration

Consider two series

$$\sum a_j, \quad \sum b_j$$

whose terms are all positive, and having the property that

$$\lim_{j \to \infty} \frac{a_j}{b_j} = 1 . \tag{8.26}$$

We already know from the limit comparison test that the two series must converge or diverge together. In a sense, the two series must converge at roughly the same rate.

However, it may happen that one series, say $\sum b_j$, is easy to sum whereas $\sum a_j$ is difficult. Let us write

$$\sum_{j=1}^{\infty} a_j = s + \sum_{j=1}^{\infty} (a_j - b_j) ,$$

where $s = \sum b_j$. From the limit condition in (8.26), we see that the series $\sum(a_j - b_j)$ has terms of smaller order than the terms of $\sum a_j$ for large j. So we may reasonably hope that it converges faster.

As an example, we may take

$$a_j = \frac{1}{j^2}, \quad b_j = \frac{1}{j\,(j+1)}.$$

The first is a p-series with $p = 2$. The second has similar convergence properties, but can be summed using partial fractions. The sum of this particular p-series

$$1 + \frac{1}{2^2} + \frac{1}{3^2} + \frac{1}{4^2} + \cdots = \frac{\pi^2}{6}$$

is the famous discovery due to Leonhard Euler, which we encountered earlier in formula (8.24). First note that

$$\begin{aligned} s &= \sum_{j=1}^{\infty} \frac{1}{j\,(j+1)} \\ &= \sum_{j=1}^{\infty} \left(\frac{1}{j} - \frac{1}{j+1}\right) \\ &= 1, \end{aligned}$$

using the telescoping series property to obtain the answer in the final step. Thus

$$\begin{aligned} \sum_{j=1}^{\infty} \frac{1}{j^2} &= 1 + \sum_{j=1}^{\infty} \left[\frac{1}{j^2} - \frac{1}{j\,(j+1)}\right] \\ &= 1 + \sum_{j=1}^{\infty} \frac{1}{j^2\,(j+1)}. \end{aligned}$$

This new series is the Kummer accelerated version of the original p-series.

Of course, there is nothing to stop us from applying this method again. Let us set

$$a'_j = \frac{1}{j^2\,(j+1)}, \quad b'_j = \frac{1}{j\,(j+1)\,(j+2)}$$

where now

$$\begin{aligned} s' &= \sum_{j=1}^{\infty} \frac{1}{j\,(j+1)\,(j+2)} \\ &= \sum_{j=1}^{\infty} \left[\frac{1}{2\,j} - \frac{1}{j+1} + \frac{1}{2\,(j+2)}\right] \end{aligned}$$

$$= \frac{1}{4},$$

again using the telescoping property. Thus

$$\begin{aligned}\sum_{j=1}^{\infty} \frac{1}{j^2} &= 1 + \frac{1}{4} + \sum_{j=1}^{\infty} (a'_j - b'_j) \\ &= 1 + \frac{1}{4} + \sum_{j=1}^{\infty} \frac{2}{j^2 (j+1)(j+2)}.\end{aligned}$$

In general, we can show that

$$\sum_{k=1}^{\infty} \frac{1}{k(k+1)(k+2)\cdots(k+n)} = \frac{1}{n^2(n-1)!}. \tag{8.27}$$

The proof of this is left as Problem 12 at the end of the chapter. Therefore, the application of Kummer acceleration n times to the original series lead us to the formula

$$\sum_{j=1}^{\infty} \frac{1}{j^2} = \sum_{j=1}^{n} \frac{1}{j^2} + n! \sum_{j=1}^{\infty} \frac{1}{j^2 (j+1)\cdots(j+n)}. \tag{8.28}$$

Again, see Problem 12.

8.6.3 Aitken acceleration

The key idea behind many forms of series acceleration is that it may be possible to determine the rate of convergence of a series even if the series sum is hard to compute. For example, suppose the following limit exists, namely

$$\rho = \lim_{n\to\infty} \frac{s - s_{n+1}}{s - s_n}, \tag{8.29}$$

where s is the sum of the series $\sum a_n$ whose partial sums are s_n. We say that the series convergence is

- *logarithmic* if $|\rho| = 1$,
- *linear* if $0 < |\rho| < 1$, and
- *hyperlinear* if $\rho = 0$.

For example, a convergent geometric series displays linear¶ convergence, while a convergent p-series is logarithmic. Series whose convergence is

¶ The term "linear" may seem odd. However, it is motivated by the idea that when a series converges linearly, then the number of initial decimal places in $s - s_n$ which are all zero increases at an approximate linear rate.

hyperlinear, such as Taylor series for the exponential function, often converge too fast for acceleration methods to be practically useful. However, logarithmic convergence often requires acceleration of the partial sums before the sum can be computed in practice.

The concept of linear convergence of series is the starting point for our next method, known as Aitken acceleration or Aitken's δ^2-method. For such series, we may suppose that

$$s_k \approx s - \alpha\,\rho^k, \quad \text{for } k = n-1,\, n,\, n+1, \text{ say,} \tag{8.30}$$

using some appropriate values of α and ρ. If this approximation holds for sufficiently large n, then we might naturally try to "solve" for s in terms of s_{n-1}, s_n and s_{n-1}. Doing so, we obtain the approximation

$$s \approx s_{n-1} - \frac{(s_n - s_{n-1})^2}{s_{n+1} - 2\,s_n + s_{n-1}} \tag{8.31}$$

$$= \frac{s_{n-1}\,s_{n+1} - s_n^2}{s_{n+1} - 2\,s_n + s_{n-1}}. \tag{8.32}$$

This approximation shows that we can accelerate linearly converging series by replacing the sequence of partial sums

$$s_1,\, s_2,\, s_3,\, s_4,\, \cdots$$

with the sequence

$$s_2',\, s_3',\, s_4',\, \cdots.$$

where

$$s_n' = \frac{s_{n-1}\,s_{n+1} - s_n^2}{s_{n+1} - 2\,s_n + s_{n-1}}. \tag{8.33}$$

This is Aitken's δ^2 transformation. See Aitken (1926). Having applied this transformation once, there is nothing to stop us from applying it again. We can define

$$s_n'' = \frac{s_{n-1}'\,s_{n+1}' - (s_n')^2}{s_{n+1}' - 2\,s_n' + s_{n-1}'} \tag{8.34}$$

and so on. Repeated applications of Aitken's δ^2 transformation produce a pyramid of values starting with any initial sequence of partial sums. For example, starting with an initial sequence of seven partial sums, we get

$$\begin{array}{llll}
s_1 & & & \\
s_2 & s_2' & & \\
s_3 & s_3' & s_3'' & \\
s_4 & s_4' & s_4'' & s_4''' \\
s_5 & s_5' & s_5'' & \\
s_6 & s_6' & & \\
s_7 & & &
\end{array}$$

In this pyramid, we may regard s_4''' as the final extrapolation of the initial sequence of seven partial sums.

Note that the partial sums at the base of this pyramid–the leftmost column–do no need to be consecutive. It is sufficient that the indices of the partial sums form an arithmetic sequence. This observation is particularly useful when summing series whose terms involve combinatorial expressions, such as the following example. In these cases, it may be the case that consecutive partial sums vary in a way that is not regular or smooth. Using more widely spaced partial sums may help restore enough smoothness to the sequence to improve the extrapolation.

The one-step Aitken δ^2 method, which transforms s_n to s'_n, is motivated by the idea of linear convergence. The repeated application of this method requires some justification, which can be given as follows. For the sake of a toy example, consider a sequence for which

$$s_n = s + c_1\,\rho_1^n + c_2\,\rho_2^n\,,$$

where $0 < |\rho_2| < |\rho_1| < 1$. Such a sequence will have linear convergence, since ρ_1 provides the dominating term. So as $n \to \infty$,

$$s_n - s \;\sim\; c_1\,\rho_1^n\,. \tag{8.35}$$

Applying one step of Aitken's δ^2 procedure gives us

$$s'_n = s + \frac{c_1\,c_2\,(\rho_1-\rho_2)^2\,\rho_1^{n-1}\,\rho_2^{n-1}}{c_1\,(1-\rho_1)^2\,\rho_1^{n-1} + c_2\,(1-\rho_2)^2\,\rho_2^{n-1}}\,. \tag{8.36}$$

As $n \to \infty$,

$$s'_n - s \;\sim\; c_2\left(\frac{\rho_1-\rho_2}{\rho_1-1}\right)^2\rho_2^{n-1}\,. \tag{8.37}$$

this can be compared with (8.35). So the effect of the δ^2 method is to eliminate the dominant term in the difference $s_n - s$.

In general, successive iterations of Aitken's δ^2 transformation will successively kill off the leading terms for series where the partial sums have a generalised asymptotic expansion of the form

$$s_n - s \;\sim\; c_1\,\rho_1^n + c_2\,\rho_2^n + c_3\,\rho_3^n + \cdots \tag{8.38}$$

where $1 > |\rho_1| > |\rho_2| > \cdots > 0$. Thus

$$\begin{aligned} s'_n - s &\;\sim\; c'_2\,\rho_2^n + c'_3\,\rho_3^n + \cdots \\ s''_n - s &\;\sim\; c''_3\,\rho_3^n + c''_4\,\rho_4^n + \cdots\,, \end{aligned}$$

and so on, in this manner.

At this stage, it is useful to point out a connection between Aitken's δ^2 method and the Padé approximants of Chapter 3. In Section 3.3.1, we noted that the calculation of a Padé approximant of relatively low degree was a better approximation than a power series of much higher degree. However, the Padé approximant was calculated from the lower order coefficients of the power series. Therefore, the Padé approximant was, in effect, accelerating the convergence of the power series.

This type of acceleration may be related to Aitken's method as follows. Suppose we consider a one-step application of the δ^2 method to a general power series

$$f(x) = a_0 + a_1 x + a_2 x^2 + a_3 x^3 + \cdots .$$

Setting

$$s_n = a_0 + a_1 x + \cdots + a_n x^n , \quad n \geq 0 ,$$

then

$$\begin{aligned} s'_1 &= \frac{s_0 s_2 - s_1^2}{s_2 - 2 s_1 + s_0} \\ &= \frac{a_0 a_1 - a_0 a_2 x + a_1^2 x}{a_2 x - a_1} , \end{aligned}$$

which is a rational function of degree (1, 1). Similarly,

$$s'_2 = \frac{-a_0 a_2 + a_0 a_3 x - a_1 a_2 x + a_1 a_3^2 x^2 - a_2^2 x^2}{-a_2 + a_3 x}$$

which is a rational function of degree (2, 1) and so on. In general, s'_n is a rational function of degree $(n, 1)$. Upon inspection, we can also observe that

$$s'_n = f_{[n, 1]}(x) ,$$

the $[n, 1]$ Padé approximant of degree $(n, 1)$ for $f(x)$. This goes some distance towards explaining the property of Padé approximants mentioned above, namely that low order Padé approximants based upon a small number of terms can achieve high accuracy. This type of series acceleration coincides with Aitken's δ^2 in the $[n, 1]$ case. However, generally the Padé approximants of degree (n, m) are distinct from the accelerants generated by repeated application of Aitken's δ^2.

We shall now turn to a type of acceleration that is even more closely linked with Padé approximation than Aitken's method. Like Aitken's δ^2, it will be most directly applicable to linear convergence, and less successful for logarithmic convergence.

8.6.4 Wynn's ϵ-transform

Wynn's ϵ-transform is defined as

$$\epsilon_n^{(-1)} = 0 , \quad \epsilon_n^{(0)} = s_n ,$$

and

$$\epsilon_n^{(k+1)} = \epsilon_{n+1}^{(k-1)} + \frac{1}{\epsilon_{n+1}^{(k)} - \epsilon_n^{(k)}}\,.$$

It will be the $\epsilon_n^{(k)}$ where k is even that will be our main concern here. The values where k is odd will be considered as intermediate quantities used in the algorithm. The algorithm proceeds iteratively based upon an initial sequence of partial sums in a way that is similar to the iterative version of Aitken's algorithm. Using the partial sums of the series as a base, we iteratively calculate $\epsilon_n^{(k)}$ for successively higher values of k, with particular attention to the cases where k is even. For example, based upon a sequence of seven partial sums, we proceed iteratively from left to right in the following array to get

$$\begin{array}{ccccccc}
\epsilon_1^{(0)} & \epsilon_1^{(1)} & \epsilon_1^{(2)} & \epsilon_1^{(3)} & \epsilon_1^{(4)} & \epsilon_1^{(5)} & \epsilon_1^{(6)} \\
\epsilon_2^{(0)} & \epsilon_2^{(1)} & \epsilon_2^{(2)} & \epsilon_2^{(3)} & \epsilon_2^{(4)} & \epsilon_2^{(5)} & \\
\epsilon_3^{(0)} & \epsilon_3^{(1)} & \epsilon_3^{(2)} & \epsilon_3^{(3)} & \epsilon_3^{(4)} & & \\
\epsilon_4^{(0)} & \epsilon_4^{(1)} & \epsilon_4^{(2)} & \epsilon_4^{(3)} & & & \\
\epsilon_5^{(0)} & \epsilon_5^{(1)} & \epsilon_5^{(2)} & & & & \\
\epsilon_6^{(0)} & \epsilon_6^{(1)} & & & & & \\
\epsilon_7^{(0)} & & & & & &
\end{array}$$

where the final step of the algorithm yields $\epsilon_1^{(6)}$ as the accelerated sum of the series based upon seven terms. Wynn (1956, 1962, 1966) examined the improvement in convergence of linearly converging series such that

$$s_n \sim s + \sum_{j=0}^{\infty} c_j\,\rho_j^n\,, \quad \text{where } 1 > \rho_0 > \rho_1 > \rho_2 > \cdots > 0\,, \tag{8.39}$$

as $n \to \infty$. Wynn showed that after an even number of steps of the ϵ-algorithm, we get

$$\epsilon_n^{(2k)} \sim s + d_k\,\rho_k^n\,, \tag{8.40}$$

where

$$d_k = c_k \left[\frac{\prod_{j=0}^{k-1}(\rho_j - \rho_k)}{\prod_{j=0}^{k-1}(\rho_j - 1)}\right]^2. \tag{8.41}$$

We conclude this brief description of Wynn's ϵ-algorithm by mentioning its connection to the theory of Padé approximants. Suppose we consider a power series

$$f(x) = a_0 + a_1\,x + a_2\,x^2 + a_3\,x^3 + \cdots \tag{8.42}$$

whose partial sums are

$$s_n = a_0 + a_1\,x + a_2\,x^2 + \cdots + a_n\,x^n\,.$$

It is not straightforward to check that in this case, $\epsilon_n^{(2k)}$ is a rational function of degree $[n+k,\, k]$. In fact, it is the Padé approximant of that degree for $f(x)$. Therefore we can write

$$f_{[n+k,\, k]}(x) = \epsilon_n^{(2k)} .$$

8.6.5 Richardson's extrapolation method

Up to this point, our methods have been motivated by linear convergence arguments. However, the series for which acceleration is important converge at a logarithmic rate. So, we shall now consider a method due to Richardson (1911, 1927).

We have seen that in a number of important series, it is possible to expand the tail of the series in an asymptotic series in n of the form

$$s_n - s \sim \frac{b_1}{n^p} + \frac{b_2}{n^{p+1}} + \frac{b_3}{n^{p+2}} + \cdots . \tag{8.43}$$

In this expansion, p need not be an integer. By an examination of the form of the series, we can hope to determine the value of p. In the argument which follows, this shall be assumed to be known.

Richardson's extrapolation method calculates a new sequence defined by

$$s'_n = s_{2n} + \frac{s_{2n} - s_n}{2^p - 1} . \tag{8.44}$$

From this definition, it easily follows from (8.43) that

$$s'_n - s \sim \frac{1}{2^p - 1}\left[\frac{b_2}{2\,n^{p+1}} + \frac{3\,b_3}{4\,n^{p+2}} + \frac{7\,b_4}{8\,n^{p+3}} + \frac{15\,b_5}{16\,n^{p+4}} + \cdots\right] . \tag{8.45}$$

This is an asymptotic series starting at order $n^{-(p+1)}$ rather than order n^{-p}. Once again, we can eliminate the leading term by defining

$$s''_n = s'_{2n} + \frac{s'_{2n} - s'_n}{2^{p+1} - 1} , \tag{8.46}$$

and so on. At each iteration of the algorithm, the leading term is eliminated. The end result is a table of values (for 16 terms, say) of the form

$$\begin{array}{ccccc}
s_1 & s'_1 & s''_1 & s'''_1 & s''''_1 \\
s_2 & s'_2 & s''_2 & s'''_2 & \\
s_4 & s'_4 & s''_4 & & \\
s_8 & s'_8 & & & \\
s_{16} & & & &
\end{array}$$

In each case, the base of the algorithm uses the partial sums whose number of terms is a power of two.

Lewis Fry Richardson (1881–1953)

Lewis Fry Richardson made diverse contributions to the applied mathematical sciences, including the theory of meteorology and the statistics of deadly quarrels between nation states. Born into a Quaker family, he chose to be a conscientious objector during World War I, and served in the Friends' Ambulance Unit during that conflict. Richardson pioneered the use of mathematical methods to deepen our understanding of the causes of war.

His work on series acceleration comes from a 1927 paper entitled "The deferred approach to the limit" published in the *Phil. Trans. Roy. Soc. London*, Ser. A., Vol. 226, pp. 299-361. This paper, jointly authored with J. Arthur Gaunt was written in two parts, the first of which was Richardson's work and contains the key acceleration technique.

Let us apply this algorithm to the series $\sum j^{-2}$. The bounds provided by the integral test give us

$$\int_n^\infty \frac{1}{x^2}\,dx \geq \sum_{j=n+1}^\infty \frac{1}{j^2} \geq \int_{n+1}^\infty \frac{1}{x^2}\,dx\,,$$

which is equivalent to $n^{-1} \geq s - s_n \geq (n+1)^{-1}$. This establishes that

$$s_n - s = \frac{-1}{n} + O\left(\frac{1}{n^2}\right).$$

We can get an asymptotic expression for $s_n - s$ using an Euler Maclaurin expansion, or directly from Maple. For example, we saw earlier that the *eulermac* command can be used to show that

$$s_n - s = -\frac{1}{n} + \frac{1}{2\,n^2} - \frac{1}{6\,n^3} + \frac{1}{30\,n^5} - \frac{1}{42\,n^7} + O\left(\frac{1}{n^9}\right).$$

Such an asymptotic expansion can be used to accelerate the convergence of s_n directly through the approximation of the difference $s - s_n$. For Richardson's algorithm, we need only note that $p = 1$. Implementing Richardson's extrapolation method on the series $\sum j^{-2}$ using a base of four partial sums, s_1, s_2, s_4, and s_8, we get

1.000000000	1.500000000	1.629629630	1.644418529
1.250000000	1.597222222	1.642569917	
1.423611111	1.631232993		
1.527422052			

The rightmost quantity in this table is s_1'''; it is our final extrapolated approximation for the series sum. This value can be compared with the exact sum $\pi^2/6$ which is approximately equal to 1.644934068.

8.6.6 *Euler acceleration*

Consider a converging alternating series $\sum(-1)^n a_n$ such that

$$\lim_{j\to\infty} \frac{a_{j+1}}{a_j} = 1\,.$$

Out in the tail of this series, any two neighbouring terms are approximately equal in absolute value, and differ only in sign. So, the contribution of any two such neighbouring terms to those partial sums which contain both will almost cancel out. This suggests that, there are large fluctuations which could be eliminated–provided the combined contributions of neighbours which almost cancel could be taken into account. Euler's transformation gives us a much more powerful tool to do this.

Suppose

$$a_1 - a_2 + a_3 - a_4 + \cdots = s$$

is a convergent alternating series, such that

$$a_1 > a_2 > a_3 > \cdots > 0\,.$$

Adding the series twice we get

$$\begin{aligned} 2\,s &= (a_1 - a_2 + a_3 - a_4 + \cdots) + (a_1 - a_2 + a_3 - a_4 + \cdots) \\ &= a_1 + (a_1 - a_2) - (a_2 - a_3) + (a_3 - a_4) - \cdots\,. \end{aligned}$$

Let $\Delta\, a_j = a_j - a_{j+1}$. Then

$$s = \frac{a_1}{2} + \frac{1}{2}(\Delta\, a_1 - \Delta\, a_2 + \Delta\, a_3 - \cdots)\,.$$

Since the terms a_j are monotone decreasing, the differences Δa_j are all positive. So the expression in parentheses is also an alternating series. Therefore, we may apply the same procedure again to this alternating series: adding it to itself and grouping terms into second differences

$$\begin{aligned} \Delta^2\, a_j &= \Delta(\Delta\, a_j) \\ &= \Delta\, a_j - \Delta\, a_{j+1} \\ &= (a_j - a_{j+1}) - (a_{j+1} - a_{j+2}) \\ &= a_j - 2\, a_{j+1} + a_{j+2}\,. \end{aligned}$$

Obviously, this can be continued. The result of the repeated grouping of terms in this fashion is a representation of the series and its sum as

$$s = \frac{a_1}{2} + \frac{\Delta\, a_1}{4} + \frac{\Delta^2\, a_1}{8} + \frac{\Delta^3\, a_1}{16} + \cdots\,, \tag{8.47}$$

where

$$\Delta^j\, a_1 = \sum_{k=0}^{j} (-1)^k \binom{j}{k} a_{1+k}\,. \tag{8.48}$$

The new series obtained in (8.47) is the *Euler transformation* of the original alternating series.

How well does it work? Looking at the denominators in (8.47), we see that they increase as powers of two. Thus the new series may be justifiably called an acceleration of the old series provided the higher order differences $\Delta^j\, a_1$ do not grow fast. To see this in action, let us apply the Euler transform to a well known series, namely

$$\ln\, 2 = 1 - \frac{1}{2} + \frac{1}{3} - \frac{1}{4} + \cdots\,.$$

This well known series is not immediate practical for computing ln 2 because its convergence is too slow. For this series, we find that $\Delta^j a_1 = 1/(j+1)$. So, the Euler transformed series for ln 2 is

$$\ln 2 = \frac{1}{2 \times 1} + \frac{1}{4 \times 2} + \frac{1}{8 \times 3} + \frac{1}{16 \times 4} + \cdots .$$

This is much better: the Euler transformation of our original series with logarithmic convergence has linear convergence. The convergence of the new transformed series can be further accelerated using Aitken's method or Wynn's ϵ-transform.

We can get some insight into the Euler transformation by considering the contribution of each term in the original series to the partials sums s'_n of the transformed series. Writing this out, we get

$$\begin{aligned} s'_n &= \frac{a_1}{2} + \frac{\Delta a_1}{4} + \cdots + \frac{\Delta^{n-1} a_1}{2^n} \\ &= \{1 - 2^{-n}\} a_1 - \{1 - (n+1)\, 2^{-n}\} a_2 \\ &\qquad + \left\{1 - \left[1 + \binom{n+1}{2}\right] 2^{-n}\right\} a_3 - \cdots . \end{aligned}$$

In this representation, we can see that the Euler transformation shrinks the oscillations of the original series. This shrinkage makes some sense if we recall that the original alternating series is enveloping, so that the partial sums of an enveloping series successively overshoot above and below the total sum of the series.

There is another representation of the Euler transformation that highlights its connection to probability theory. For each nonnegative integer value $m \geq 0$, let K_m be a nonnegative integer-valued random variable with binomial distribution $\mathcal{B}(m, \frac{1}{2})$. (For the case $m = 0$, we shall understand that $K_0 = 0$ with probability one.) Then the Euler transformation of the alternating sum $s = \sum (-1)^{j-1} a_j$ can be written as

$$s = \frac{1}{2} \sum_{m=0}^{\infty} E\left[(-1)^{K_m} a_{1+K_m}\right] .$$

So, the Euler transformation is determined by a family of binomial smoothings of the original series.

8.7 Applications of acceleration methods

8.7.1 Ties in multiple random walks

Consider a game in which each of r people independently and simultaneously tosses a coin. Over successive trials, each individual records the total number of heads obtained. After n trials, say, a tie is said to occur if all r individuals have the same number of heads. We shall consider the problem of computing the expected number of heads as $n \to \infty$.

It is well known that for $r \leq 3$, the ties for such a random walk are persistent events.|| Therefore there will be infinitely many such events as $n \to \infty$. So we shall restrict attention to the case where $r > 3$, where the number T of ties is finite with probability one. After exactly n tosses, the probability of a tie is

$$u_n = \sum_{j=0}^{n} \left[\binom{n}{j} \frac{1}{2^n} \right]^r .$$

So

$$E(T) = \sum_{n=0}^{\infty} u_n \,. \tag{8.49}$$

Note that the $n = 0$ term is included, and that $u_0 = 1$. Using Stirling's approximation it is possible to show that

$$u_n \sim \frac{1}{\sqrt{r}} \left(\frac{2}{\pi\, n} \right)^{(r-1)/2} \qquad \text{as } n \to \infty \,. \tag{8.50}$$

See Problem 13. While this approximation is satisfactory for establishing the convergence properties of $\sum u_n$, it is inadequate for evaluating $E(T)$ precisely. Direct numerical evaluation of the series by summing a large number of terms is not very practical. The sum can be approximated to high accuracy by analytical methods beyond the scope of the discussion here. However, such a procedure is only available to us for this particular series, and not for all series converging slowly in this way. Our purpose here will be to use this series as a test of acceleration.

Let us try Richardson's method. In this case, we shall choose a base of partial sums for the left-hand column consisting of values s_n where $n = 2^m$, for $m = 1, 2, \ldots, 9$. To choose the value of p, we may refer to the asymptotic approximation obtained in formula (8.50), where we found that the n term of the sum goes to zero at rate $O(n^{-(r-1)/2})$. For $r = 4$, this reduces to $O(n^{-3/2})$, which implies that the tail of the series is of order

$$O\left(\int_n^{\infty} \frac{dx}{x^{3/2}} \right) = O(n^{-1/2}) \,.$$

So we may set $p = 1/2$ for the iteration which produces the second column of the table. The third column will require $p = 3/2$, and so on,

|| See Feller (1968, p. 316). The problem has some similarity to the calculation of the random walk constants associated with random walks on lattices.

s_{2m}	s'_{2m}							s_{2m}^{viii}
1.195312500	1.426309552	□	□	□	□	□	□	1.492446184
1.262969970	1.464747165	□	□	□	□	□	□	
1.322069142	1.481688293	□	□	□	□	□		
1.368820509	1.488454289	□	□	□	□			
1.403860432	1.491000029	□	□	□				
1.429383029	1.491928581	□	□					
1.447702197	1.492262010	□						
1.460753464	1.492380850							
1.470016911								

Table 8.2 *Accelerated partial sums for the expected number of ties using Richardson's method (with intermediate values omitted for brevity)*

with each column incrementing the value of p by one. Table 8.2 shows the accelerated values. In fact, the value obtained by Richardson acceleration is accurate to six significant figures.

Of course, the series could also be accelerated by Aitken's or Wynn's method. These methods are designed for series which converge linearly. So the longer tail of this series makes Aitken's and Wynn's methods underestimate the sum of the series. Nevertheless, we may reasonably hope that these acceleration methods improve the numerical result. However, the reader is warned that applying an acceleration method when it has no asymptotic justification is a dangerous game.

8.7.2 Geometric means and Khintchine's constant

Let N be a strictly positive integer-valued random variable. If the distribution of N is heavy tailed, say $P(N > n) = O(n^{-1})$, or heavier, then the mean and higher moments will all be infinite. In such cases, the *geometric mean* of the distribution may be finite, and if so, may provide a suitable measure of centrality for the distribution of N. If we let ω be the geometric mean of N, then

$$\ln \omega = \sum_{j=1}^{\infty} (\ln j) \cdot P(N = j)\,. \tag{8.51}$$

Unfortunately, if N has a heavy-tailed distribution, then the convergence of this series is likely to be very slow. If the problem is simply that of computation, then a series acceleration method is appropriate.

An important example of this kind is to be found in the computation of *Khintchine's constant.* This constant is the geometric mean arising in the continued fraction expansions of normal numbers. A brief explanation is as follows. Suppose X is a strictly positive random variable having some continuous distribution on the positive reals. Then there exists a unique continued fraction representation of X as

$$X = N_0 + \cfrac{1}{N_1 + \cfrac{1}{N_2 + \cfrac{1}{N_3 + \cfrac{1}{\ddots}}}}$$

where N_0, N_1, N_2, and so on, are positive integers. Khintchine (1964) discovered that there exists a real number ω such that

$$P\left(\lim_{m\to\infty} \sqrt[m]{N_1\, N_2 \cdots N_m} = \omega\right) = 1\,. \tag{8.52}$$

In fact, ω is the geometric mean of the limiting distribution of N_m as $m \to \infty$. The probability function of this limiting distribution can be shown to be

$$\begin{aligned}\lim_{m\to\infty} P(N_m = j) &= \frac{1}{\ln 2} \ln\left[1 + \frac{1}{j\,(j+2)}\right] \\ &\sim O(j^{-2}) \text{ as } j \to \infty\,,\end{aligned}$$

again, with probability one. Therefore, an infinite series representation for the value of Khintchine's constant is given by

$$\ln\omega = \sum_{j=1}^{\infty} \frac{\ln j}{\ln 2} \ln\left[1 + \frac{1}{j\,(j+2)}\right]. \tag{8.53}$$

Unfortunately, the convergence of this series is even slower than $\sum j^{-2}$, due to the presence of the additional factor $\ln j$. However, the logarithm is a fairly flat function, and so a conservative acceleration of the series can be achieved using Richardson's method under the rough assumption that $s_n - s = O(n^{-1})$. Since the difference $s_n - s$ is really of greater order than $O(n^{-1})$, Richardson's acceleration method will tend to underestimate the sum if we set $p = 1$ for the first iteration and increment p by one for each iteration thereafter.

For Richardson's algorithm, let us take a set of partial sums s_n, where $n = 2^m \times 100$ for $m = 0, \ldots, 6$. Setting $p = 1$ in (8.44), and successively incrementing p by one, we get the values shown in Table 8.3. The estimate on the far right is our accelerated sum for the series. So, based on Richardson's algorithm and the given partial sums, our estimate for ω is

$$\omega \approx \exp(0.9876902347)$$

0.9080365395	□	□	□	□	0.9875263696	0.9876902347
0.9427186946	□	□	□	□	0.9876876743	
0.9627178807	□	□	□	□		
0.9740145133	□	□	□			
0.9803014212	□	□				
0.9837612607	□					
0.9856484896						

Table 8.3 *Accelerated sums for Khintchine's constant using Richardson's algorithm (with intermediate values omitted for brevity)*

$$\approx \quad 2.685025526\,.$$

This value can be compared with more precise estimates for ω based upon more extensive computations. It is known that

$$\omega = 2.685452001\ldots.$$

Khintchine's constant is known to many more decimal places than this. For example, Bailey *et al.* (1997) calculated the constant to more than 7000 decimal places. However, calculations of such precision use series for ω which converge much faster than the one given here.

8.8 Comparing acceleration techniques

We have now considered a number of methods for accelerating series. A comparison among these methods is rather difficult because very few series will satisfy tidy regularity conditions necessary to make a particular acceleration method more effective than others. While the series calculated in Section 8.7.1 fits the assumptions for Richardson's method better than Wynn's or Aitken's, the series calculated in Section 8.7.2 fits none of the assumptions that well. Nevertheless, the corrections obtained by these methods were all in the right direction, even if the order of the error was not reduced.

In all cases, the success of an acceleration method must be judged in terms of the computational efficiencies obtained from it. Typically, the computational costs of summing a series are roughly proportional to the number of terms in the sum. More generally, we might suppose that the computational cost of summing n terms of a series is some function $\mathbf{C}(n)$. We might also let $\mathbf{C}_0(n)$ denote the computational cost of accelerating the partial sums out to n terms. Suppose that $m > n$ are two positive integers such that $s'_n - s \approx s_m - s$. It seems reasonable to measure the computational efficiency obtained by accelerating the first n terms as

$$\mathbf{Eff}(n) = \frac{\mathbf{C}(m)}{\mathbf{C}(n) + \mathbf{C}_0(n)} .$$

Typically, acceleration will not be worthwhile unless m is much larger than n.

8.9 Divergent series

8.9.1 Antilimits

Divergent series have arisen in numerous places in earlier sections of this book. In particular, asymptotic series are often divergent. Although these series are of importance both theoretically and computationally, so far we have not had a unified way of summing such series.

Some intriguing clues about the summability of divergent series can be found among the series acceleration techniques that we have considered so far. In some cases, the series transformations such as Aitken's and Euler's can be applied to divergent series so as to make them convergent. It is worth formalising this idea. When it exists, let us define an *antilimit* s of a divergent series $\sum a_j$ to be the sum of the series after it has been transformed by some standard acceleration procedure. For example, the series

$$1 + \rho + \rho^2 + \rho^3 + \cdots$$

has limit $1/(1-\rho)$ for $|\rho| < 1$. When $|\rho| > 1$, the series diverges. However, we can still say that $1/(1-\rho)$ is the antilimit of the series, because after a one-step application of Aitken's δ^2 method, we see that

$$s_n = \frac{1-\rho^n}{1-\rho}, \ \forall n \geq 1 \ \text{ and } \ s'_n = \frac{1}{1-\rho}, \ \forall n \geq 2 .$$

In qualitative terms, the partial sums of series converge towards limits and diverge away from antilimits.

8.9.2 Why should we sum divergent series?

Sometimes the properties of functions can be proved simply through their expansions. A very simple example of this principle is that the identity

$$\frac{1}{1-x} + \frac{1}{1+x^2} - \frac{x}{1-x^2} = \frac{2}{1-x^4} ,$$

while easily verified algebraically, can be derived immediately from series expansions of the rational functions. Thus

$$\begin{aligned}\frac{1}{1-x}+\frac{1}{1+x^2}-\frac{x}{1-x^2} &= (1+x+x^2+\cdots)+(1-x^2+x^4-\cdots)\\ &\quad -(x+x^3+x^5+\cdots)\\ &= 2+2\,x^4+2\,x^8+\cdots\\ &= \frac{2}{1-x^4}\,.\end{aligned}$$

While the original identity is valid for all $|x| \neq 1$, the proof by series expansion is only valid for $|x| < 1$. However, the proof remains valid if the series sums are interpreted as limits for $|x| < 1$ and antilimits for $|x| > 1$.

- A unified theory for the summation of divergent and convergent series should allow the free manipulation of functions through their series expansions. This should lead to the construction of simpler proofs of some identities without restrictions on the domains of the functions or the need for the extensions of those domains by analytic continuation.

A minimal condition for a unified methodology for summing both convergent and divergent series is the following.

- A summation method will be said to be *regular* provided that it sums all convergent series to their usual sums. Thus a summation method should be an extension of the usual concept of series convergence to a more general class of series.

At this stage, we should clarify an ambiguity in the definition of regularity. Consider the sequence $1+2+4+8+\cdots$. Understood in the usual sense, this is a divergent series. However, as a series in the extended real numbers (including $\pm\infty$), the sum of this series exists and is $+\infty$. That is, the partial sums of the series converge in the extended reals to $+\infty$. In this respect, this series is quite different from $1-2+4-8+\cdots$ whose oscillations prevent the partial sums of the series from converging even among the extended reals. The latter series is therefore divergent in a stricter sense than the former. So, we may interpret the definition of regularity given above in two senses, depending on whether series $\sum 2^j$ is understood as divergent or as convergent in the extended reals.

- A summation method is said to be *totally regular* if it sums series to s where
$$\liminf s_n \leq s \leq \limsup s_n\,,$$
the limits superior and inferior being defined for the extended real numbers.

For example, a summation method which sums diverging geometric series to the antilimit using Aitken's δ^2 method is not totally regular because it sums $1+2+4+\cdots$ to -1 rather than $+\infty$. While regularity is clearly a desirable property of any summation method, the desirability of total regularity will depend on the context in which the series is studied.

What other properties should summation methods have? The following laws must be satisfied if we are to be able to manipulate series with standard algebraic techniques.

1. If $\sum a_j = s$, then $\sum c\, a_j = c\, s$.
2. If $\sum a_j = s$ and $\sum b_j = t$, then $\sum (a_j + b_j) = s + t$.
3. We must have $\sum_{j \geq 1} a_j = s$, if and only if $\sum_{j \geq 2} a_j = s - a_1$.

The reader who is interested in learning more about the fascinating subject of divergent series is encouraged to read G. H. Hardy's classic work, *Divergent Series*, which is available as a reprint from the American Mathematical Society. See Hardy (1991). While the subject has not become mainstream with statistics, it continues to fascinate. The existence of asymptotic series, which have great power for the calculation of functions but which do not usually converge, is a reminder of other paradigms for series summation, most notably that of Euler.

Such generalised summation methods can be used to provide formal definitions of the expectations of some discrete non-integrable random variables. This is of some value in discussing examples such as the St. Petersburg paradox involving finite random variables with infinite expectations. A generalisation of the law of large numbers was proposed by Feller (1968, 1971). See Problem 14, which explores this further.

We finish with one final formula, the merits of which are left to the reader to judge, to wit:

$$0! - 1! + 2! - 3! + 4! - \cdots = 0.5963\ldots.$$

8.10 Problems

1. Use Bertrand's test to prove that the harmonic series

$$1 + \frac{1}{2} + \frac{1}{3} + \frac{1}{4} + \frac{1}{5} + \cdots$$

diverges.

2. Does the series

$$\sum_{j=1}^{\infty} j^{-(j+1)/j}$$

converge?

3. Prove that if

$$\sum_{j=1}^{\infty} a_j$$

is a convergent series whose terms are positive, then the series

$$\sum_{j=1}^{\infty} a_j^{j/(j+1)}$$

is also convergent. (This was problem B4 from the 1988 William Lowell Putnam competition.)

4. Prove that if $\sum_{j=1}^{\infty} a_j$ is a convergent series with positive terms, then the series

$$\sum_{j=1}^{\infty} (a_1\, a_2 \, \cdots \, a_j)^{1/j}$$

is also convergent. (This is a toughie!)

5. Let $X_1,\, X_2,\, X_3,\, \ldots$ be independent identically distributed $\mathcal{N}(0,\, 1)$. Show that for all values of θ the series

$$\sum_{j=1}^{\infty} \frac{X_j\, \sin(j\,\theta)}{n}$$

converges with probability one.

6. Let X_j, $j \geq 1$ be any sequence of random variables. Prove that there exists a sequence of positive constants a_j such that

$$\lim_{j\to\infty} \frac{X_j}{a_j} = 0$$

with probability one.

7. Suppose $X_1,\, X_2,\, X_3,\, \ldots$ are independent identically distributed Poisson random variables. Let $a_j = \ln j / \ln \ln j$. Prove that

$$P\left(\limsup_{j\to\infty} \frac{X_j}{a_j} = 1\right) = 1\,.$$

8. Use the Euler-Maclaurin sum formula to prove the following identities.

   (a) The sum of positive integers:

   $$1 + 2 + 3 + \cdots + n = n\,(n+1)/2\,.$$

   (b) The sum of squares of positive integers:

   $$1^2 + 2^2 + 3^2 + \cdots + n^2 = n\,(n+1)\,(2\,n+1)/6\,.$$

   (c) The sum of the odd positive integers:

   $$1 + 3 + 5 + \cdots + (2\,n - 1) = n^2\,.$$

   In each case, don't forget to evaluate the constant c using special cases of n.

9. It is well known that the harmonic series diverges. What about

   $$\frac{1}{n+1} + \frac{1}{n+2} + \frac{1}{n+3} + \cdots + \frac{1}{2\,n}\,?$$

   Does this go to infinity in n? If not, find the limit.

10. Little Tommy wants to collect toy cars from boxes of cereal. Tommy knows that each box contains a car, and that there are n different types of cars available to be collected. Tommy is good at probability, and assumes that each type of car is equally likely in any box, and that the toy cars are inserted independently into each box. The family buys one box of cereal each week. Prove that the expected number of weeks until Tommy has all n types of cars is asymptotically

    $$n\,\ln n + \gamma\,n + \frac{1}{2}$$

    for large n, where γ is Euler's constant.

11. An urn contains n balls, each individually numbered from 1 to n. Two random numbers N and M are obtained by drawing a ball from the urn two times with replacement.

    (a) Let $n = k\,p$, where p is a prime number and k is any positive integer. Find an expression for the probability that the greatest common divisor of N and M is *not* divisible by p.

    (b) Let $n = k\,p\,q$, where p and q are distinct prime numbers. Find an expression for the probability that the greatest common divisor of N and M is not divisible by p or q.

(c) As $n \to \infty$ show that the probability N and M are relatively prime is given by

$$\prod_{p \text{ prime}} \left(1 - \frac{1}{p^2}\right)$$

where the product is over all primes.

(d) Take the reciprocal of this product and expand to show that for large n, the probability N and M are relatively prime is approximately $6/\pi^2$.

12. The following two formulas were useful in an application of Kummer acceleration. Prove that

$$\sum_{k=1}^{\infty} \frac{1}{k\,(k+1)\,(k+2)\,\cdots\,(k+n)} = \frac{1}{n^2\,(n-1)!},$$

and

$$\sum_{j=1}^{\infty} \frac{1}{j^2} = \sum_{j=1}^{n} \frac{1}{j^2} + n! \sum_{j=1}^{\infty} \frac{1}{j^2\,(j+1)\,\cdots\,(j+n)}.$$

13. Prove (8.50), namely that

$$\sum_{j=0}^{n} \left[\binom{n}{j} \frac{1}{2^n}\right]^r \sim \frac{1}{\sqrt{r}} \left(\frac{2}{\pi\, n}\right)^{(r-1)/2} \qquad \text{as } n \to \infty.$$

14. Alice and Bob decide to play the following game. A fair coin is tossed independently until it lands heads. Let N be the number of tosses required. If N is odd, then Alice pays Bob 2^N dollars. If N is even, Bob must pay Alice 2^N dollars. Let X be the total amount of money that Alice pays Bob in this game, which may be positive or negative. Is the game favourable to either player? The reader will recognise that this game is a thinly disguised version of the St. Petersburg paradox. It is well known that random variables with infinite expectations do not obey a law of large numbers. However, a generalisation of the law of large numbers to the St. Petersburg paradox has been given by Feller (1968, p. 252) for games with variable "entrance fees." Discuss the following.

(a) The game would be favourable to Bob if $E(X) > 0$, favourable to Alice if $E(X) < 0$ and fair if $E(X) = 0$. Formally,

$$\begin{aligned} E(X) &= 2 \cdot 2^{-1} - 2^2 \cdot 2^{-2} + 2^3 \cdot 2^{-3} - \cdots \\ &= 1 - 1 + 1 - \cdots \end{aligned}$$

which is undefined. Therefore, the concept of fairness is not meaningful here.

(b) In practice, the number of tosses must be bounded. Any partial sum of the series $1 - 1 + 1 - 1 + \cdots$ will be greater than or equal to zero, with many of them strictly positive. Therefore, under any finite termination rule after n tosses

$$E(X \mid N \leq n) \geq 0$$

with strict inequality when n is odd. Therefore, the game is in Bob's favour.

(c) Suppose we let v be the value of the game, as measured by some method yet to be determined. In the conditional model where the first toss is a head, then clearly $v = 2$. Conditionally on the first toss being a tail, the roles of Alice and Bob are reversed, with all the payoffs based on subsequent tosses being doubled. So the value of the game conditional of an initial tail is $-2\,v$. Thus

$$\begin{aligned} v &= \frac{1}{2} \cdot 2 + \frac{1}{2} \cdot (-2v) \\ &= 1 - v\,, \end{aligned}$$

which is solved by $v = 1/2$. Thus the game is slightly in Bob's favour by 50 cents!

(d) The game is fair in the classical sense.** Suppose this coin tossing game is repeated many times under independent and identical conditions. After k repetitions, let S_k be the total amount of money that Alice has paid Bob over those trials where N was odd, and let S'_k be the total amount of money that Bob has paid Alice over those trials where N was even. Equivalently,

$$S_k - \max(X_1,\,0) + \max(X_2,\,0) + \cdots + \max(X_k,\,0)$$

and

$$S'_k = \max(-X_1,\,0) + \max(-X_2,\,0) + \cdots + \max(-X_k,\,0)\,,$$

where $X_1, \ldots, X_k$ are the payments (positive or negative) from Alice to Bob on each of the k trials. The game is fair in the classical sense because, for all $\epsilon > 0$,

$$P\left(\left|\frac{S_k}{S'_k} - 1\right| > \epsilon\right) \to 0\,,$$

as $k \to \infty$.

** This is Feller's terminology. See Feller (1968, p. 252) and Feller (1971, p. 236).

CHAPTER 9

Glossary of symbols

Calculus

$f^{(n)}(x)$	The n-th derivative of $f(x)$.	p. 25.
$\partial^n : f \mapsto f^{(n)}$	Derivative operator acting on f.	p. 37.
$\int_{t_0-i\infty}^{t_0+i\infty} f(z)\,dz$	Contour integral in complex plane along the vertical line through t_0 on real axis.	p. 239.
$\int_C f(z)\,dz$	Contour integral in complex plane along the directed closed contour C.	p. 229.
$\oint_C f(z)\,dz$	Contour integral in complex plane around the closed contour C taken in counterclockwise sense.	p. 264.

Special constants, functions and sets

π	3.14159	p. 3.
e	2.71828... .	p. 4.
γ	Euler's constant 0.57721... .	p. 294.
$\ln(x)$	Natural logarithm of x.	p. 2.
i	Usually $\sqrt{-1}$ and not an integer.	p. 46.
$\Re(z)$	Real part of complex number z.	p. 60.
$\Im(z)$	Imaginary part of complex number z.	p. 229.
$E_1(y)$	Exponential integral.	p. 73.
$\Gamma(x)$	Gamma function at x.	p. 32.

$\Phi(x)$	Standard normal distribution function.	p. 43.
$\phi(x)$	Standard normal density function.	p. 43.
$\zeta(x)$	The difference in the natural logarithms of $\Gamma(x)$ and Stirling's approximation.	p. 72.
$Z^+ \times Z^+$	The lattice of nonnegative (i.e., zero and positive) integers in the plane.	p. 93.
$\#(B)$	The number of elements in the (finite) set B.	p. 93.
$\binom{n}{j}$	The number of subsets of j items from a set of n. "n choose j."	p. 1.
$n!$	The number of ways to order a set of n items. "n factorial."	p. 4.
B_n	The n-th Bernoulli number (Abramowitz & Stegun).	p. 73.
B_n^*	The n-th Bernoulli number (Whittaker & Watson).	p. 73.
$\det(H)$	Determinant of the matrix H.	p. 218.

Order symbols

$f(n) = O(n^k)$	Function $f(n)$ is of order n^k as $n \to \infty$.	p. 9.
$f(n) = o(n^k)$	Function $f(n)$ is of smaller order than n^k as $n \to \infty$.	p. 12.
$f(n) = O_p(n^k)$	Function $f(n)$ is of order n^k in probability as $n \to \infty$.	p. 13.
$f(n) = o_p(n^k)$	Function $f(n)$ is of smaller order than n^k in probability as $n \to \infty$.	p. 13.
$f(n) \sim g(n)$	Functions $f(n)$ and $g(n)$ are asymptotically equivalent as $n \to \infty$.	p. 11.

Series and sequences

$\limsup x_n$	The limit superior of sequence x_n.	p. 7.

$\liminf x_n$	The limit inferior of sequence x_n.	p. 7.
$f(x) \sim \sum a_n x^{-n}$	$f(x)$ has asymptotic expansion $\sum_n a_n x^{-n}$ as $x \to \infty$.	p. 48.
$f(x) \sim h(x) \sum a_n x^{-n}$	$f(x)/h(x)$ has asymptotic expansion $\sum_n a_n x^{-n}$ as $x \to \infty$.	p. 48.
$\sum_n a_n \rightsquigarrow s$	The sum $\sum_n a_n$ envelops s.	p. 40.

Probability distributions and statistics

$E(T)$ or $E\,T$	Expectation of random variable T.	p. 15.
$E_\theta(T)$ or $E_F(T)$	Expectation of T assuming parameter θ or distribution F.	p. 126.
$\mathrm{Var}(T)$	Variance of random variable T.	p. 15.
$\mathrm{Var}_\theta(T)$ or $\mathrm{Var}_F(T)$	Variance of T assuming parameter θ or distribution F.	p. 134.
$X \stackrel{d}{=} Y$	Random variables X and Y identically distributed.	p. 56.
$X_n \stackrel{d}{\Longrightarrow} F$	Random variables X_n converge in distribution to F.	p. 14.
$X_n \stackrel{P}{\to} c$	Random variables X_n converge in probability to c.	p. 14.
$X_n \stackrel{a.s.}{\to} c$	Random variables X_n converge almost surely to c.	p. 14.
μ_j	The j-th moment of a random variable $E(X^j)$.	p. 34.
κ_j	The j-th cumulant of a random variable.	p. 34.
$\mathcal{B}(n, \theta)$	Binomial for n independent trials with probability weight θ for each trial.	p. 3.
$\mathcal{C}(\theta, a)$	Cauchy distribution centred at θ with scale parameter a.	p. 225.
$\mathcal{E}(\lambda)$	Exponential distribution with mean λ^{-1}.	p. 64.
$\mathcal{X}(k)$	Chi-square distribution with k degrees if freedom.	p. 32.

$\mathcal{X}(k, \nu)$	Non-central chi square with k degrees of freedom and non-centrality parameter ν.	p. 32.
$\mathcal{G}(\alpha, \lambda)$	Gamma distribution with shape α, scale λ and mean $\alpha\lambda^{-1}$.	p. 70.
$\mathcal{N}(\mu, \sigma^2)$	Normal with mean μ and variance σ^2.	p. 3.
$\mathcal{P}(\mu)$	Poisson with mean μ.	p. 20.
$\mathcal{U}(a, b)$	Continuous uniform distribution on $[a, b]$.	p. 67.
$M(t)$	Moment generating function $M(t) = E(e^{t\,X})$.	p. 19.
$A(t)$	Probability generating function $A(t) = E(t^X)$.	p. 20.
$K(t)$	Cumulant generating function $K(t) = \ln M(t)$.	p. 34.
$\chi(t)$	Characteristic function $\chi(t) = E(e^{i\,X\,t})$.	p. 57.

Likelihood statistics

$L_n(\theta)$	Log-likelihood function of a parameter θ for sample size n.	p. 143.
$\ell_n(\theta)$	Log-likelihood function $\ell_n(\theta) = \ln L_n(\theta)$.	p. 143.
$\widehat{\theta}_n$	Maximum likelihood estimator $\widehat{\theta}_n = \arg\max_\theta L(\theta)$.	p. 143.
$u_n(\theta)$	Score function $u_n(\theta) = \ell_n'(\theta)$.	p. 153.
$i_n(\theta)$	Observed information function $i_n(\theta) = -\ell_n''(\theta)$.	p. 153.
$i_n(\widehat{\theta}_n)$	Observed information.	p. 153.
$I(\theta)$	Expected information function $I(\theta) = n^{-1}\, E_\theta\, i_n(\theta)$.	p. 153.

CHAPTER 10

Useful limits, series and products

Limits

Unless otherwise specified, limits are as $x \to \infty$ or $n \to \infty$.

$\frac{(\ln x)^k}{x} \to 0$

$\frac{x^k}{e^x} \to 0$

$(1 + n^{-1})^n \to e$

$\frac{\sin x}{x} \to 1$

$n^{1/n} \to 1$

$\sqrt[n]{\frac{n^n}{n!}} \to e$

Series

$1 + 2 + 3 + \cdots + n = \frac{n\,(n+1)}{2}$

$a + (a + d) + (a + 2\,d) + (a + 3\,d) + \cdots + [a + (n-1)\,d] = n\,a + \frac{n\,(n-1)\,d}{2}$

$1 + 3 + 5 + \cdots + (2\,n - 1) = n^2$

$2 + 4 + 6 + \cdots + (2\,n) = n\,(n+1)$

$1^2 + 2^2 + 3^2 + \cdots + n^2 = \frac{n\,(n+1)\,(2\,n+1)}{6}$

$$1^3 + 2^3 + 3^3 + \cdots + n^3 = (1 + 2 + 3 + \cdots + n)^2$$

$$1^2 + 3^2 + 5^2 + \cdots + (2\,n - 1)^2 = \tfrac{n\,(4\,n^2-1)}{3}$$

$$a + a\,r + a\,r^2 + a\,r^3 + \cdots + a\,r^{n-1} = \tfrac{a\,(1-r^n)}{1-r}$$

$$\tbinom{n}{0} + \tbinom{n}{1} + \tbinom{n}{2} + \tbinom{n}{3} + \cdots + \tbinom{n}{n} = 2^n$$

$$\tbinom{n}{0} - \tbinom{n}{1} + \tbinom{n}{2} - \tbinom{n}{3} + \cdots + (-1)^n \tbinom{n}{n} = 0$$

$$\tbinom{n}{0}^2 + \tbinom{n}{1}^2 + \tbinom{n}{2}^2 + \tbinom{n}{3}^2 + \cdots + \tbinom{n}{n}^2 = \tbinom{2\,n}{n}$$

$$\tbinom{n}{0} + \tbinom{n}{2} + \tbinom{n}{4} + \tbinom{n}{6} + \cdots + \tbinom{n}{2\lfloor \frac{n}{2} \rfloor} = 2^{n-1}$$

$$\tfrac{1}{1\times 2} + \tfrac{1}{2\times 3} + \tfrac{1}{3\times 4} + \cdots + \tfrac{1}{(n-1)\times n} = 1 - \tfrac{1}{n}$$

Products

$$\left(1 - \tfrac{1}{2}\right)\left(1 - \tfrac{1}{3}\right)\left(1 - \tfrac{1}{4}\right)\cdots\left(1 - \tfrac{1}{n}\right) = \tfrac{1}{n}$$

$$\left(1 - \tfrac{1}{2^2}\right)\left(1 - \tfrac{1}{3^2}\right)\left(1 - \tfrac{1}{4^2}\right)\cdots\left(1 - \tfrac{1}{n^2}\right) = \tfrac{n+1}{2\,n}$$

$$\left(1 - \tfrac{1}{1\times 3}\right)\left(1 - \tfrac{1}{2\times 4}\right)\left(1 - \tfrac{1}{3\times 5}\right)\cdots\left(1 - \tfrac{1}{n\times(n+2)}\right) = \tfrac{2\,(n+1)}{n+2}$$

$$\left(1 + \tfrac{1}{1}\right)\left(1 - \tfrac{1}{2}\right)\left(1 + \tfrac{1}{3}\right)\cdots\left(1 + \tfrac{(-1)^{n+1}}{n}\right) = \tfrac{2\,n+1+(-1)^{n+1}}{2\,n}$$

$$\sin\tfrac{x}{k}\,\sin\tfrac{x+\pi}{k}\,\sin\tfrac{x+2\,\pi}{k}\,\cdots\,\sin\tfrac{x+(k-1)\,\pi}{k} = 2^{1-k}\,\sin x$$

$$\sin\tfrac{x+\pi/2}{k}\,\sin\tfrac{x+3\,\pi/2}{k}\,\sin\tfrac{x+5\,\pi/2}{k}\,\cdots\,\sin\tfrac{x+(2\,k-1)\,\pi/2}{k} = 2^{1-k}\,\cos x$$

References

Abramowitz, M. & Stegun, I. A. editors (1972). *Handbook of Mathematical Functions.* Dover, New York.

Aitken, A. C. (1926). On Bernoulli's numerical solution of algebraic equations. *Proc. Roy. Soc. Edin.* 46, 289–305.

Aitken, A. C. & Silverstone, H. (1942). On the estimation of statistical parameters. *Proc. Roy. Soc. Edinburgh, Series A* 61, 186–194.

Amari, S.-I. (1985). *Differential-Geometrical Methods in Statistics.* Springer Lecture Notes in Statistics 28. Springer, Berlin.

Bahadur, R. R. (1964). On Fisher's bound for asymptotic variances. *Ann. Math. Statist.* 35, 1545–1552.

Bailey, D. H., Borwein, J. M. & Crandall, R. (1997). On the Khintchine constant. *Mathematics of Computation* 66, 417–431.

Baker, G. A. & Graves-Morris, P. R. (1996). *Padé Approximants.* Encyclopaedia of Mathematics and Its Applications. Cambridge University, Cambridge, UK.

Barndorff-Nielsen, O. (1980). Conditionality resolutions. *Biometrika* 67, 293–310.

Barndorff-Nielsen, O. (1983). On a formula for the distribution of the maximum likelihood estimator. *Biometrika* 70, 343–365.

Barndorff-Nielsen, O. E. & Cox, D. R. (1989). *Asymptotic Techniques for Use in Statistics.* Chapman and Hall, London.

Beran, R. J. (1999). Hájek-Inagaki convolution theorem. *Encyclopedia of Statistical Sciences,* Update Volume 3. Wiley, New York, 293–297.

Bickel, P. J. & Doksum, K. A. (2001). *Mathematical Statistics: Basic Ideas and Selected Topics Vol. I.* Second Edition. Prentice Hall, Upper Saddle River, New Jersey.

Billingsley, P. (1995). *Probability and Measure.* Third Edition. Wiley, New York.

Billingsley, P. (1999). *Convergence of Probability Measures.* Second Edition. Wiley, New York.

Breiman, L. (1968). *Probability.* Addison-Wesley, Reading, Massachusetts.

Butler, R. W. (2007). *Saddlepoint Approximations with Applications.* Cambrige University Press, Cambridge, UK.

Chow, Y. S. & Teicher, H. (1988). *Probability Theory: Independence, Interchangeability, Martingales.* Second Edition. Springer Texts in Statistics. Springer, New York.

Cox, D. R. & Reid, N. (1987). Parameter orthogonality and approximate conditional inference. *J. Roy. Statist. Soc. B* 49, 1–39.

Cramér, H. (1946a). *Mathematical Methods of Statistics.* Princeton University, Princeton, NJ.

Cramér, H. (1946b). A contribution to the theory of statistical estimation. *Skand. Akt. Tidskr.* 29, 85–94.

Daniels, H. E. (1954). Saddlepoint approximations in statistics. *Ann. Math. Statist.* 25, 631–650.

Darmois, G. (1945). Sur les lois limites de la dispersion de certains estimations. *Rev. Inst. Int. Statist.* 13, 9–15.

de Bruijn (1981). *Asymptotic Methods in Analysis.* Dover, New York.

Debye, P. (1909). Näherungsformelm für die Zylinderfunktionen für grosse Werte des Arguments und unbeschränkt veränderliche Werte des Index. *Math. Ann.* 67, 535–558.

Durrett, R. (1996). *Probability: Theory and Examples,* Second Edition. Duxbury, Belmont.

Erdélyi, A. (1956). *Asymptotic Expansions.* Dover, New York.

Feller, W. (1968). *An Introduction to Probability Theory and Its Applications, Vol. I.* Wiley, New York.

Feller, W. (1971). *An Introduction to Probability Theory and Its Applications, Vol. II.* Wiley, New York.

Ferguson, T. S. (1982). An inconsistent maximum likelihood estimate. *J. Amer. Statist. Assoc.* 77, 831–834.

Fisher, R. A. (1922). On the mathematical foundations of theoretical statistics. *Phil. Trans. Roy. Soc. London, Series A* 222, 309–368.

Fisher, R. A. (1925). Theory of statistical estimation. *Proc. Cam. Phil. Soc.* 22, 700–725.

Fisher, R. A. (1934). Two new properties of mathematical likelihood. *Proc. Roy. Soc. Ser. A* 144, 285–307.

Fraser, D. A. S. (1968). *The Structure of Inference.* Wiley Series in Probability and Mathematical Statistics. Wiley, New York.

Fréchet, M. (1943). Sur l'extension de certaines evaluations statistiques de petits echantillons. *Rev. Int. Statist.* 11, 182–205.

Gibson, G. A. (1927). Sketch of the History of Mathematics in Scotland to the end of the 18th Century. *Proc. Edinburgh Math. Soc. Ser. 2,* 1–18, 71–93.

Gurland, J. (1948). Inversion formulae for the distribution of ratios. *Ann. Math. Statist.* 19, 228–237.

Hájek, J. (1970). A characterization of limiting distributions of regular estimates. *Zeit. Wahrsch. verw. Geb.* 14, 323–330.

Haldane, J. B. S. (1942). Mode and median of a nearly normal distribution with given cumulants. *Biometrika* 32, 294.

Hampel, F. R. (1968). *Contributions to the theory of robust estimation.* Ph. D. Thesis, University of California, Berkeley.

Hardy, G. H. (1991). *Divergent Series.* AMS Chelsea, Providence, Rhode Island.

Hayman, W. K. (1956). A generalization of Stirling's formula. *J. Reine Angew. Math.* 196, 67–95.

Hougaard, P. (1982). Parametrizations of non-linear models. *J. Roy. Statist. Soc. Ser. B* 44, 244–252.

Huzurbazar, V. S. (1948). The likelihood equation, consistency and the maxima of the likelihood function. *Ann. Eugen.* 14, 185–200.

Inagaki, N. (1970). On the limiting distribution of a sequence of estimators with uniform property. *Ann. Inst. Statist. Math.* 22, 1–13.

Inagaki, N. (1973). Asymptotic relations between the likelihood estimating function and the maximum likelihood estimator. *Ann. Inst. Statist. Math.* 25. 1–26.

James, W. and Stein, C. (1961). Estimation with quadratic loss. *Proc. Fourth Berkeley Symp. Math. Statist. Prob.* 1, University of California Press, 311–319.

Johnson, R. A. (1967). An asymptotic expansion for posterior distributions. *Ann. Math. Statist.* 38, 1899–1907.

Johnson, R. A. (1970). Asymptotic expansions associated with posterior distributions. *Ann. Math. Statist.* 41, 851–864.

Kass, R. E., Tierney, L. & Kadane, J. B. (1988). Asymptotics in Bayesian computation (with discussion). In *Bayesian Statistics 3*, edited by J. M. Bernardo, M. H. DeGroot, D. V. Lindley & A. F. M. Smith. Clarendon Press, Oxford, 261–278.

Kass, R. E., Tierney, L. & Kadane, J. B. (1990). The validity of posterior expansions based on Laplace's method. In *Bayesian and Likelihood Methods in Statistics and Econometrics*, edited by S. Geisser, J. S. Hodges, S. J. Press & A. Zellner, North-Holland Amsterdam, 473–488.

Khintchine, A. (1924). Über einen Satz der Wahrscheinlichkeitsrechnung. *Fundamenta Mathematicae* 6, 9–20.

Khintchine, A. (1964). *Continued Fractions.* University of Chicago Press, Chicago.

Kolmogorov, A. N. (1929). Über das Gesetz des iterierten Logarithmus. *Math. Ann.* 101, 126-135, 1929.

Le Cam, L. (1953). On some asymptotic properties of maximum likelihood estimates and related Bayes' estimates. *University of California Publ. in Statist.* 1, 277–330.

Le Cam, L. (1960). Locally Asymptotically Normal Families of Distributions. *Univ. of California Publications in Statistics Vol 3, no. 2.* University of California, Berkeley and Los Angeles, 37–98.

Le Cam, L. & Yang, G. L. (2000). *Asymptotics in Statistics: Some Basic Concepts.* Second Edition. Springer Series in Statistics. Springer, New York.

Lehmann, E. L. (1983). *Theory of Point Estimation.* Wiley, New York.

Lehmann, E. L. & Casella, G. (1998). *Theory of Point Estimation.* Springer, New York.

Lugannani, R. & Rice, S. (1980). Saddle point approximation for the distribution of the sum of independent random variables. *Adv. Appl. Prob.* 12, 475–490.

Neyman, J. & Pearson, E. S. (1933). On the problem of the most efficient tests of statistical hypotheses. *Phil. Trans. Roy. Soc. Ser A* 231, 289–337.

Perron, O. (1917). Über die näherungsweise Berechnung von Funktionen großer Zahlen. *Sitzungsber. Bayr. Akad. Wissensch. (Münch. Ber.)*, 191–219.

Poincaré, H. (1886). Sur les integrales irregulières des equations linéaires. *Acta Mathematica* 8, 295–344.

Pólya, G. and Szegö, G. (1978). *Problems and Theorems in Analysis I.* Springer Classics in Mathematics. Springer, Berlin.

Rao, C. R. (1945). Information and the accuracy attainable in the estimation of statistical parameters. *Bull. Calcutta Math. Soc.* 37, 81–91.

Rao, C. R. (1962). Apparent anomalies and irregularities in maximum likelihood estimation (with discussion). *Sankhya Ser. A*, 24, 73–101.

Richardson, L. F. (1911). The approximate arithmetical solution by finite differences of physical problems including differential equations, with an application to the stresses in a masonry dam. *Phil. Tran. Roy. Soc. London, Ser. A* 210, 307–357.

Richardson, L. F. (1927). The deferred approach to the limit. *Phil. Tran. Roy. Soc. London, Ser. A* 226, 299–349.

Rudin, W. (1987). *Real and Complex Analysis*, Third edition. McGraw-Hill, New York.

Sheppard, W. F. (1939). *The Probability Integral.* British Ass. Math. Tables, Vol 7. Cambridge University, Cambridge, UK.

Spivak, M. (1994). *Calculus.* Publish or Perish, Houston, Texas.

Temme, N. M. (1982). The uniform asymptotic expansion of a class of integrals related to cumulative distribution functions. *SIAM J. Math. Anal.* 13, 239–253.

Wald, A. (1949). Note on the consistency of the maximum likelihood estimate. *Ann. Math. Statist.* 20, 595–601.

Wall, H. S. (1973). *Analytic Theory of Continued Fractions.* Chelsea, Bronx, N. Y.

Whittaker, E. T. & Watson, G. N. (1962). *A Course of Modern Analysis: An Introduction to the General Theory of Infinite Processes and of Analytic Functions with an Account of the Principal Transcendental Functions*, Fourth Edition. Cambridge University, Cambridge, UK.

Wilks, S. S. (1938). The large-sample distribution of the likelihood ratio for testing composite hypotheses. *Ann. Math. Statist.* 9, 60–62.

Wong, R. (2001). *Asymptotic Approximations of Integrals.* SIAM Classics in Applied Mathematics. SIAM, Philadelphia.

Wynn, P. (1956). On a procrustean technique for the numerical transformation of slowly convergent sequences and series. *Proc. Camb. Phil. Soc.* 52, 663–671.

Wynn, P. (1962). Acceleration techniques in numerical analysis, with particular reference to problems in one independent variable. *Proc. IFIPS, Munich*, Munich, pp. 149–156.

Wynn, P. (1966). On the convergence and stability of the epsilon algorithm. *SIAM J. Num. An.* 3, 91–122.

Index

Accelerating convergence, xiii
computational efficiency, 313
Ahmed, F., xiv
Airfoil design, 124
Aitken acceleration, 300
and Padé approximant, 302
Algebraic functional, 126, 128
"polynomial-like", 128
as expectation, 126
as local approx., 126
of order k, 126
Algorithm, xiii
Almost lower bound, 7, 18
Almost upper bound, 7, 18
Alternating series, 41, 42, 47, 72
Alternative hypothesis, 33
Amari, S.-I., 118
Amplitude, 229
Analysis, 16, 18
Analytic continuation, 315
Analytic function, 26, 54, 56, 264
at infinity, 51
at zero, 51
Ancillary statistic, 243, 249
maximal, 243, 249, 250
Antilimit, 314
Applied mathematicians, xiii
Applied mathematics, xiii
Approximate confidence interval, 114
Approximate normal pivotal, 112–114
Approximately symmetric, 66
Arc sine law, 74
Arithmetic series, 1, 2
Asymptotic analyst, 17
Asymptotic bias, 109, 146, 165, 202
Asymptotic comparison, 164
Asymptotic distribution, 32
Asymptotic efficiency, xii, 145, 146, 148, 150, 168
Asymptotic equivalence, 10–12
Asymptotic formula, 21
Asymptotic normality, xii, 3, 111, 114, 161, 165
Asymptotic optimality, 3
Asymptotic property, 16
Asymptotic rate of convergence, 67
Asymptotic relative efficiency, 167
Asymptotic series, xii, 3, 24, 45, 47–49, 51–57, 62–64, 68, 75, 97, 199
derivative term by term, 50
divergent, 63
for Mills ratio, 50
for stable d., 62
for stable density, 57
for tail density, 82
function not unique, 49
generalised, 291
in $\sqrt{n}$, 66
in Maple, 53
initial "convergence", 63
integration term by term, 50
large samples, 66
paradoxical behaviour, 63
product of, 49
sum of, 49
truncated, 64
unique expansion, 49
Asymptotic sufficiency, 145
Asymptotic unbiasedness, 164
Asymptotic variance, 111, 146, 204
Asymptotically normal estimator, 146
Asymptotics, xi, 3
Atomic number, 68

Auxiliary estimator, 150
Averaged log-likelihood, 219

Bachmann, P., 8
Baker, G. A., 93
Barndorff-Nielsen formula, 241, 246, 277
  for exponential model, 248
  for location model, 251
  for normal model, 248
  for transformation model, 251
Basu's theorem, 249
Bayes estimator, 151, 215, 222
Bayesian statistics, 219
Bell curve, 193
  approx. for integrand, 194, 215
Bernoulli distribution, 113
Bernoulli number, 72, 293
  notation for, 293
Bernoulli trials, 113, 272
Bertrand's series, 281, 288
Bertrand's test, 282, 284, 316
Bessel function, 74
Best unbiased estimator, 113
Bhattacharyya function, 156, 187
  regularity for, 155
  score function as, 158
Bhattacharyya functions, 154
Bias, 146, 202
Bias of estimator, 110
Bias reduction, xii, 109, 110
Bickel, P., 21
Bilinear form, 118
Binet, J. P. M., 72
Binomial
  coefficient, 2
  distribution, 1, 20, 21
  expansion, 1, 2, 64
  theorem, *see* Binomial expansion
Binomial coefficient, 101
Binomial distribution, 138, 309
  and Hayman's method, 271
  saddle-point approx, 275
Biometrika tables, 90
Birthday problem, 124
Borel-Cantelli lemma, 287, 288
Bounded above, 6
Bounded below, 6
Bounded derivative, 106, 110
Bounded function, 12
Bounded in probability, 12, 13, 18
  at infinity, 13, 18
Bounded set, 7
Bounding constant, 55

C-fraction, 86, 97
  finite, 86
Calculus, 2, 5
Cardinality of set, 93
Carrier's rule, 63
Carrier, G. F., 63
CAS, 17, 18
Cauchy condensation theorem, 280, 281, 289
Cauchy distribution, 56, 62, 81, 92, 225
  as stable dist., 57
  Pitman estimator for, 225
Cauchy residue, 268
Cauchy residue theorem, 264
Cauchy's integral theorem, 229
Cauchy, A.-L., 4
Cauchy-Hadamard formula, 25, 28–30, 69
Cauchy-Riemann equations, 234
  and conjugacy, 236
Central limit theorem, xiii, 2, 3, 15, 19, 99, 123, 130, 133, 164, 206
  Lindeberg-Feller, 15, 16, 20, 141
  local, 206
Central moment, 70, 115
Centred likelihood ratio, 154, 156
Centred variable
  from one-step iteration, 150
Centred variables, 150
Change of variables, 31
Characteristic function, xi, 46, 52, 57, 58, 74, 90, 205, 238
  asymptotic series for, 52
Chebyshev's inequality, 19, 285
Chernoff, H., 172
Chi-square distribution, 30–33, 123

degrees of freedom, 31, 32
noncentral, 30, 32
Closed contour, 229
Clumping index, 271
Cluster point, 7
Coefficient, 10, 35, 76, 103
Coefficient matrix, 94
Comparison test, 11, 279, 280
Complete sufficient statistic, 247
Completeness of reals, 6
Complex analysis, xi
Complex analytic function, 58
Complex conjugate, 91
Complex number, 23
Complex plane, xiii, 58, 60, 229
Complex series, 47
Complex valued function, 55
Computer, 16, 17
Computer algebra, 17, 18
Conditional density, 277
Conditional distribution, 249
Conditionality principle, 250
Confidence interval, 82, 111, 114
Configuration of sample, 250
Conjugate function, 236
Conjugate functions, 234
Consistency, xii, 143, 144, 159
Fisher, 143
redefined by Fisher, 144
Constant
of integration, 31, 112, 113
of proportionality, 32
Contiguity, 148
model, 149
neighbourhood, 148, 172
parameter, 148
Contiguous sequence, 149, 172
for integer parameter, 191
Continued fraction, xii, 75, 85
C-fraction, 86, 97
compact notation, 86
expansion of function, 85
from asymptotic series, 87
from power series, 86
normal distribution, 88
S-fraction, 97
Continuous function, 15
Continuous measurements, 68
Contour, 229
closed, 229
oriented, 229
perturbed, 239
simple closed, 264
Contour integral, xiii, 57, 58, 91, 229
Contour of integration, 58, 60
with common endpoints, 60
Convergence, 11, 14, 40, 62
accelerating, xiii, 297
almost sure, 14, 15
almost surely, 14, 285
hyperlinear, 300
in L_p, 14
in distribution, 14, 15, 19, 21, 100, 121, 122
in probability, 13–15, 19, 285
linear, 300
logarithmic, 300
of mean, 100
slow, 28
superexponential, 68
to a constant, 14
Convergent series, 23, 62, 63, 72, 279, 287, 297
and LIL, 289
Convex function, 160
Convex set, 129
Correlation coefficient, 112, 114
sample, 114
Coupon collector problem, 318
Covariance matrix, 217
Cramér assumptions for m.l.e.
imply l.a.n., 150
Le Cam's criticisms, 149
Cramér's assumptions for m.l.e., 147
Cramér, H., xii, 135, 146
Cramér-Rao inequality, 159, 167
Cramér-von Mises statistic, 123
Cui, Z., xiv
Cumulant, xii, 33–35, 38–40, 70, 74, 108, 110, 115, 116
from moments, 34
in Maple, 35
recursion for, 35
second, 70

standardised, 241
third, 115
tilted, 241
Cumulant generating function, 34, 39, 238
Curved exponential family, 117
Curvilinear integral, 229

Daniels' saddle-point method, 238
Daniels, H. E., 242
and concertina, 242
mechanical watches, 242
Darboux's method, 268
approximant, 268
for negative binomial, 273
De Bruijn, N. G., 17
De Moivre, A., 2, 3
Decimal places, 17
Degrees of freedom, 31–33
Delta method, xii, 35, 99, 123
coefficients, 104
for distributions, xii, 100, 106, 121, 122
for functionals, 133
for general distributions, 121
for moments, xii, 100, 103, 105, 108–110
in Maple, 108
paired terms, 107
ratios, 119
several variables, 119
variance formula, 107
Density function, xii, 31–33, 190
Dependent random variables, 15
Differentiable function, 37
Differential equation, 89, 115–117, 139
Differentiation operator, 37, 38
Differentiation term by term, 29
Dirac delta, 135
Dirac delta function, 24
Discrete component, 125
Dispersion parameter, 166
Distribution, 99
limit, 99
Distribution function, xii, 3, 14, 16, 28, 43, 125, 129
empirical, 123, 125, 127
normal, 28
Divergent series, 23, 40, 41, 45, 46, 52, 62, 63, 206, 287, 314
asymptotic, 63
Doksum, K., 21
Domain of convergence, 23, 71
Domain of function, 10
Double exponential distribution, 189

Economics, 260
Edgeworth expansion, xiii, 40, 242, 258
and cumulants, 259
and Gram-Charlier series, 261
compared to saddle-point, 262
for mean, median, mode, 262
Hermite polynomials for, 261
sharpens central limit approx., 259
tilted, xiii, 242
Edgeworth, F. Y., 40, 260
Efficiency, 143, 144
Efficient likelihood estimator, 150
Egyptians, 1, 2
Empirical distribution function, 123, 127, 129, 130
Empty set, 6, 7
infimum of, 6
supremum of, 6
Entire function, 26
Enveloping, 73
Enveloping series, xii, 40–42, 44 47, 50, 72
definition, 40
Equivalence transformation, 97
Equivariance, 109, 221, 248
Equivariant estimator, 221
Error, 17
Error function, 64
Error mean square, 33
Error term, 8, 17
Estimator, xii, 17
Euclidean algorithm, 94
Euler acceleration, 307
Euler's constant, 294, 318
Euler, L., 4, 48, 292, 295, 299

Euler-Maclaurin formula, 291, 318
in Maple, 294
Expected information, 147
Expected information function, 153
Experimental design, 33
Exponential decreasing error, 64
Exponential distribution, 42, 64, 73, 84, 116, 138, 188, 243, 297
asymptotic expansion for, 52
density function, 42
with lag parameter, 174
Exponential family, 116, 246, 249
k-dimensional, 249
curved, 117
generating functions of, 193
one-parameter, 145
Exponential function, 12, 38, 58, 67, 79
continued fraction for, 87
Exponential growth, 45
Exponential integral, 73
Exponential operator, 38
Exponential tilt, 37, 242
Exponential transformation model, 251
Extended real numbers, 23, 315

F distribution, 30, 33
Faà di Bruno's formula, 35
Faà di Bruno, F., 35
Factorial, 2, 72, 200, 203
Factorial growth, 45
Factorial moment, 71
Fibonnaci number, 37
Finite differences, 27
Finite parameter space, 160
Fisher consistency, 143, 144
Fisher information, 147
Fisher, R. A., xii, 143
1922 paper, 143
Fluctation, 286
Fluctuation
ϵ-fluctuation, 287
Formal power series, 37
Foundations of statistics, 250
Fourier inversion, 205, 238
Fourier series, 59
Fourier, J., 59
Fractional power, 53
Function series, 23
Functional, xii, 17, 125
algebraic, 126, 128
statistical, 123

Gâteaux derivative, 130
examples, 132
idempotent, 132
Gamma distribution, 70, 73, 97, 113, 244
estimation of shape, 257
saddle-point approx. for density, 274
Gamma function, 60, 72, 96, 201
Gaussian elimination, 94
Generalised asymptotic series, 68, 291
of Poincaré type, 68
Generalised series, 53
Generating function, xi, 16
Generation function, 29
Geometric distribution, 30, 272
Geometric mean, 30, 311
Gram-Charlier A series, 261
Grattan-Guinness, I., 59
Graves-Morris, P. R., 93
Greatest lower bound, *see* Infimum
Group of transformations, 251

Hölder's inequality, 102
Hj́ek projection, 141
Hammersley-Chapman-Robbins inequality, 158
Hampel, F. R., 136
Hankel structure, 94
Hardy, G. H., 316
Harmonic function, 234
Harmonic series, 294, 316
Hayman admissible function, 265
Heqat, 1
Hermite polynomial, 261, 276
Hessian matrix, 213

Hodges' superefficient estimator, 169, 171, 173, 176, 188
Hodges, J. L., 169
Hougaard, P., 116–118
Huang, R., xiv
Huzurbazar, H., 147
Hyperasymptotic series, 63
Hyperasymptotics, 65
  approximation by, 65
Hyperbolic cosine, 234
Hyperlinear convergence, 300
Hypothesis testing, 32
  equality of variances, 33

Identically distributed, 16, 56
Imaginary axis, 60, 238, 239
Imaginary part, 47, 58, 62
  of series, 60
Infimum, 4, 6
  of sequence, 6
  of set, 6
Infinite expectation, 207
Infinite sequence, 6
Infinite series, 23
Influence curve, 136, 140
  and robustness, 136
  for mean, 136
  for median, 136
Influence function, 130, 133
  "almost" unique, 130
Information function, 118, 153
  Fisher, 147
Information inequality, 157
Integer lattice, 93
Integer-valued parameter, 67, 191
Integrability, 196
Integral equation, 276
Integral remainder, 44
Integration by parts, 43, 46, 47, 51, 52, 54, 73, 74, 97, 212, 252
Interarrival time, 73
Interval of convergence, 25, 69
Inverse Fourier transform, 57, 58, 205
Inversion, 97
Iterated logarithm, law of, 288
  and CLT, 288
  Feller's extension, 289

Jacobian, 31
James-Stein estimator, 187
Jensen's inequality, 137, 160
Joint density function, 31
Jones, P. S., 27

Kernel density estimator, 144
Khintchine's constant, 311, 312
Kolmogorov-Smirnov statistic, 132
Kummer acceleration, 298, 319
Kummer test, 282
Kurtosis, 70

Lagrange remainder, 41, 163
Landau, E., 8
Laplace approximation, xiii, 193
  applications in statistics, 219
  basic formulas, 195
  for location model, 220
  for multiple integration, 212
  for product integrand, 215
  in Maple, 201
  maximum at boundary, 211
  proofs, 207
Laplace method, xiii
Laplace series, 199
Laplace transform, 54
Laplace's diff. equation, 234
Laplace, P.-S., 88, 197
Lattice, 93
Law of iterated logarithm, 288
  and CLT, 288
  Feller's extension, 289
Law of large numbers, 15, 239, 319
  strong, 15, 160
  weak, 15, 19
Law of unconscious statistician, 214
Le Cam asymptotics, xiii, 148
Le Cam, L., xii, 148
Least probable error, 145
Lebesgue measure zero, 170
Leibniz, G., 2
Likelihood asymptotics, 143

historical overview, 143
Likelihood function, xii, 82, 83, 152, 247
logarithm of, xii
approximately normal, 117, 118
global maximum, 147
logarithm of, 143
power series expansion, 152
properties of, 152
stable variance, 117, 118
Likelihood ratio, 253
Likelihood region, 83
interval, 83–85
Limit, xii, 4, 20, 23
Limit comparison test, 280
Limit distribution, 67
Limit inferior, xii, 6–8, 18, 25
of sequence, 7
of set, 7
Limit superior, xii, 6–8, 18, 25
of sequence, 7
of set, 7
Lindeberg condition, 16, 20
Line integral, 229
Linear approximation, 12
Linear convergence, 300
Linear statistical functional, 127, 131
Linear transformation, 31
Linearised Poisson distribution, 112
LMVUE, 156
Local asymptotic admissibility, 175, 176
Local asymptotic minimaxity, xiii, 176, 181, 186
Local asymptotic normality, xiii, 149, 177, 179, 181, 186, 191
and l. a. quadratic, 179
Local asymptotic risk, 171–173, 175, 191
constant, 188
Local asymptotics, 148, 171
technical meaning, 148
Local central limit theorem, 206
Local linearity, 99
Locally asymptotically quadratic, 150, 177, 178
Location equivariance, 221
Location invariant, 277
Location parameter, 220, 250
Log-likelihood, xii, 83, 85, 143, 152
averaged, 219
stable variance, 117
Log-relative likelihood, 84, 190
Logarithm, 2
Logarithmic convergence, 300
Lower bound, 7, 45
of sequence, 4
of set, 4
Lugannani, R., 253
Lugannani-Rice formula, 253, 257

M-estimating equation, 256
M-estimator, 134, 256
M-functional, 134, 147, 161
Maclaurin, C., 48, 292
Maple, xiii, 18, 20, 26, 35, 53, 54, 82, 88, 94, 108, 109, 139, 257
asymptotic series, 53
continued fractions in, 88
delta method, 108
Euler-Maclaurin formula in, 294
expansions in, 20
power series, 35
simplification in, 202
Maple command
LambertW, 272
Order, 10
asympt, 20, 53, 201, 202
coeftayl, 26
collect, 109
confrac, 88
convert, 88
eulermac, 294
expand, 201, 202
inverse, 37
mtaylor, 109
numapprox, 79, 82
pade, 79
powcos, 37
powcreate, 108
powerseries, 35
powexp, 37, 108
powseries, 26, 35, 37, 108, 202

powsin, 37
series, 53
simplify, 201, 202
solve, 245
subs, 109
taylor, 10, 26, 53
tpsform, 36, 37, 109
Marcinkiewicz-Zygmund ineq., 102
Marginal density function, 31
Marginal distribution, 215, 249
Markov's inequality, 19
Maroufy, V., xiv
Marriott, P., 118
Martingale convergence theorem, 286
Mathematical statistics, xi
Maximal invariant, 277
Maximum likelihood estimator, xii, xiii, 67, 83, 84, 109, 111, 116, 134, 143, 153
primus inter pares, 150
as global maximum, 143
as local maximum, 143
as M-estimator, 134, 256
asymptotically normal, 135, 147, 161
asymptotically sufficient, 183
backlash against, 148
bias adjusted, 117, 118, 175
consistency of, 159, 163
constant l. a. risk, 183
early optimism for, 145, 146, 148
efficiency of, 145, 148
equivariance, 109
Fisher consistent, 144
l. a. admissible, 184
l. a. minimax, 185
normalised, 117, 118
regularity for, 146, 147
McIsaac, M., xiv
Mean, 21, 30, 42, 67, 262
Mean functional, 127, 132, 136
Median, 123, 146, 202, 262
Method of Darboux, 268
Method of maximum likelihood, 143
Method of moments, 145
Mills' ratio, 46, 48, 50, 88, 257
as asymptotic series, 50, 68, 88
continued fraction for, 88
Mixed partial derivative, 217
Mixture distribution, 135
Mode, 30, 262
Modern computing, 16
Modes of convergence, 14
Modular structure, xi
Moment, xii, 33–35, 66, 67, 70, 74, 99, 101, 103, 104, 107, 115, 119, 203
from cumulants, 34
in Maple, 35
limiting, 99
product, 119
recursion for, 35, 70
Moment generating function, 3, 19, 33, 38, 39, 47, 54, 108, 238
differentiable at zero, 19
logarithm of, 20
multivariate, 217
Monte Carlo, xiii
Monte Carlo methods, 193

Natural numbers, 10, 11
Negative binomial distribution, 272
Neighbourhood, 9–11
Neighbourhood of origin, 54
Newton, I., 2, 4, 5
Newton-Raphson, 83
Newton-Raphson iteration, 150
Neyman type A distribution, 271
clumping index, 271
saddle-point approx. for tail, 272
Neyman, J., 146
Neyman-Pearson lemma, 146
Noguchi, K., xiv
Noncentral chi-square, 33
Noncentral chi-square distribution, 30, 32, 33
Noncentral F distribution, 33
noncentrality parameters, 33
Noncentrality parameter, 32, 33
Nondecreasing function, 30
Nondegenerate distribution, 165
Nonparametric maximum likelihood, 190